BARRIER CONTRACEPTIVES

Current Status and Future Prospects

BARRIER CONTRACEPTIVES

Current Status and Future Prospects

Proceedings of the Fourth Contraceptive Research and Development Program International Workshop, Santo Domingo, Dominican Republic, March 22–25, 1993

Held in Collaboration with the Family Health International, Profamilia of the Dominican Republic, and World Health Organization, and with Major Financial Support from the United States Agency for International Development

EDITORS

CHRISTINE K. MAUCK
Contraceptive Research and Development Program
Arlington, Virginia

MILTON CORDERO
Profamilia, Santo Domingo, Dominican Republic

HENRY L. GABELNICK
Contraceptive Research and Development Program
Arlington, Virginia

JEFFREY M. SPIELER
Agency for International Development
Washington, D.C.

ROBERTO RIVERA
Family Health International
Durham, North Carolina

WILEY-LISS

A JOHN WILEY & SONS, INC., PUBLICATION
New York • Chichester • Brisbane • Toronto • Singapore

Address All Inquiries to the Publisher
Wiley-Liss, Inc., 605 Third Avenue, New York, NY 10158-0012

Printed in the United States of America

The publication of this volume was facilitated by the authors and editors who submitted the text in a form suitable for direct reproduction without subsequent editing or proofreading by the publisher.

Library of Congress Cataloging-in-Publication Data

Contraceptive Research and Development Program. International
Workshop (4th : 1993 : Santo Domingo, Dominican Republic)
Barrier contraceptives : current status and future prospects /
editors, Christine K. Mauck ... [et al.].
p. cm.
"Proceedings of the Fourth Contraceptive Research and Development (CONRAD) Program International Workshop, in collaboration with Family Health International (FHI), Profamilia of the Dominican Republic, and the World Health Organization, and with major financial support from the United States Agency for International Development (A.I.D.), held In Santo Domingo, Dominican Republic, March 22–25, 1993."
Includes index.
ISBN 0-471-30440-9
1. Contraceptives, Vaginal—Congresses. 2. Condoms—Congresses
3. Contraceptives, Vaginal—Research—Congresses. 4. Sexually transmitted diseases—Prevention—Equipment and supplies--Congresses. 5. Sexually transmitted diseases—Prevention--Congresses. I. Mauck, Christine K. II. Title.
[DNLM: 1. Contraception—methods—congresses. 2. Contraceptive Devices—congresses. WP 640 C7635 1993]
RG137.2.C66 1993
613.9'435—dc20
DNLM/DLC
for Library of Congress 93-34517

The text of this book is printed on acid-free paper.

CONTENTS

Contributors ix

Preface
Christine Mauck, Milton Cordero, Henry Gabelnick, Jeffrey Spieler, and Roberto Rivera xiii

Keynote Address: The Last Barrier
Malcolm Potts xv

PART I: BARRIER USE

1. Demographics of Barrier Contraceptive Use
Jacqueline Darroch Forrest 3
DISCUSSION

2. Contraceptive Efficacy of Barrier Contraceptives
James Trussell 17
DISCUSSION

3. Acceptability of Barrier Methods for Prevention of Unwanted Pregnancy and Infection
Nancy E. Williamson and Carol Joanis 53
DISCUSSION

4. Feminist Perspective on Barrier Use
Judy Norsigian 69
DISCUSSION

PART II: BARRIERS RESEARCH

5. Vaginal Anatomy and Physiology as They Relate to Sexual Intercourse
Subir Roy 77
DISCUSSION

6. Microbiology of the Vagina and Cervix
Peter Mason and Lovemore Gwanzura 91
DISCUSSION

7. Vaginal Barrier Contraceptives and Infection Risk
Felicia H. Stewart 105

8. A Contraceptive and Prophylactic Barrier: Current Concepts for the Male Condom
Howard Charles Price 123
DISCUSSION

PART III: RESEARCH METHODOLOGIES

9. Vaginal Contraceptive Drugs and Devices: Regulatory Requirements
Lisa D. Rarick and Lillian Yin 137
DISCUSSION

10. Chemical Vaginal Contraceptives: Preclinical Evaluation
Gustavo F. Doncel 147
DISCUSSION

11. Vaginal Contraceptive Efficacy: Animal Models
Lourens J.D. Zaneveld 163
DISCUSSION

12. Postcoital Tests
Kamran S. Moghissi 173
DISCUSSION

13. Considerations for the Clinical Evaluation of Chemical and Physical Barrier Contraceptive Efficacy
Roberto Rivera, Gaston Farr, Lanetta Dorflinger, Carol Connell, Rosalie Dominik, Deborah Menius, Marilyn Metcalf-Whittaker, and Carolyn Reusché 185
DISCUSSION

PART IV: BARRIERS AND STDs

14. HIV, Other STDs, and Barriers
Katherine M. Stone 203
DISCUSSION

15. Use of the SIV/Rhesus Macaque System to Test Virucides Designed to Prevent the Sexual Transmission of HIV
Christopher J. Miller 213
DISCUSSION

16. Barrier Methods for the Prevention of Gonococcal and Chlamydial Infection
Penelope J. Hitchcock 225
DISCUSSION

PART V: CLINICAL STUDIES ON THE EFFECTIVENESS OF BARRIERS IN PREVENTING STDs

17. Four Family Health International Studies on Barrier Methods and Sexually Transmitted Infections
Ronald E. Roddy 245

18. Prevention of Vaginal Trichomoniasis by Compliant Use of the Female Condom
David E. Soper, Donna Shoupe, Gary A. Shangold, Mona M. Shangold, Jacqueline Gutmann, and Lane Mercer 249

19. Condom Use Evaluation in Hotel and Motel Rooms
Ernesto Guerrero, Santo Rosario, and Lydia Bond.................. 255

20. Efficacy of Nonoxynol-9 Contraceptive Sponge Use in Preventing Heterosexual Acquisition of HIV in Nairobi Prostitutes
Elizabeth Ngugi, Joan Kreiss, King Holmes, Jackoniah Ndinya, Peter Waiyaki, Pacita L. Roberts, Irene Rumijo, Rose Sajabi, Joyce Kimata, Thomas R. Fleming, Aggrey Anzala, Donna Holton, and Francis Plummer 261

21. PROTECTAID, a New Vaginal Contraceptive Sponge With Anti-STD Properties
Alexandre Psychoyos 265

22. Development of Vaginal Microbicides
Marc Karam.. 271

23. Praneem Polyherbal Cream and Suppositories
G.P. Talwar, Sanjay Garg, Ritu Singh, P.L. Sharma, Veneeta Dhar, Vibha Taluja, Suman Dhawan, and S.N. Upadhyay................. 273

24. South to South Experience With Polyherbal Cream in the Treatment of Vaginal Infection
O.A. Ladipo.. 277

25. Praneem Polyherbal Cream: Phase I Clinical Trials in India
Sanjay Garg, Rajinder Kaur, S.N. Upadhyay, and G.P. Talwar 283

General Discussion 287

PART VI: CONTRACEPTIVE APPLICATION OF BARRIER METHODS WORKING GROUPS

26. Preclinical Study Design Working Group Report
Moderator: David Katz
Rapporteur: Douglas S. Colvard................................. 295
DISCUSSION

27. Clinical Study Design Working Group Report
Moderator: Milton Cordero
Rapporteur: Lisa Rarick 311
DISCUSSION

28. Issues of Acceptability and Behavior Working Group Report
Moderator: Virginia S. Cain
Rapporteur: Nancy E. Williamson 323
DISCUSSION

29. Product Development Working Group Report
Moderator: Gabriel Bialy
Rapporteur: Jeffrey Spieler 339
DISCUSSION

30. HIV and Other STDs Working Group Report
Moderator: Penelope Hitchcock
Rapporteur: Lee Claypool 353
DISCUSSION

PART VII: WORKSHOP PARTICIPANTS

Workshop Participants. 365

Index 375

CONTRIBUTORS

Aggrey Anzala, University of Nairobi, Nairobi, Kenya **[261]**

Gabriel Bialy, National Institutes of Health, Bethesda, MD 20892 **[339]**

Lydia Bond, Pan American Health Organization, Washington, DC 20037 **[255]**

Virginia S. Cain, National Institutes of Health, Bethesda, MD 20892 **[323]**

Lee Claypool, Contraceptive Research and Development Program, Arlington, VA 22209 **[353]**

Douglas S. Colvard, Contraceptive Research and Development Program, Arlington, VA 22209 **[295]**

Carol Connell, Family Health International, Research Triangle Park, NC 27713 **[185]**

Milton Cordero, Profamilia, Santo Domingo, Dominican Republic **[311]**

Veneeta Dhar, National Institute of Immunology, 110067 New Delhi, India **[273]**

Suman Dhawan, National Institute of Immunology, 110067 New Delhi, India **[273]**

Rosalie Dominik, Family Health International, Research Triangle Park, NC 27713 **[185]**

Gustavo F. Doncel, Eastern Virginia Medical School, Norfolk, VA 23507 **[147]**

Lanetta Dorflinger, Family Health International, Research Triangle Park, NC 27713 **[185]**

Gaston Farr, Family Health International, Research Triangle Park, NC 27713 **[185]**

Thomas R. Fleming, University of Washington, Seattle, WA 98195 **[261]**

Jacqueline Darroch Forrest, The Alan Guttmacher Institute, New York, NY 10003 **[3]**

Sanjay Garg, National Institute of Immunology, 110067 New Delhi, India **[273, 283]**

Ernesto Guerrero, Union Latino Americana Contra de Enfermedades de Transmission Sexual (ULACETS), Santo Domingo, Dominican Republic **[255]**

Jacqueline Gutmann, Yale University, New Haven, CT 60637 **[249]**

Lovemore Gwanzura, University of Zimbabwe Medical School, Harare, Zimbabwe **[91]**

Penelope J. Hitchcock, National Institutes of Health, Bethesda, MD 20892 **[225, 353]**

King Holmes, University of Washington, Seattle, WA 98195 **[261]**

Donna Holton, University of Manitoba, Winnipeg, Manitoba, Canada **[261]**

Carol Joanis, Family Health International, Research Triangle Park, NC 27713 **[53]**

The numbers in brackets are the opening page numbers of the contributors' articles.

Marc Karam, Global Programme on AIDS, World Health Organization, CH-1211 Geneva 27, Switzerland **[271]**

David Katz, University of California, Davis, CA 95616 **[295]**

Rajinder Kaur, National Institute of Immunology, 110067 New Delhi, India **[283]**

Joyce Kimata, Kenya Medical Research Institute, Nairobi, Kenya **[261]**

Joan Kreiss, University of Washington, Seattle, WA 98195 **[261]**

O.A. Ladipo, South to South Cooperation in Reproductive Health, 40210-350 Salvador, Bahia, Brazil **[277]**

Peter Mason, University of Zimbabwe Medical School, Harare, Zimbabwe **[91]**

Deborah Menius, Family Health International, Research Triangle Park, NC 27713 **[185]**

Lane Mercer, University of Chicago, Chicago, IL 60637 **[249]**

Marilyn Metcalf-Whittaker, Family Health International, Research Triangle Park, NC 27713 **[185]**

Christopher J. Miller, California Regional Primate Research Center, University of California, Davis, CA 95616 **[213]**

Kamran S. Moghissi, Wayne State University School of Medicine, Detroit, MI 48201 **[173]**

Jackoniah Ndinya, University of Nairobi, Nairobi, Kenya **[261]**

Elizabeth Ngugi, University of Nairobi, Nairobi, Kenya **[261]**

Judy Norsigian, Boston Women's Health Book Collective, Somerville, MA 02144 **[69]**

Francis Plummer, University of Manitoba, Winnipeg, Manitoba, Canada **[261]**

Malcolm Potts, University of California, Berkeley, CA 94704 **[xv]**

Howard Charles Price, Family Health International, Research Triangle Park, NC 27709 **[123]**

Alexandre Psychoyos, Laboratoire de Physiologie de la Reproduction, Hôpital Bicêtre, Bâtiment INSERM, Kremlin-Bicêtre 94276, France **[265]**

Lisa D. Rarick, US Food and Drug Administration, Rockville, MD 20857 **[137, 311]**

Carolyn Reusché, Family Health International, Research Triangle Park, NC 27713 **[185]**

Roberto Rivera, Family Health International, Research Triangle Park, NC 27713 **[185]**

Pacita L. Roberts, University of Washington, Seattle, WA 98195 **[261]**

Ronald E. Roddy, Family Health International, Research Triangle Park, NC 27709 **[245]**

Santo Rosario, Centro de Orientacion e Investigacion Integral (COIN), Santo Domingo, Dominican Republic **[255]**

Subir Roy, USC School of Medicine, Women's Hospital, Los Angeles, CA 90033 **[77]**

Irene Rumijo, University of Nairobi, Nairobi, Kenya **[261]**

Rose Sajabi, University of Nairobi, Nairobi, Kenya **[261]**

Gary A. Shangold, University of Chicago, Chicago, IL 60637 **[249]**

Mona M. Shangold, Hahnemann University, Philadelphia, PA 19102 **[249]**

P.L. Sharma, Postgraduate Institute of Medical Education and Research, 160012 Chandigarh, India **[273]**

Donna Shoupe, University of Southern California, Los Angeles, CA 90007 **[249]**

Ritu Singh, National Institute of Immunology, 110067 New Delhi, India **[273]**

David E. Soper, Medical College of Virginia, Richmond, VA 23298 **[249]**

Jeffrey Spieler, Office of Population, US Agency for International Development, Washington, DC 20523 **[339]**

Felicia H. Stewart, Valley Center for Women's Health, Sacramento, CA 95825 **[105]**

Katherine M. Stone, Centers for Disease Control and Prevention, Atlanta, GA 30333 **[203]**

Vibha Taluja, National Institute of Immunology, 110067 New Delhi, India **[273]**

G.P. Talwar, National Institute of Immunology, 110067 New Delhi, India **[273, 283]**

James Trussell, Princeton University, Princeton, NJ 08544 **[17]**

S.N. Upadhyay, National Institute of Immunology, 110061 New Delhi, India **[273, 283]**

Peter Waiyaki, Kenya Medical Research Institute, Nairobi, Kenya **[261]**

Nancy E. Williamson, R&D/Office of Health/HIV-AIDS Division, Agency for International Development, Washington, DC **[53, 323]**

Lillian Yin, US Food and Drug Administration, Rockville, MD 20857 **[137]**

Lourens J. D. Zaneveld, Rush-Presbyterian-St. Luke's Medical Center, Chicago, IL 60612 **[163]**

PREFACE

On March 22-25, 1993 the Contraceptive Research and Development (CONRAD) Program of Eastern Virginia Medical School, in collaboration with Family Health International (FHI), Profamilia of the Dominican Republic and the World Health Organization, and with major financial support from the United States Agency for International Development (A.I.D.), convened the "International Workshop on Barrier Contraceptives: Current Status and Future Prospects." The Workshop brought together 165 participants from 20 countries, including basic and applied biomedical and social science researchers, clinical researchers, public health specialists, policy-makers, and women's health advocates from academia and the public and private sectors. The format for the workshop included plenary sessions, special working groups on pre-clinical study design, clinical study design, issues of acceptability and behavior, product development, and human immunodeficiency virus (HIV) and other sexually transmitted diseases (STDs), with ample opportunity for formal and informal discussion both during and after the sessions.

The goals of the workshop were to:

- Present the state-of-the-art of barrier methods in preventing pregnancy and STDs;
- Review and update technical data in related areas;
- Identify future research priorities in contraception and prevention of STDs;
- Facilitate networking among and between researchers and other participants;
- Educate relevant policy-makers and public health workers;
- Address obstacles to barrier method development, use and acceptability; and
- Review current and potential products.

Although the primary mission of the CONRAD Program is the development of improved and totally new, safe, effective, and acceptable methods of fertility regulation suitable for use in developing countries, CONRAD, the co-sponsors and the funding agencies have recognized the vital importance of considering how contraceptive technology can have a positive, negative or neutral influence on the spread of STDs, especially AIDS. Thus, one cannot undertake a contraceptive development research program without giving some priority to the control and prevention of STDs. With direct support from A.I.D., and with critical support from the U.S. National Institutes of Health (specifically the National Institute of Child Health and Human Development and the National Institute of Allergy and Infectious Disease) and Centers for Disease Control, provided through interagency agreements between these agencies and A.I.D., the CONRAD Program has developed a research program investigating the mechanisms controlling, and prevention of, the sexual transmission of HIV and other pathogens. This work complements and supplements the work conducted by the other co-sponsors of the workshop, all of whom either support, work on or are concerned about the interface of contraceptive technology and STD control.

Science is propelled forward by the exchange of information. These proceedings include the presentations made at the plenary sessions, the discussion following the presentations and the results of the five working groups. They summarize the current state of knowledge, research and development of barrier methods of contraception and STD prevention. This book is the output of a lively meeting and, although the printed word can never capture the dynamics of the event nor the formal and informal exchanges that occurred, it is being published so that a wider audience might benefit from the crucial information shared on two interrelated topics critical to the health and welfare of women, men and families worldwide.

We are greatly indebted to Ms. Marlene LaMont and Mr. Nash Herndon for their tireless efforts to produce this book while the conference is still fresh in the minds of the participants and enthusiasm for implementing its recommendations is high.

Christine Mauck
Milton Cordero
Henry Gabelnick
Jeffrey Spieler
Roberto Rivera

KEYNOTE ADDRESS: THE LAST BARRIER

Malcolm Potts

University of California, Berkeley
2408 Parker Street
Berkeley, CA 94704

Twenty five years ago, barrier methods seemed destined to go the way of the steam locomotive -- historically important, much relied on by those that used them, but about to be overtaken by more effective technologies. That did not happen and today there is a renaissance of interest in barrier methods. This is partly because they are a logical way to interrupt the processes leading to conception. In addition, what has kept inexpensive barrier methods center stage, more specifically, is their unique ability to slow the spread of sexually transmitted diseases (STDs) and of the human immunodeficiency virus (HIV).

The basic scientific aspects of the field are covered in this volume from a practical perspective, i.e. what are the opportunities and what can be expeditiously done to develop new barrier methods of contraception? Attention is also given to strategies for the clinical testing of new methods and to issue of drug regulatory approval (see chapter 9).

The once almost moribund field of barrier method research now offers a number of genuine opportunities. The female condom (Reality/Femidom) is now marketed in western Europe and the United States. Two new vaginal barriers (Lea's Shield and Femcap) are being evaluated in clinical trials and will, hopefully, obtain marketing approval in the near future.

But the message that runs most strongly through this book is the need for a barrier or chemical method to slow the spread of HIV. It is, perhaps, the single most important technology we could offer the world at the present moment. Reality/Femidom is welcome but we also need something that cannot be seen or felt by the man, and that a woman can use to protect herself against HIV and other STDs when she cannot compel her partner to use a condom. There are hundreds of millions of such sexual partnerships around the world, in both developed and developing countries. Sexual intercourse, tragically, is not always something that takes place in a loving and equal partnership. Women are sometimes given no choice by drunken or abusive partners. They may suspect their partner is unfaithful, yet have no way to negotiate any protection. We should also note that the relative risk of transmission of HIV to women from men is significantly greater than the other way around.

Modern scientific medicine has never faced a new worldwide epidemic prior to AIDS. If a health care program in a developing country fails to implement its measles vaccination program on time, approximately as many children will die this year as died last year, or the year before. If we fail in the case of HIV prevention, then we will have twice as many cases in core groups to deal with in twelve months time, and probably twice as many in the general population in three or four years.

We have already lost a great deal of time in responding to this dreadful epidemic. The number of HIV-positive people is projected to jump from six million in 1988 to 40 million in the year 2000. Forty million is more people than died as combatants and civilians in World War II. Most of us who make decisions about health care, or conduct research and development on barrier methods of contraception, are healthy. How do we turn cold statistics into individual suffering and death? And what of the tens of millions of loved ones, children, wives, husbands, lovers, who will watch in anguish and suffer themselves - and they will worry over money, over their own future and over the next generation.

Barrier methods are part of the triad of actions -- education, treating STDs and offering barrier methods -- that can slow the spread of AIDS. A recent analysis by the World Bank suggests that STD treatment may be one of the most cost-effective and powerful interventions, particularly when directed at groups with high-risk behaviors [World Bank, 1992]. Unfortunately, we are not hearing what epidemiology has to tell us. I know of an African country where ten to thirty percent of urban women attending some antenatal clinics are HIV positive. The STD clinics have acetaminophen and cough syrup but up country they do not even have the cough syrup! We are not doing very well. But, at least the people of East Africa would use antibiotics and condoms if they had them. In 1992, a Dublin court found the Irish Family Planning Association guilty of the crime of selling a condom! Mr. Richard Branson, who owns Virgin Records and Virgin Atlantic Airline, had the courage to permit the Irish Family Planning Association to sell condoms in his record stores. The case was lost and the Irish Family Planning Association was fined £500. Nevertheless, the FPA continues its "criminal" activities.

Obviously, there is an imperative to use what exists but a plausible case can also be made that a virucide could be formulated for vaginal use which would kill HIV without damaging the vaginal epithelium. A number of FDA-approved compounds exist, although not for contraception, including benzalkonium chloride and chlorhexidine, and they might become the basis of a new method. Other chemical entities can be investigated. The formulation of the product is likely to be crucial. A practical way of testing it, as Roddy of FHI has shown (See Chapter 17), is to use informed volunteers who were not exposed to STDs or HIV, but who are willing to have vaginal examination after a week or so of use, in order to demonstrate whether any damage to the vaginal epithelium has occurred. It is quite possible that the dose required to kill the virus might be very low. Different formulations for regular sexual partnerships and commercial sex workers might need to be developed.

The experience of family planning is that successful programs depend on a range of choices delivered through a variety of channels of distribution. It is impossible to raise contraceptive prevalence beyond a rather low level with a single method of family planning. When several methods are offered, a higher prevalence can be reached. For example, IUDs are more effective contraceptives than barrier methods, but they are neither appropriate for all couples nor wanted by all couples. This is equally true for oral contraceptives. The same probably applies to preventing HIV; methods are also likely to vary in their acceptability and efficacy. Recent research in the United States by Rosenberg and colleagues demonstrated that when women at an STD clinic were followed up, those who used diaphragms or spermicides had a lower re-infection rate than those whose partners used condoms [Rosenberg et al., 1992]. The likely explanation is that although condoms may be more effective in preventing the transmission of pathogens, women are more likely to use a vaginal method consistently than men are to use a condom on a regular basis. Computer models suggest that even if a vaginal method were less effective than a condom in slowing HIV transmission, and even if some condom users switched to a method easier to use, the use of the vaginal method would still add to successful prevention.

Personally, I would advocate:

- **Revising the regulations controlling the introduction of barrier methods, taking into account the new imperative of slowing the spread of HIV.** Some of our current requirements are rather like those of a bureaucrat who insists on recording the weight of every sack of food going into starving Somali's to 10 decimal places; it is not science, it is pedantry.
- **Interpreting existing regulations to take into account the new imperative of HIV.** We are caught between an awareness of the exponential spread of the virus and the difficulty in moving regulations of national drug and device regulatory authorities. Yet the epidemiology of the epidemic tells us we cannot conduct business as usual; new problems require new solutions. We must accept the responsibility that has always fallen on the good physician or public health official to do the best for those we serve, even if it exposes us to risk or potential criticism.
- **Avoiding the expense and time of over-testing a drug or device with the potential to slow HIV in every country.** If a method has been responsibly tested in one country, then use it in another. For example, there can be little scientific justification for repeating all previous work with Femidom or diaphragms in every country.

- **Recognizing that regulations devised to test contraceptive efficacy may not be appropriate for testing the potential to slow HIV.** With a disease where numbers double as rapidly as HIV, a poor method today is better than a perfect one tomorrow. I would suggest that such a virucidal agent should be distributed if it can be shown (a) that it kills HIV in vitro and (b) if it can be shown that there is no evidence of damage to the vaginal epithelium or gross disruption of the normal vaginal environment. The clinical demonstration that it slows HIV transmission will be important, but time consuming, difficult, expensive and ethically challenging; it is possible such a demonstration will only become possible **after** widespread marketing.
- **Pursuing priorities.** The need for such a product has been apparent for some time. It has been discussed at NIH and A.I.D. in America, and at the Medical Research Council in the United Kingdom. Nothing practical, however, has happened.

If the world were invaded by little green men from outer space who had already sentenced tens of millions to death, we would surely react in a rational way to defend ourselves. We have failed to do that in the case of the AIDS epidemic. It is exceedingly urgent that funding agencies (either nationally or through international cooperation), both governmental and non-governmental, work out a strategy for developing new barrier methods that women can use to protect themselves against HIV transmission.

We need to make a commitment to prove whether the development of new methods is possible or not. If a vaginal virucide can be developed, it will provide a most important missing weapon against the spread of HIV. We have already lost a great deal of time waiting for each other to act. This is not a normal situation. Millions of lives are at stake. The last barrier to progress is not in the method, but in the management of human skills and the allotment of funds. This book provides the most up-to-date scientific basis possible for making decisions about the support of research in a key preventive area. It is hoped it will excite the interest of researchers and ensure that a small percentage of the large sums of money currently going into AIDS treatments will be assigned to prevention.

REFERENCES

World Bank, 1992. Tanzania: AIDS Assessment and Planning Study. World Bank, Washington.

Rosenberg MJ, Davidson AJ, Chen JH, Judson FN, Douglas JM (1992): Barrier contraceptives and sexually transmitted diseases in women: a comparison of female dependent methods and condoms. Am J Public Health 82 (5):660-74.

PART I: BARRIER USE

1

DEMOGRAPHICS OF BARRIER CONTRACEPTIVE USE

Jacqueline Darroch Forrest

Vice President for Research
The Alan Guttmacher Institute
111 Fifth Avenue
New York, NY 10003

INTRODUCTION

Barrier contraceptives include both the male condom, which functions solely as a barrier method, and a variety of female vaginal methods such as the diaphragm, cervical cap and sponge, which can act as barriers but are also used with spermicide, and spermicidal foams, jellies and creams that may be used alone. These include the earliest modern contraceptive, the condom, and one of the newest, the contraceptive sponge. Work is currently under way to reformulate the shape and materials used for condoms into new options for men, as well as for women, and to develop new spermicide and microbicides for vaginal use.

Information on the current levels and patterns of barrier method use can help inform both the development of newer methods and the distribution of current ones. This paper will use available survey information to describe what is known about current levels of knowledge, accessibility and use of barrier contraceptive methods.

SOURCES AND LIMITATIONS OF DATA

Most of the information presented here comes from summary reports of country surveys carried out in the second half of the 1980s as part of the Demographic and Health Surveys or under other auspices [Liskin et al., 1990; Rutenberg et al., 1991; United Nations, 1989]. Newer barrier methods are not yet represented in available surveys, however. In addition, most of the country-wide information on barrier method use comes from surveys focused on women's fertility and reproductive health. In such surveys, barrier methods have been asked about only in the contraceptive context, with

Barrier Contraceptives: Current Status and Future Prospects, pages 3–15

no querying about possible exposure to sexually transmitted diseases, including HIV, or the use of barrier methods for the purpose of disease prevention.

KNOWLEDGE OF BARRIER CONTRACEPTIVE METHODS

Knowledge of a method is a prerequisite to its use and is related to the amount of information made available to potential users. Levels of knowledge about modern contraceptive methods are high across most of the countries included in the DHS, but knowledge levels for the condom and, especially, for vaginal methods are relatively low (Table 1). The proportion of married women of reproductive age who knew at least one modern contraceptive method ranged from 28.6 percent to 99.9 percent and in half of the countries in which there were DHS surveys, at least 93.2 percent of married women knew about modern methods. Knowledge levels for vaginal methods ranged from 4.0 percent to 85 percent, but the median value for all DHS countries was 25.9 percent. Condom knowledge levels were generally higher than those for vaginal methods: a range of 8.2 percent to 96 percent and a median value of 55.7 percent.

These methods have often been introduced into countries through family planning programs that give heavy emphasis to other methods such as the pill, which is most widely known, with a median knowledge level of 91.1 percent (not shown). The level of knowledge about condoms is strongly related to the knowledge level for oral contraceptives (r=0.88). Knowledge of vaginal methods is less strongly related to the knowledge level of the pill (r=0.52) or to the level for the condom itself (r=0.56), suggesting that vaginal methods have not been treated the same way as the condom in informational efforts.

ACCESS TO BARRIER CONTRACEPTIVES

Attempts have been made to characterize the degree of access to contraceptive methods through judgments of the proportion of couples with ready and easy geographic and financial access to these methods. The most recent such assessment estimated that the average proportion of couples with ready and easy access to the condom was 61 percent of couples in developing countries and 89 percent of those in developed countries [Population Crisis Committee, 1992]. In 44 of the 95 developing countries, at least 80 percent of couples had ready and easy access to condoms, while 20 percent or fewer had access in 19 other developing countries. At least 80 percent of couples had ready and easy access to condoms in 24 of the 29 developed countries.

In both developing and developed countries, the average level of access to condoms was eight percentage points higher than the level of access to pills and IUDs. In developing countries, where condoms have often been introduced along with other methods by organized family planning programs, the proportions of couples with ready and easy access to pills and IUDs and to condoms are closely related (r=0.86). The relationship is not so strong in developed countries, where condom availability usually preceded the growth of organized family planning programs (r=0.56).

LEVELS OF BARRIER METHOD USE

Worldwide estimates of levels of contraceptive use must be pieced together from surveys in some countries and assumptions about use levels where there have not been recent surveys. Such estimates suggest that somewhat more than half of the married couples of reproductive age in the world today use contraceptives, with the level approaching three-quarters of married couples in more developed countries and probably just exceeding half of couples in the developing countries [United Nations, 1989: 42, 62; Maulden and Ross, 1992:6; Liskin et al., 1990:4].

About 5 percent of all married couples of reproductive age in the world are estimated to use condoms for contraception — 12 to 13 percent of the couples in developed countries and 3 to 4 percent of those in developing countries. These condom users account for 9 to 10 percent of all contraceptive methods used by married couples — for about 19 percent of total use in the more developed regions and for some 5 to 6 percent of use in the less developed countries. Estimated levels of condom use among married couples of reproductive age have changed little since the early 1980s although there have been increases in some individual countries, but there are indications of increases on overall levels of condom use, perhaps indicating greater use outside of marriage [Liskin et al., 1990:3; United Nations, 1989:42].

Use of vaginal barrier and spermicide methods is much lower in both developed and developing countries. Only 2 percent of all married couples using contraceptives are estimated to use vaginal methods — 3 percent of all married contraceptive users in developed countries and 1 percent of those living in developing regions [United Nations, 1989:42].

Estimates of barrier contraceptive use among married couples of reproductive age undoubtedly understate the total levels of use, however, because they include only married couples, ignoring sexual activity among unmarried women and men and extramarital activity, and because they are based on surveys asking about contraceptive use and not about prevention of sexually transmitted diseases. In addition, women tend to underreport condom use by 1 to 2 percentage points or more, whether because they are asked only about contraceptive use, because they don't realize their partner used a condom or because they don't consider it a method they are relying on [Liskin et al., 1990:3; United Nations, 1989:5]. Liskin and colleagues [1990:4] combined data on worldwide condom production, numbers of married couples using condoms and coital frequency to estimate that at least 40-50 percent, and perhaps as many as 80 percent, of all condoms used annually are used outside of marriage. While surveys in Canada, the United States and the Federal Republic of Germany in the early 1980s showed the condom accounting for lower proportions of contraceptive use among unmarried than among married couples, a 1988 survey in the United States found the relationship had switched. By then, sexually active unmarried women aged 15 to 44 were more than twice as likely as currently married women to be relying on condoms [Kost and Forrest, 1992; Kost, 1992].

Table 1. Percentage of currently married women of reproductive age who said they knew about any modern method, any vaginal method and the condom and the percentage who ever used, and were currently using a method, a vaginal method or condom, 1985–1989.

	Knowledge			Ever use			Current use				% of couples with ready and easy access to condoms
	Modern methods	Vaginal methods	Condom	Any method	Vaginal method	Condom	Any method	Vaginal method	Condom	% of users using barrier methods	
SUB–SAHARAN AFRICA											
Botswana	94.4	51.3	87.1	63.0	1.3	10.1	33.0	0.0	1.3	3.9	100
Burundi	63.8	6.9	14.9	30.1	0.1	0.3	8.7	0.0	0.1	1.1	80
Ghana	76.5	38.0	49.5	37.0	8.3	4.5	12.9	0.3	0.3	4.7	50
Kenya	91.3	26.7	55.7	45.0	2.1	4.3	26.9	0.4	0.5	3.3	80
Liberia	68.0	9.8	26.2	18.8	1.0	1.7	6.4	0.2	0.0	3.1	0
Mali	28.6	6.0	8.2	19.0	0.4	0.4	4.7	0.1	0.0	2.1	60
Senegal	67.8	8.6	26.5	37.9	0.8	1.4	11.3	0.1	0.1	1.8	30
Togo	81.4	25.9	35.9	73.5	3.7	3.3	33.9	0.6	0.4	2.9	80
Uganda	77.9	11.6	31.1	21.5	0.2	0.7	4.9	0.0	0.0	0.0	80
Zimbabwe	97.8	14.5	80.3	79.0	0.3	17.0	43.1	0.0	1.2	2.8	80
NORTH AFRICA & NEAR EAST											
Egypt	98.2	41.0	44.6	59.5	5.6	9.1	37.8	0.4	2.4	7.4	90
Morocco	97.5	23.7	59.8	58.9	2.3	6.5	35.8	0.1	0.5	1.7	70
Tunisia	99.0	61.3	66.3	68.2	11.7	9.2	49.8	1.0	1.3	4.6	90
Turkey	na	na	na	na	na	na	63	na	7	11.1	90
ASIA											
Bangladesh	na	na	na	na	na	na	25.2	0.2	1.8	7.9	90
China	na	na	na	na	na	na	77	1	2	3.9	100
Hong Kong	na	na	na	na	na	na	81	na	26	32.1	na
Indonesia	94.2	4.0	65.2	65.0	0.2	6.3	47.7	0.0	1.6	3.4	100
Korea, Rep. of	na	na	na	na	na	na	70.4	na	7.2	10.2	100
Pakistan	na	na	na	na	na	na	7.6	0.1	1.7	23.7	50
Philippines	na	na	na	na	na	na	45.3	0	1.2	2.6	80
Sri Lanka	99.1	14.6	73.3	73.9	0.2	9.9	61.7	0.0	1.9	3.1	100
Taiwan	na	na	na	na	na	na	78	na	14	17.9	100
Thailand	99.5	16.8	88.5	83.6	0.5	12.8	65.5	0.0	1.1	1.7	100

LATIN AMERICA/CARIBBEAN											
Bolivia	67.5	23.0	28.5	45.8	2.3	3.4	30.3	0.1	0.3	1	70
Brazil	99.9	36.1	86.5	86.3	6.0	23.4	66.2	0.5	1.7	3	100
Colombia	99.4	85.0	71.0	83.0	19.1	10.4	64.8	2.3	1.7	6	100
Costa Rica	na	na	na	na	na	na	69.5	0.7	13.4	20	100
Dom. Rep.	99.3	68.2	87.8	73.0	6.8	17.1	49.8	0.2	1.4	3	100
Ecuador	90.0	56.3	51.0	62.9	11.4	9.1	44.3	1.2	0.6	4	100
El Salvador	na	na	na	na	na	na	47.3	0.2	1.2	3	100
Grenada	na	na	na	na	na	na	31	3.8	8	38	na
Guatemala	71.6	21.5	36.5	34.0	0.9	6.4	23.2	0.4	1.2	7	80
Honduras	na	na	na	na	na	na	41	na	2	5	100
Jamaica	na	na	na	na	na	na	55	na	9	16	100
Mexico	93.2	64.1	66.6	70.2	9.2	11.2	52.7	0.6	1.9	5	100
Paraguay	na	na	na	na	na	na	44.8	0.5	2.3	6	70
Peru	86.6	46.3	51.4	65.1	8.0	9.7	45.8	1.0	0.7	4	100
Trin. & Tob.	98.9	80.5	96.0	83.1	22.6	49.1	52.7	5.0	11.8	32	100
DEVELOPED COUNTRIES											
France	na	na	na	na	na	na	72	na	4	6	90
Hungary	na	na	na	na	na	na	73.1	0.9	3.5	6	100
Japan	na	na	na	na	na	na	64	na	45	70	100
Netherlands	na	na	na	na	na	na	72	na	6	8	100
New Zealand	na	na	na	na	na	na	74	na	5	7	100
Spain	na	na	na	na	na	na	59.4	na	12.2	21	100
United States	na	na	na	na	na	na	74	7	11	24	90
Fed. Rep. of Germany	na	na	na	na	na	na	77.9	1.2	5.7	9	100
Minimum	28.6	4.0	8.2	18.8	0.1	0.3	0.0	0.0	0.0	0	0.0
Maximum	99.9	85.0	96.0	86.3	22.6	66.2	81.0	7.0	45.0	70	100.0
Median	93.2	25.9	55.7	63.0	2.3	37.8	47.7	0.4	1.7	5	100.0

Note: For DHS survey countries, the minimum, maximum and median values for the columns in current use are: Any method – 4.7, 66.2, 37.8; vaginal method – 0.0, 5.0, 0.2; condom – 0.0, 11.8, 1.1; % of users using barrier methods – 0.0, 31.9, 3.3.

Source: Countries with full information, in which DHS surveys were carried out: Rutenberg et al., 1991.
Countries with only current use of any method and of condom: Liskin, et al., 1990; United Nations, 1989; Mosher, 1990.
Access score: Population Crisis Committee, 1992.

USE AMONG MARRIED COUPLES

There are wide differences within areas of the world in the levels of barrier contraceptive use (Table 2). Levels of condom use among married couples of reproductive age are highest in Japan, where the condom accounts for about 70 percent of all contraceptive use, and in Denmark, Finland and Sweden, where almost a quarter of married couples use condoms [Liskin et al., 1990]. In contrast, only 5 percent of couples in Australia and New Zealand rely on condoms.

Few married couples in developing countries have ever used barrier contraceptives (Table 1). While the median level of ever use of any method in the 25 DHS countries was 63.0 percent, it was only 9.1 percent for the condom and 2.3 percent for vaginal methods. However, at least 10 percent of married women in eight countries had used condoms and in four countries, at least 10 percent had ever used vaginal methods.

Highest documented levels of current condom use are in Japan and Hong Kong, where 45 percent and 26 percent, respectively, of married women of reproductive age rely on condoms. Use is much less common in other countries, however. Of the countries

TABLE 2. Estimated use of condoms for contraception by married couples of reproductive age, 1990

Region	Number using condoms (millions)	Percentage distribution of condom users	Condom users as percentage of all married couples of reproductive age
World Total	45	100	5
DEVELOPED AREAS			
Total	21	47	12
Australia & New Zealand	0.2	0.4	5
Europe			
East (includes USSR)	4	9	6
Scandinavia	0.7	2	23
South	2	4	10
West	3	7	9
Japan	8	18	45
North America	4	9	10
DEVELOPING AREAS			
Total	24	53	4
Asia & Pacific			
China	4	9	2
Indian subcontinent	16	36	7
Other Asia & Pacific	2	4	3
Latin America & Caribbean	1	2	2
Near East & North Africa	1	2	3
Sub-Saharan Africa	0.3	1	0.3

Source: Liskin et al., 1990:4.

for which data are available from the late 1980s, 10 to 14 percent of couples were using condoms in only Costa Rica, Spain, Taiwan, Trinidad and Tobago and the United States. In all the DHS countries but Trinidad and Tobago, no more than 2.4 percent of married couples were using condoms for contraception. In the developing countries, the level of condom use among married couples is moderately related to the estimated proportion of couples with ready and easy access to condoms (r=.45).

Current use of vaginal methods is less common than condom use. The highest levels reported in available data from the late 1980s are in the United States and in Trinidad and Tobago, 7.0 percent and 5.0 percent of currently married women, respectively.

In the DHS countries, the proportion of married women using condoms or vaginal methods is weakly related to the overall level of use (r=0.40 for condom and any method, and r=0.31 for vaginal and any method). In general, however, in DHS countries where condom use is higher, use of vaginal methods is also likely to be relatively high (r=0.82).

The differences across sociodemographic subgroups in barrier contraceptive use levels are similar to the overall patterns of reversible contraceptive use, but since barrier use levels are quite low in most of the DHS countries, the differences across subgroups are small (Table 3). The proportion of married women using condoms or vaginal methods tends to be lowest among women aged 15 to 24 and those who have not yet had a child and highest among those 25 to 34 and those who have had one to four children. In general, barrier contraceptive use is higher in urban than in rural areas and increases with higher levels of women's education.

USE AMONG UNMARRIED WOMEN AND MEN

Very little is known about use of barrier methods by unmarried women and men, especially those in developing countries. Liskin et al. [1990] reported that one-third to two-thirds of unmarried men in their adolescence and early 20s in Latin America, the Caribbean and Africa had had sex during the prior month. The proportion of those who had used a condom ranged from 9 to 11 percent in Mexico City and Santiago, Chile; from 20 to 28 percent in Monrovia, Liberia, Quito and Guayaquil, Equador, and Sao Paulo, Brazil; from 47 to 48 percent in Ibadan, Nigeria and Jamaica; and to 64 percent or higher in Banjul, The Gambia and Harare, Zimbabwe.

Surveys of unmarried adolescents in the late 1970s and early 1980s found 5 to 6 percent of those who had had intercourse were using condoms in Canada and Hungary; 21 to 22 percent were using them in Denmark, the United Kingdom and in the United States; and, that 43 percent of sexually active unmarried adolescents in Spain were using condoms [United Nations, 1988]. In the United States in 1988, 51 percent of sexually active unmarried women ages 15 to 17 used condoms, as did 40 percent of those ages 18 to 19 [Kost, 1992]. The proportion of all unmarried women ages 15 to 44 who were sexually active, fecund and not intending to become pregnant who used condoms increased from 7 percent in 1982 to 13 percent in 1988 [Forrest, 1993].

Table 3. Percentage of currently married women 15–49 reporting current use of barrier contraceptive methods, by age, parity, residence education, Demographic and Health Surveys, 1986–1989

	Age			Living children				Residence		Education		
Region	15–24	25–34	35–49	0	1–2	3–4	5+	Urban	Rural	None	Prim.	Sec.+
SUB–SAHARAN AFRICA												
Botswana	1.1	1.4	1.2	1.0	1.0	1.0	1.2	2.5	0.7	0.6	1.0	3.1
Burundi	0.2	0.0	0.1	0.0	0.0	0.2	0.2	0.3	0.1	0.1	0.3	0.4
Ghana	1.0	0.6	0.3	0.8	0.6	0.2	0.8	1.0	0.3	0.3	0.7	1.7
Kenya	0.9	0.9	0.9	0.0	1.0	0.7	1.1	1.3	0.8	0.8	0.7	2.7
Liberia	0.3	0.0	0.3	0.5	0.1	0.3	0.0	0.5	0.0	0.6	0.6	0.6
Mali	0.1	0.1	0.1	0.0	0.1	0.1	0.2	0.4	0.0	0.0	0.2	3.3
Senegal	0.0	0.4	0.0	0.0	0.1	0.2	0.1	0.3	0.0	0.3	0.3	0.7
Togo	1.0	1.0	1.0	0.0	1.1	1.5	0.8	2.3	0.5	1.3	1.2	4.5
Uganda	0.0	0.0	0.0	0.0	0.1	0.0	0.0	0.2	0.0	0.0	0.0	0.3
Zimbabwe	1.0	1.4	1.0	0.0	1.5	1.2	1.1	1.5	1.0	0.8	1.1	1.8
NORTH AFRICA												
Egypt	0.5	3.1	3.8	0.0	3.5	3.3	2.6	4.9	0.8	1.1	3.3	6.5
Morocco	0.1	0.9	0.7	0.0	0.6	1.0	0.7	1.2	0.3	0.5	0.7	2.8
Tunisia	2.2	2.0	2.7	0.0	2.6	2.5	2.3	3.0	1.2	1.5	3.7	2.6
ASIA												
Indonesia	0.8	1.9	1.7	0.2	1.3	2.4	1.6	4.2	0.6	0.4	1.0	5.3
Sri Lanka	1.4	2.1	2.0	0.3	2.5	2.2	0.7	3.0	1.7	0.2	1.0	2.7
Thailand	1.1	1.2	1.0	1.0	1.3	1.0	0.3	2.4	0.8	1.0	0.7	3.5
LATIN AMERICA/CARIBBEAN												
Bolivia	0.3	0.8	0.2	0.0	0.7	0.6	0.0	0.7	0.2	0.0	0.1	1.2
Brazil	1.1	2.9	1.9	1.8	3.0	1.6	1.0	2.3	1.6	1.7	1.9	3.4
Colombia	2.7	5.1	3.5	2.1	4.7	4.4	2.8	4.5	3.0	2.2	3.8	4.8
Dom. Rep.	1.2	2.2	1.2	0.2	2.6	1.8	0.7	1.8	1.2	0.9	1.1	3.2
Ecuador	1.3	2.0	2.0	0.9	2.4	1.9	1.2	2.2	1.4	0.7	1.4	2.8
Guatemala	1.4	2.0	0.9	0.5	2.5	1.6	0.6	2.8	0.9	0.1	1.9	6.9
Mexico	1.8	2.7	2.8	1.1	3.2	2.5	2.0	2.9	1.6	0.1	2.4	3.9
Peru	1.9	1.8	1.5	0.0	2.5	2.3	0.6	2.4	0.4	0.2	0.9	3.1
Trin. & Tob.	14.8	19.2	15.3	8.4	19.5	18.5	15.2	15.8	17.6	7.4	15.6	18.3
Minimum	0.0	0.0	0.0	0.0	0.0	0.0	0.0	0.2	0.0	0.0	0.0	0.3
Maximum	14.8	19.2	15.3	8.4	19.5	18.5	15.2	15.8	17.6	7.4	15.6	18.3
Median	1.0	1.4	1.0	0.2	1.3	1.5	0.8	2.3	0.8	0.6	1.0	3.1
Ave	1.5	2.2	1.8	0.8	2.3	2.1	1.5	2.6	1.5	0.9	1.8	3.6

Note: Data for Brazil and Guatemala are for women 15–44.
Source: Rutenberg, et al., 1991.

In a number of developed countries, condoms play an important role at first intercourse. Surveys of adolescent women in the late 1970s and early 1980s showed that condoms were the most commonly used method at first intercourse in the Federal Republic of Germany (used by 32 percent at first intercourse), Norway (48 percent), Sweden (52 percent) and the United Kingdom (36 percent) [United Nations, 1988]. In the Netherlands, 29 percent of women had used condoms at first intercourse, but even more (46 percent) had used the pill. The proportion of adolescent women in the United States reporting condoms had been used at first intercourse increased from 23 percent in 1982 to 47 percent in 1988 [Forrest and Singh, 1990].

CHARACTERISTICS OF BARRIER METHOD USE

Barrier methods, especially the condom, are promoted both for contraceptive protection and for prevention of sexually transmitted disease. Data from a recent national survey in the United States give some insight into the degree of use of condoms for disease prevention [Kost and Forrest, 1992]. (Ninety-five percent of those using any method for prevention of sexually transmitted disease used condoms [Forrest, 1992]). Among all those using condoms, 45 percent used them solely for pregnancy prevention, 33 percent for prevention of STDs, including HIV, and 22 percent used them for both purposes. Patterns differed sharply by marital status, however: 72 percent of married women, but only 15 percent of unmarried women using condoms did so solely for pregnancy prevention, compared with 11 percent of married and 31 percent of unmarried women using them for disease prevention and 17 percent of married and 54 percent of unmarried women using them for both purposes [Kost, 1992].

Forty-three percent of those using the condom were also using another contraceptive method, either in addition or in alternation with the condom [Kost and Forrest, 1992]. Only 81 percent of condom users had used the method at last intercourse, however, and those with more than one sexual partner in the last three months were significantly less likely to have done so than were other women. Across all age groups, those using the condom for contraceptive purposes, either solely or in addition to STD prevention, were much more likely to have used the condom at last intercourse (68 to 100 percent) than were those using condoms only to prevent STDs (7 to 44 percent) [Kost, 1992].

Inconsistent method use is not limited to the United States. Only about 60 percent of men using condoms in Bangladesh, and 30 percent of those in Barbados, reported using them at every act of intercourse [Liskin et al., 1990].

DISCUSSION

While there are wide differences between countries, levels of barrier method use are generally higher in developed than in developing countries. Some of this difference may reflect more widespread geographic and financial access to the methods in developed countries. However, while barrier methods, especially the condom, were introduced and used in many developed countries long in advance of organized family planning service programs, the opposite is the case in most developing countries where barrier methods were introduced through the family planning programs along with other methods such as pills, IUDs and sterilization.

Condoms are the leading method of contraception only in Japan, and use levels in almost all developing countries are quite low, with even fewer married women using vaginal methods. It is only in recent years, however, that condoms, and sometimes vaginal methods as well, have been promoted not only as contraceptives but also for prevention of STDs, particularly HIV. Disease prevention accounts for most of the condom use among unmarried couples in the United States, but those using condoms only for STD prevention are much less likely than contraceptors to use them consistently. The next round of DHS surveys will provide information about whether attention to both the disease and pregnancy prevention aspects of condom use is leading to increased levels of use in developing countries as well.

REFERENCES

Forrest JD (1993): Unpublished tabulation from the 1982 and 1988 National Survey of Family Growth.

Forrest JD (1992): Statement before the Obstetrics-Gynecology Devices Panel of the Food and Drug Administration, Washington, D.C., June.

Forrest JD, Singh S (1990): The sexual and reproductive behavior of American women, 1982-1988. Fam Plan Persp 22:206-214.

Kost K (1992): Condom use in the United States: Who uses them and how well? Paper presented at the annual meeting of the American Public Health Association, Washington, D.C., November.

Kost K, Forrest JD (1992): American women's sexual behavior and exposure to risk of sexually transmitted diseases. Fam Plan Persp 24:244-254.

Liskin L, Wharton C, Blackburn R (1990): Condoms—Now more than ever. Pop Rep Series H, No 8:1-35.

Mauldin WP, Ross JA (1992): Contraceptive use and commodity costs in developing countries, 1990-2000. Int Fam Plan Persp 18:4-9.

Mosher, WD (1990): Contraceptive practice in the United States, 1982-1988. Fam Plan Persp 22:198-205.

Population Crisis Committee (1992): "World Access to Birth Control." Washington, D.C.: Population Crisis Committee.

United Nations (1988): "Adolescent Reproductive Behavior." ST/ESA/SER.A/109. New York: United Nations.

United Nations (1989): "Levels and Trends of Contraceptive Use As Assessed in 1988." ST/ESA/SER.A/110. New York: United Nations.

DISCUSSION

Jeffrey Spieler: Perhaps one of the most startling pieces of information presented was the chart showing use of condom at last active intercourse based on motivation, i.e. those who used condoms for contraception and also those who used condoms for contraception and STD prevention. In the groups who said they were only using it for STD prevention and the youngest group had 40 percent use of condom as a last act properly, and the older groups got down to almost none at all. I had this perhaps naive feeling that if one were using a condom to prevent HIV you would be a much more reliable user because the outcome of not using could be a life sentence and most women don't see pregnancy, even when maternal mortality is very high, as a life sentence. So, it seems conter-intuitive that there should be such a low use at last act when the person declares they were using it for STDs.

Jacqueline Forrest: I find it fascinating, too, and I think there are a couple of things happening. First, let me just mention what you noted about the difference in age, that younger condom users who were using condoms for infection prevention were more likely to use them at last intercourse than the older users. This isn't controlled for any other factors in relationships or whatever, but there are a number of indications that adolescents as well as some of those in their early twenties, but especially adolescents, may be doing better than older women. The other part, though, about why it's so low compared to those who are using condoms for pregnancy prevention, is that I have a different way of looking at it. You said if they are thinking they could be killed by what they might get infected with, then they would have a high motivation to use. The point is that by telling themselves and a partner it's not that I'm using this because we may both want to prevent pregnancy, which is socially acceptable, but rather I'm using it because I think you're the kind of person who can give me an infection that is interpersonally and socially a much more difficult thing to do.

Jeffrey Spieler: The piece that's missing is how many people wanted to use it at the last act and there you might see a discrepancy in those who would have wanted to but were not able to negotiate it, perhaps.

Jacqueline Forrest: That's a good point. It would be interesting to study because you'd have to do it prospectively. After the fact, whether they had wanted to use it might be very much related to whether they got an infection or not.

Laneta Dorflinger: I just want a clarification on the projections for need by the year 2000. In the number of couples using barrier methods, was the projection based strictly on the number of couples, married women of reproductive age using condoms?

Jacqueline Forest: The projection comes from an article by Parker Maulden and John Ross and it is first of all estimating, looking only at married couples, how many need to be using contraception to meet the UN mid-level population projections for the year 2000. What it is saying is that the population of the world is going to grow by so much and therefore we are going to need some more contraception for that growth and then it includes an estimate that overall use of contraceptives would increase from 51 to 59 percent of married women of reproductive age, so that would be an additional growth.

It then simply allocates all those hopeful users in the year 2000 by the current patterns of contraceptive use, but it is just among married couples and it is coming from surveys that are only asking about condom use for contraceptive use not STD prevention only. So, it may be a very low estimate of how many couples are going to be needing to use condoms.

Alfred Shihata: How can you accurately estimate the number of people who use condoms properly and the people who improperly use them?

Jacqueline Forrest: That is a good question. I hope that will be a bit closer to the answer at the end of this meeting and hopefully also James Trussell is going to touch on that a little bit. Certainly the data that I presented just on the proportion who had used the method at last intercourse gives us great pause. We often say a user is a user and we know in terms of these methods that that is not the case so that's a key question.

Roberto Rivera: You said that even though contraceptive use prevalence and the total number of users of condoms is likely to increase, the proportion of couples using condoms is not going to increase which is not good considering the need for protection against STDs. Why is that? I don't think we should be content with that.

Jacqueline Forrest: No, that is a good point, but I was looking at the implications even if we essentially stayed with the current methods, and the current kind of acceptability of those methods, compared to others and delivery of those methods. We still are facing a large number of couples across the world who would be newly using in this case condoms. That, in and of itself, has important implications for service delivery.

Erica Gollub: This is just to confirm one of your conclusions about vaginal methods. It seems to me that this is very, very low usage of vaginal contraceptives other than condoms in both Europe and developing countries. Only one or two percent of women of reproductive age are using vaginal methods worldwide other than condoms. Is that correct?

Jacqueline Forrest: Yes, yes, I think it is about 1 percent or fewer world-wide married women using vaginal methods and about 5 percent who are using condoms. So it's less than a 1 to 5 ratio.

Erica Gollub: And that includes all kinds--spermicides and sponges with spermicide?

Jacqueline Forrest: Yes, diaphragms are included in that as well.

Nancy Williamson: I just wanted to comment that there is some evidence that married women under-report condom use. You had pointed out that a lot of the condoms world-wide are being used in Asia, India, Bangladesh, for example, and there are studies in both India and Bangladesh showing that married women under-report. If you do a husband and wife survey, and if either the husband or the wife report use, the couple is counted, you get considerably larger proportions of couples using condoms.

Jacqueline Forrest: That is a good point, although I would note that in what I was looking at, which was a secondary source review, the estimate of the under-reporting was generally at the level of 1 to 2 percent, although in some studies it has been higher.

Virginia Cain: You talked mainly about women of reproductive age and a little bit about unmarried women and adolescents. Clearly the risk for HIV and STDs does not end with reproductive years. Are you aware of any work that's going on looking at barrier methods in women who are post-reproductive age?

Jacqueline Forrest: No, that is a really good question. In the United States data, with which I am most familiar, among formerly married women we have extremely high levels of contraceptive failure rates compared to married women and even never married women for the same method. Formerly married women may be less aware of their susceptibility and they may be less able to negotiate some of the risks. They grew up, at least in the United States, in an era when all they had to worry about was pregnancy, many of them in the early days of oral contraceptives, so that they never had much experience with barrier method use, and it may be even more difficult for them than for adolescents to move into using some of these methods.

Judy Norsigian: I just wanted to underscore one of the problems we have with these data and that has to do with the fact that prevalence is so closely tied to availability and with something like the diaphragm, although technically it may be available in a region it often is not, so that of course your prevalence rates might even be lower than they would otherwise be and that comes up all the time as we look at these issues, not just with the diaphragm.

Jacqueline Forrest: I appreciate that. I tried to look at it in terms of correlations between condom or vaginal method use and the pill but certainly in my mind was that point that I think we have a developed country situation where these methods, condom, diaphragm and even other spermicides, were available before and outside of the family planning programs and the advent of oral contraceptives, and I think that gives you a very different context than developing countries where the methods have been introduced primarily in family planning programs where I would suggest they have not been a key focus in terms of either education or in terms of availability.

Kamran Moghissi: You did not comment on cultural prohibition or inhibition and whether or not that would influence the use of barrier contraceptives.

Jacqueline Forrest: It is a good point. However, there were not strong regional patterns looking at large geographic regions of the world in either knowledge, ever use, or current use of contraception so that at least from the data that I had there was no pattern that said Latin America and the Caribbean are here and North Africa is here at a different level as a group. There were countries with high and low levels of knowledge and use spread across the world.

2

CONTRACEPTIVE EFFICACY OF BARRIER CONTRACEPTIVES

James Trussell

Office of Population Research
Princeton University
21 Prospect Avenue
Princeton, NJ 08544

INTRODUCTION

Four pieces of information about contraceptive effectiveness would help couples to make an informed decision when choosing a contraceptive method.

- *Failure rates during typical use* show how effective the different methods are during actual use (including inconsistent *or* incorrect use).
- *Failure rates during perfect use* show how effective methods can be, where perfect use is defined as following the directions for use — correctly *using* the method at every act of intercourse.
- *Failure rates during imperfect use* show how ineffective methods will be if they are used incorrectly or inconsistently.
- *The percentage of perfect users* or percentage of months during which a method is used perfectly reveals how hard it is to use a method correctly and consistently.

The difference between failure rates during imperfect use and failure rates during perfect use reveals how forgiving of imperfect use a method is. The difference between failure rates during typical use and failure rates during perfect use reveals the conse-

Barrier Contraceptives: Current Status and Future Prospects, pages 17–51

quences of imperfect use. This difference depends both on how unforgiving of imperfect use a method is and on how hard it is to use that method perfectly.

The barrier contraceptive methods reviewed in this paper are the male condom, spermicides, the sponge, the cervical cap, the diaphragm, the female condom and Vaginal Contraceptive Film (VCF). All studies reviewed here were conducted in the United States unless otherwise noted in the text or in Tables 1-6. As will be demonstrated in the following sections, the information on the contraceptive efficacy of Vaginal Contraceptive Film is of questionable reliability. Information on contraceptive efficacy during perfect use is available only for the sponge, the cervical cap, the diaphragm, and the female condom. There is very good information from well-designed, well-executed, and well-analyzed clinical trials of contraceptive efficacy during typical use of the sponge, the cervical cap, the diaphragm, and the female condom. Information on efficacy during typical use of the male condom, spermicides, the sponge, and the diaphragm is available from survey data, particularly the National Surveys of Family Growth (NSFG).

Data from nationally representative surveys have the advantage of pertaining (in theory) to the actual population of users of each method, but they suffer from the disadvantage of reporting errors, especially since efficacy information is obtained from retrospective reports of contraceptive use and contraceptive failures (in the NSFG, the reporting period extends at least three years prior to the date of interview). Failure rates based on survey data are thought to be too low because induced abortions are not fully reported in the surveys [Jones and Forrest 1992b]. Jones and Forrest [1992a] add estimated contraceptive failures that were not reported in the 1988 NSFG to compute corrected contraceptive failure rates for the male condom, the diaphragm and spermicides (as well as the pill and periodic abstinence). However, there are reasons to doubt whether these corrected failure rates more accurately reflect reality. It is likely that women in abortion clinics (surveys of whom provided the information for the correction) overreport use of a contraceptive at the time they became pregnant, since any perceived responsibility for the pregnancy would be shifted from the woman to the method. Moreover, it seems likely that women in personal interviews for the NSFG also tend to overreport use of a contraceptive at the time of a conception leading to a live birth. Evidence for this suspicion is provided by a 6 percent first-year probability of failure for the IUD — a method with little scope for user error — among married women in the 1976 and 1982 NSFGs. This probability is much higher than probabilities obtained from clinical trials of IUDs [Trussell and Kost 1987]. Thus, failure rates based on the NSFG would tend to be too low because induced abortions (and contraceptive failures leading to induced abortions) are underreported but would tend to be too high because contraceptive failures leading to live births are overreported. These two sources of bias would tend to cancel, whereas adjustment for underreporting of induced abortion would make the failure rates too high.

Information from well-supervised clinical trials is also subject to potential bias because it is plausible that subjects may alter their behavior while they know that they are being observed, so that efficacy results during typical use may not be generalizable beyond the trial conditions. Specifically, subjects may be more likely to use a contraceptive method perfectly either to please the investigator or to avoid embarrassment or even possible censure when the investigator discovers imperfect use. However, given the high levels of imperfect use reported in well-supervised trials [Richwald et al. 1989; Trussell and Grummer-Strawn 1990; Trussell, Strickler and Vaughan 1993; Trussell et al. 1993], it is not at all evident that the Hawthorne effect is important.

MEASURING CONTRACEPTIVE EFFICACY DURING PERFECT USE

Many investigators have recognized the importance of distinguishing between contraceptive failure associated with user error and failure attributable to the inherent inadequacy of the method itself. By convention, all failures that occur during a month (or menstrual cycle) in which a barrier method is either not used at every act of intercourse or used improperly are labeled user failures. Only those pregnancies that occur during a month (cycle) during which a method is used perfectly are classified as method failures; *perfect use* does not imply that the method did not fail, only that the method was used consistently and correctly according to a well-specified set of rules. There is no problem with this convention (indeed, it is necessary), but it must be understood that not all pregnancies classified as user failures will in fact be user failures. If, for example, in one month (or cycle) a method is not used or is used incorrectly during two acts of intercourse but used correctly during eight other acts, then a resulting pregnancy cannot be unambiguously causally linked to user error instead of method failure.

The standard procedure is next to compute separate *method* and *user* failure rates (pregnancies divided by exposure). In this standard procedure, all exposure from both perfect and imperfect use is included in the denominator of *both* method *and* user failure rates. Under this convention, method and user failures are treated as two of many possible *causes* of method discontinuation. The sum of all cause-specific discontinuation rates is the overall or total discontinuation rate, and the sum of method and user failure rates is the overall or total failure rate. The common misinterpretation is that the resulting method rate yields information about the inherent efficacy of the method. That this interpretation is incorrect is obvious when one realizes that inherent method efficacy can be measured only when the numerator (method failures) is assessed in relation to the proper risk set (exposure only when the method is used perfectly). Method failure rates computed by the standard procedure confound inherent method efficacy with the proportion of exposure that is characterized by perfect use.

Consider the following example that illustrates just how misleading the conventional interpretation can become. We decide to evaluate an innovative contraceptive regime that is claimed not only to prevent pregnancy but also to enhance health. It requires a woman to stand on her head and drink orange juice during each act of intercourse. We conduct a clinical trial, which yields the amazing result that the method failure rate is 0 percent, even though common sense would dictate that the method is absolutely worthless (except insofar as perfect use would drastically reduce coital frequency). Unbeknown to us, however, this 0 percent failure rate is due entirely to the fact that, quite sensibly, no woman ever achieved perfect use in any month. By convention, there can be no pregnancies classified as method failures if all exposure is characterized by imperfect use. Such reasoning leads to the conclusion that it makes no sense to treat method and user failures as *causes* of discontinuation. To be treated as a cause, an event must be a possible outcome for those in the risk set.[1] In an IUD trial, removal for pain/bleeding, expulsion and accidental pregnancy are proper competing risks; in contrast, user *versus* method failures make no more sense as competing risks than do white *versus* non-white pregnancies or teenage *versus* adult pregnancies, because each of these alternative events is conditional on a separate risk set.

The analytical error is the failure to realize that, by convention, method failures can occur only when the method is being used correctly. Logically, as discussed above, method failures can occur during both imperfect and perfect use. During perfect use, however, the only failures that can occur are method failures, and method failure rates can be measured only when the method is used correctly and consistently. The *difference* between the failure rate during imperfect use and the failure rate during perfect use provides a measure of the contribution of imperfect use to the propensity to fail. Because the experimental design of most studies probes for risk-taking or imperfect use only when there is a pregnancy (with the World Health Organization study of the ovulation method of periodic abstinence constituting an important exception [Trussell and Grummer-Strawn 1990]), method failure rates could not be computed even if the investigators recognized the error in the standard procedure. As a consequence of this computational error, failure rates that are reported as method failure rates in the literature are too low, with the extent of the downward bias unknowable. User failure rates reported in the literature are also downward biased for the same reason. Whether conventionally computed method failure rates are more or less severely biased than user failure rates depends on the proportion of total exposure characterized by perfect use.

Suppose that information on perfect and imperfect use is not available for each month (or cycle) of use. Instead, at follow-up visits scheduled at three-month intervals, women report whether they used their method correctly at every act of intercourse. Therefore, information on consistency and correctness of use is available only for three-month intervals, not for each menstrual cycle. The analyst will not be able to provide estimates of failure rates during always-imperfect use, since three-month segments that are not perfect-use-only segments will contain a mix of months of perfect use and months of imperfect use. However, estimates of failure rates during perfect use can be computed by using only those intervals of exposure during which the method was used perfectly.

Unfortunately, having only three-month intervals of perfect-use exposure creates a potentially insoluble problem for analysis. If month-by-month (preferably cycle-by-cycle) information were available, then there would be both more exposure classified as perfect

---

[1] For example, death rates by cause do add to the overall death rate because all persons are assumed to be at risk of dying from any cause. Clearly, one might argue that all persons are not exposed to every cause of death, even when analysis is age- and gender-specific. For example, a woman is not at risk of maternal mortality if she is sterile (or not having intercourse). However, such examples illustrate problems with the cause-of-death methodology rather than providing support for the standard procedure of computing method failure rates.

use (because some three-month segments of imperfect use contain cycles of perfect use) and more failures during perfect use. It might appear that failure rates during perfect use would still be unbiased since only intervals of exposure during which the method is used consistently are employed in the analysis.

Under certain assumptions, this reasoning is incorrect. If the likelihood of imperfect use each month is the same for all women and if women who become pregnant stop using contraception in that month, then failure rates during perfect use when three-month segments are employed will be biased upward compared with failure rates during perfect use based on one-month segments. By extension, the probability of failure during a year among women who always used a method perfectly will exceed the probability of failure during a year of perfect use computed from all perfect-use months, including those contributed by women who sometimes used the method imperfectly [Dominik, Trussell and Walsh 1993]. To understand why, consider the perfect-use failure rate in the first month. The numerator is the number of women who become pregnant during the first month of perfect use and the denominator is the number of women who use the method perfectly in the first month. When the analysis is confined to always-perfect users, then the first-month perfect-use failure rate is too high: the numerator is the same but the denominator is too small because those who used the method perfectly in the first month but who were imperfect users later are removed. Similar reasoning confirms that the bias is highest in the first month and declines thereafter; there is no bias in the last month of the analysis interval (for example, the third month of each three-month interval if three-month segments are employed).

In fact, many women do continue to use contraception after becoming pregnant, and some women are more likely to be imperfect users than others. What happens if these two assumptions of the model are violated? If women who become pregnant continue to use contraception because they do not know that they are pregnant, then the bias will be reduced. To understand why, consider again the perfect-use failure rate in the first month. The numerator will be smaller than the number of women who become pregnant during the first month of perfect use because some of these will use the method imperfectly thereafter. When three-month segments are employed, if pregnant women continue to use contraception for two months following the month they become pregnant, then the problem vanishes. If the propensity for imperfect use varies among women, then the bias is also reduced;[2] in the extreme case in which there are two distinct groups of always-perfect users and always-imperfect users, the problem disappears.

The empirical significance of this potential bias is unclear. In an analysis of contraceptive failure of the ovulation method (OM) of periodic abstinence, the probability of failure during a year of perfect use (3.2 percent) was higher (not lower) than the 12-month probability of failure based only on women who always used OM perfectly (3.1 percent) [Trussell and Grummer-Strawn 1990]. Nevertheless, estimates of the probability of failure during perfect use of the cap [Richwald et al. 1989], the Reality female condom

---

[2] For example, consider two scenarios. In the first, the probability of imperfect use each month is 25% for all women. In the second, half the women have a 50% probability and half have a 0% probability of imperfect use each month. The bias is smaller in the second scenario.

[Trussell et al. 1993], and the diaphragm, sponge and cervical cap [Trussell, Strickler and Vaughan 1993] may be biased upward because information on perfect use during each cycle was not available.

MALE CONDOM

Of the 13 studies in Table 1, eight [Peel 1972; Jones and Forrest 1992a; Bracher and Santow 1992; Schirm et al. 1982; Grady et al. 1983; Vaughan et al. 1977; Grady et al. 1986; Westoff et al. 1961] are based on retrospective reports of women in sample surveys. Results from these population-based surveys are simply not comparable with results from a carefully designed and executed prospective clinical trial because the research designs are so different.

Two of the five remaining studies are not based on clinical trials. John [1973] interviewed three groups in Shepshead, England — (1) a random sample of married women in the town's only clinical practice, (2) all women who attended the family planning clinic in the town during a one-year period, and (3) all women in the town whose pregnancy ended in a birth or an induced abortion during that year — to obtain their retrospective reports of contraceptive experience. Potts and McDevitt [1975] assessed the efficacy of spermicidally-lubricated condoms by recruiting a sample of condom users from the regular mail-order customers at the Marie Stokes Memorial Centre. Prospective male participants were asked to use the new condom exclusively for a period of two years and to complete a questionnaire after two years of use. The results are very difficult to generalize to the larger population of condom users, for several reasons. The sample of males was rather old — 77 percent were aged 40 or over, 51 percent were aged 45 or over, and 24 percent were aged 50 or older — and so might be expected to have low coital frequency, given the decline in coital frequency with age [Trussell and Westoff 1980]. Many of their female partners — being about two years younger — would be well past the age of peak fecundity. Finally, only pregnancies leading to a live birth were counted as failures [Potts 1993].

The remaining three studies are all based on the same prospective clinical trial (the Oxford/FPA study). At recruitment, all women were required to be married and aged 25-39 and to have been using the diaphragm, pill or IUD successfully for at least five months. The last criterion ensures that the sample is self-selected for better than average contraceptive use or lower than average fecundity or coital frequency. Combined with the age criterion, the sample of eventual condom users also would be expected to have lower than average coital frequency.

SPERMICIDES

As is evident from Table 2, the proportion of women becoming pregnant within the first year of typical use of spermicides varies widely. There are six clinical trials with first-year probabilities of failure or Pearl indexes below 2 percent and three additional trials with results between 2 percent and 5 percent. Yet there are 12 studies, eight of these being clinical trials, with results greater than 15 percent. The problem is that there is no clinical trial meeting modern standards of design, execution and analysis.

While our intuition dictates that female barrier methods (with spermicides) and spermicides used alone should have similar inherent efficacy, we may be understating the efficacy of spermicides by ignoring the fact that the spermicidal products used alone differ from those used in conjunction with barriers. The modern efficacy literature on spermicides is dominated by studies of foams and suppositories, and high spermicide efficacy is documented almost exclusively in these studies. There are few studies of creams and gels used alone, and those with the lowest failure rates are more than 30 years old. It is plausible that the biophysical properties of foams and suppositories result in better dispersion both in the vagina and in cervical secretions. Evidence for this hypothesis can be found in an early randomized trial with a crossover design (so that all subjects used each product); however, even in this study the failure rates for foam are high [Mears 1962]. We consider it more likely that many of the spermicide studies suffer from flaws in analysis or design that are not apparent in the brief published descriptions.

It is interesting to note that the first clinical trial of Emko vaginal foam is also one of the few studies to provide separate failure rates for cycles in which the product was used at every act of intercourse and cycles in which unprotected intercourse occurred [Mears 1962]. The design of that trial was also quite sophisticated. Women were randomly assigned to six groups. Each group used three different spermicidal products for three cycles each. The six groups represented all possible permutations of orders of use of the three products. If the failure rate for three cycles of consistent use of Emko vaginal foam is extrapolated, then the implied proportion failing in the first year of consistent use is 8.9 percent.

SPONGE, CERVICAL CAP, AND DIAPHRAGM

As can be seen in Tables 3, 4 and 5, the proportion of women who become pregnant accidentally during the first year of use of the sponge, cervical cap and diaphragm exceeds 10 percent in most studies reported in the literature. The extreme outlier is a study of the diaphragm reported by Lane et al. [1976]. Among 2,168 women (only 1.1 percent of whom were lost to follow-up in the first year), the proportion becoming pregnant in the first year of use was 2.1 percent. Since 29 percent of women in that study were under 20 years of age and 71 percent were not married, one might suspect that the pregnancy rate was so low because women had intercourse infrequently. However, the first-year probability of failure among the 139 married women aged 18-34 with at least one child was only 2.7 percent. The authors' explanation for the unusual success stresses (1) careful, thorough and objective explanation of all methods so that self-selection of the diaphragm was made with full knowledge; (2) exceptionally careful instruction that bolstered the patients' self-confidence; and, (3) a mechanism for continuing supervision.

Diaphragm advocates frequently mention the "Mary Lane effect" as evidence that thorough and supportive counseling by clinicians can instill motivation and compliance among clients. The 2 percent probability of failure reported in this study is cited by many family planning practitioners and women's health activists as evidence that the inherent efficacy of the diaphragm must be high. Nevertheless, conversations with some of the investigators in subsequent clinical trials in which women were randomly assigned to the sponge, cervical cap or diaphragm reveal that they too devoted exceptional attention to patient training and motivation. Yet the fractions of diaphragm users failing within the first year of use in these trials are much higher (16.7 percent [Bernstein *et al.* 1986] and 12.9 percent [McIntyre and Higgins 1986]). Why is the "Mary Lane effect" not found in these other studies?

The most likely explanation is that the analysis by Lane et al. [1976] was flawed. The cutoff date for the analysis coincided with the closing of the Margaret Sanger clinic in August 1973. Recruitment of women into the study began in January 1971 and ended in December 1972. The protocols called for a follow-up visit to the clinic after one week and annually thereafter. The analysts followed the recommendations of Tietze and Lewit [1973, 1974], whereby all women who are two or fewer months overdue for an appointment at the end of a study are assumed to be continuing users until that date. Therefore, all women who kept the one-week appointment, but who were not seen at their next appointment scheduled to occur any time after two months before the close of the study, were credited with up to 14 months of successful diaphragm use. It is incorrect, however, to assume periods of successful use of any method unless the user actually informs the investigator that she is still using the method and has not become pregnant [Trussell and Kost 1987]. The use of the (incorrect) Tietze-Lewit convention in this study undoubtedly led to a sizeable underestimation of the failure rate.

Data on the relative efficacy of the female barrier methods during *typical* use (both perfect and imperfect use) are exceptionally good because they come from two clinical trials conducted in the United States in which participating women were randomly assigned to use either the sponge or the diaphragm or to use either the cervical cap or the diaphragm.

Sponge-Diaphragm Trial. Data on 1,439 women from 13 clinics yielded a first-year probability of failure of 17.0 percent for the sponge and 12.5 percent for the diaphragm [Edelman et al. 1984].[3] Reanalysis of these data revealed, however, that the first-year probability of failure for the sponge among women who had borne children (28.3 percent) was much higher than among women who had never borne a child (13.9 percent). No difference in probabilities of failure by parity was found for diaphragm users. These results show that the sponge is as effective as the diaphragm among nulliparous women but much less effective than the diaphragm among parous women, perhaps because the sponge comes in only one size (too small for parous women).

Cap-Diaphragm Trial. Data on 1,394 women in Los Angeles, San Diego and San Francisco yielded a first-year probability of failure of 17.4 percent for the cervical cap and 16.7 percent for the diaphragm. Probabilities of failure among parous women during

---

[3] In a reanalysis of these data [McIntyre and Higgins 1986], the first-year probabilities of failure for the sponge and diaphragm were revised to 17.4% and 12.9%, respectively.

the first year were 29.9 percent for the cervical cap and 29.4 percent for the diaphragm. The first-year probabilities of failure among nulliparous women were 14.9 percent for the cervical cap and 14.5 percent for the diaphragm [Bernstein et al. 1986]. Hence the cervical cap, like the sponge, appears to have lower efficacy among women who have borne children. However, unlike the results from the sponge-diaphragm trial, results from the cap-diaphragm trial show higher failure rates among parous than among nulliparous diaphragm users.

Trussell, Strickler and Vaughan [1993] reanalyzed the data from these two trials in order to discover the efficacy of the female barrier methods when used perfectly and the effect of parity on the risk of failure of these methods. They employed multivariate life-table analysis to identify the effects of parity during perfect use, net of the effects of other confounding factors such as frequency of intercourse and age of the woman. They found that the risk of failure was substantially reduced when the analysis is limited to intervals in which women reported consistent use of the method. The percentage of women experiencing failure in the first 12 months of use was 17.2 percent for all sponge use, but only 12.0 percent for consistent use. The corresponding percentages were 17.7 percent and 12.5 percent for the cervical cap, and 12.5 percent to 17.1 percent and 6.1 percent to 8.4 percent for the diaphragm. The risk of pregnancy during perfect (both consistent and correct) use of these methods would be lower. They computed lower-bound estimates of failure rates during perfect use by removing those pregnancies (and corresponding exposure) that were classified by clinicians as user errors during periods of consistent use. Lower-bound estimates of the fractions failing during the first year of perfect use were 11.4 percent for the sponge, 10.3 percent for the cervical cap and 4.3 percent to 5.3 percent for the diaphragm. The risk of failure during consistent use was significantly higher for women who had carried a full-term pregnancy than for those who had not among users of the sponge (20.5 percent *versus* 9.5 percent) and the cervical cap (27.0 percent *versus* 9.9 percent), but not among users of the diaphragm. Lower-bound estimates of the fractions failing during the first year of perfect use of the sponge were 19.4 percent for parous and 9.0 percent for nulliparous women. The corresponding estimates for the cervical cap were 25.7 percent for parous and 7.6 percent for nulliparous women. The conclusions are that efficacy during perfect use is higher for the diaphragm than for the sponge or the cervical cap, especially for women who have borne children, and that efficacy during perfect use of the sponge or the cervical cap is far lower among women who have borne a child than among women who have not.

Richwald et al. [1989] found that among women who reported at the conclusion of a clinical investigation that they had always used the cervical cap during every act of intercourse, had always used it with a spermicide and had never worn the cap for longer than 72 hours, 6.1 percent became pregnant within their first year of use. There are two reasons for suspecting that this estimate of the proportion failing during perfect use would be too low. First, women were instructed "to insert extra spermicide in the vagina or use condoms during the first two months of cap use" [Richwald et al. 1989, page 147]. Second, it is likely that women who do not become pregnant during a clinical trial will — when questioned at the exit interview — overreport perfect use. Comparison of data collected at the end of the trial with the data from interval visits in the cap-diaphragm trial shows that 24.2 percent of women who claimed at their exit interview that they were consistent users throughout the trial had reported at a previous interval visit that they were inconsistent users during that interval [Trussell, Strickler and Vaughan 1993].

FEMALE CONDOM

In order to test the contraceptive efficacy of the Reality female condom, Family Health International (FHI) and the Contraceptive Research and Development Program (CONRAD) conducted a clinical trial in nine centers, six in the United States and three in Latin America. Altogether, 353 women (240 in the United States and 113 in Latin America) entered the six-month trial, of whom 328 (221 in the United States and 107 in Latin America) contributed data on efficacy. The final audited data showed a six-month life-table probability of failure of 15.1 percent. Probabilities of failure were lower among women in the United States (12.4 percent) than among women in Latin America (22.2 percent), although the difference is not statistically significant [Trussell *et al.* 1993]. Because the Reality female condom is intended to be used for protection against both pregnancy and STD transmission, the most natural efficacy comparison would be with the male condom. Given the differences in research design, however, comparisons between the contraceptive efficacy of the Reality female condom and the contraceptive efficacy of male condoms cannot be made with any confidence.

How does the contraceptive efficacy of Reality compare with that of the sponge, cervical cap and diaphragm? As described above, data on the relative efficacy of the sponge, cervical cap and diaphragm during typical use are exceptionally good because they come from two clinical trials in which participating women were randomly assigned to use either the sponge or the diaphragm [FHI 1985] or to use either the cervical cap or the diaphragm [Bernstein et al. 1986]. Data from the Reality trial should be comparable to data from these two prospective trials, with three important caveats, all of which, *ceteris paribus*, would make failure rates for Reality higher. First, the sponge-diaphragm and cap-diaphragm trials counted only clinically confirmed pregnancies. In contrast, pregnancy tests were given to women in the Reality trial at their time of exit and two weeks later, and pregnancies were counted if they were chemically confirmed. Second, the Reality protocols called for exclusion of women whose usual coital frequency was fewer than eight times per month and discontinuation of women if their coital frequency fell below four times per month. Third, the sponge-diaphragm and cap-diaphragm trials seem to have been less well supervised — as evidenced by six-month LFU probabilities that were considerably higher: 19.1 percent for the sponge and 23.3 percent for the diaphragm in the sponge-diaphragm trial, 20.0 percent for the cervical cap and 24.5 percent for the diaphragm in the cap-diaphragm trial, *versus* 8.2 percent for all centers and 8.7 percent for domestic centers in the Reality trial [Trussell *et al.* 1993]. All analyses of these data employ the standard assumption that the subsequent experience of those lost to follow-up is the same as the experience of those observed. However, LFU rates as high as those observed in the sponge-diaphragm and cap-diaphragm trials inevitably raise concern about the quality of the results. Rare attempts to locate those LFU in order to validate the universal assumption that while using the study product they became pregnant at the same rate as those who remained under observation (and that none were pregnant at the time they were last observed) produce sobering results. For example, Tietze, Poliakoff, and Rock [1951] found that the Pearl index for calendar rhythm rose from 9.4 to 14.4 as a result of resolution of cases LFU.

In the sponge-diaphragm trial, data on 1,437 women from 13 clinics in the United States yielded six-month life-table probabilities of failure of 11.6 percent for the sponge and 7.7 percent for the diaphragm. In the cap-diaphragm trial, data on 1,394 women in Los Angeles, San Diego and San Francisco yielded six-month probabilities of failure of 10.2 percent for the cervical cap and 8.3 percent for the diaphragm. The overall probability of failure for Reality for the United States sample (12.4 percent) is not statistically different from probabilities of failure for the sponge, cervical cap or diaphragm during typical use [Trussell et al. 1993].

The probabilities of failure during six months of perfect use of Reality by women who met the compliance criteria stipulated in the study protocol were 2.6 percent in the United States (n = 135) and 4.3 percent in the entire sample (n = 201). Women who reported fewer than four acts of intercourse during the month prior to any follow-up visit, who did not use Reality at every act of intercourse, who ever reported not following the Reality instructions, or who used another method of contraception were censored at the beginning of the first interval where non-compliance was noted. In the sponge-diaphragm trial, the fractions of women becoming pregnant during six months of perfect use lie between 8.5 percent and 8.8 percent for the sponge and between 1.7 percent and 2.7 percent for the diaphragm. In the cap-diaphragm trial, the fractions of women becoming pregnant during six months of perfect use lie between 4.3 percent and 5.8 percent for the cervical cap and between 3.1 percent and 4.7 percent for the diaphragm. These comparisons reveal that the probability of failure during perfect use of Reality is very similar to that for the diaphragm, slightly lower than that for the cervical cap, and considerably lower than that for the sponge; these rates statistically significantly differ only when the sponge and Reality are compared. The fact that the differences in the protocols for the three trials would lead one to expect higher failure rates for Reality, if the contraceptive efficacy of the methods were equal, only reinforces the conclusion that the efficacy of Reality during perfect use is similar to the efficacy of the sponge, cervical cap and diaphragm during perfect use [Trussell et al. 1993].

VAGINAL CONTRACEPTIVE FILM (VCF)

Information on the contraceptive efficacy of VCF is summarized in the product brochure produced by the manufacturer [Apothecus Pharmaceutical Corporation 1992] and in Table 6. A multicenter trial was conducted by 18 investigators in nine countries. A total of 824 women contributed 6,695 months of exposure and 44 pregnancies. The resulting Pearl index is 7.9. The results of this entire multicenter trial have not been subjected to peer review or published. However, the investigators at one of the study sites terminated their trial after observing nine pregnancies in 194 months of exposure contributed by 54 women; 63 women were enrolled in the trial, six (10 percent) of whom were LFU and contributed no data and three of whom never used the product [Smith et al. 1974]. The observed 4.6 percent risk of pregnancy per month implies a 43 percent probability of pregnancy within one year of use. Another published study based on the data from the multicenter study sites in Sweden reported a Pearl index of 9.0; 237 women contributed 1,866 months of exposure and 14 pregnancies [Frankman, Raabe and Ingemansson 1975]. Two other studies not connected with the multicenter trial have been published. An Italian trial followed 37 women who contributed 924 months of exposure and three pregnancies over a period of 31 months. The resulting Pearl index is 3.9 [Brigato et al. 1982]. A Japanese trial followed 168 women who contributed 2,161 months of exposure and only one pregnancy (in month 2); 118 women (70 percent of the total) used the product for more than 12 months. All women had been pregnant at least once prior to entering the trial. The resulting Pearl index is 0.56 [Iizuka et al. 1980]. It is very difficult to evaluate the quality of these studies, given the paucity of information provided; for example, there are no data concerning loss to follow-up except in the report by Smith et al. [1974]. VCF does not seem to have been subjected to a clinical trial with rigorous current standards regarding design, execution, and analysis.

However, a reasonably well-controlled trial of VCF used in conjunction with a diaphragm was conducted in Scotland. A total of 269 women in 10 family planning clinics agreed to use the diaphragm and a spermicide as their sole method of contraception and were randomly assigned to use either VCF (n = 136) or spermicidal cream or jelly (n = 133). Of the 269 women, 156 (78 in each group) completed the 12-month trial. The Pearl index for the VCF group (10.3, ten pregnancies and 1,162 months of exposure) is higher than that for the control group (7.1, seven pregnancies and 1,188 months of exposure) [Loudon et al. 1991]. Although the Pearl indexes for the two groups are not statistically different, the power of the test is small given the small sample sizes. For example, if the Pearl indexes for the two groups were really 11.0 and 6.0, a two-sided test of size 0.05 would detect a significant difference only about a quarter of the time. Unfortunately, there is no mention of loss to follow-up in the published report; we can assume that at least 10 women were LFU since they never returned for a follow-up visit, but there must have been others as well.

CONCLUSION

In this brief review of studies of contraceptive efficacy of barrier methods, we have identified several glaring gaps in the literature. First, clinical trials that are well-designed, well-executed and well-analyzed exist only for the sponge, cervical cap, diaphragm and female condom [Bernstein et al. 1986; Richwald et al. 1989; FHI 1985; FHI and CONRAD 1993]. Spermicides, the male condom and Vaginal Contraceptive Film simply have never been subjected to the same level of scrutiny. Second, information on efficacy during perfect use exists only for those four methods (sponge, cervical cap, diaphragm and female condom) that have been evaluated in clinical trials with current standards of design and execution [Richwald et al. 1989; Trussell, Strickler and Vaughan 1993; Trussell et al. 1993]. An important caveat is that none of these trials was actually designed to determine contraceptive efficacy during perfect use, so that even these data are not ideal for computing perfect-use failure rates. Indeed, estimates of the probability of failure during perfect use of these methods may be biased upward because information on perfect use during each cycle was not available. Third, comparative efficacy results from trials in which methods have been randomly assigned exist only for the sponge and diaphragm [FHI 1985] and for the cervical cap and diaphragm [Bernstein et al. 1986]. Results of these two trials can be spliced together to infer the relative efficacy of the three methods, since the diaphragm was common to both. Inferring the relative efficacy of the sponge, cervical cap, diaphragm and female condom is made easier by the fact that a well-designed and well-executed clinical trial of the female condom does exist [Trussell et al. 1993]. However, assessing the efficacy of these methods relative to the efficacy of spermicides, the male condom and Vaginal Contraceptive Film is very much more difficult, if not impossible. In short, our knowledge base is thin, especially for spermicides and male condoms.

ACKNOWLEDGMENTS

Tables 1-6 will appear in Hatcher RA, Trussell J, Stewart F, Stewart GK, Kowal D, Guest F, Cates W, Policar M. Contraceptive Technology 1994-1996. New York NY: Irvington Publishers, 1994 (In press). They are printed here with the permission of Contraceptive Technology Communications, Inc.

Charlotte Ellertson, Monica Selinger and Judith Tilton provided helpful comments on an earlier draft of this paper.

TABLE 1. Summary of studies of contraceptive failure: condom[a]

		Failure Rate			
			Pearl Index		
Reference	*N For Analysis*	*Life-table 12 Month*	*Index*	*Total Exposure*	*Maximum Exposure*
Potts and McDevitt [1975]	397	2.1[b]			
Peel [1972]	96		3.9	3698 mo	60 mo
Glass *et al.* [1974]	2057	4.2			
Vessey *et al.* [1988]	?		4.4	10000[c] yr	24 mo
John [1973]	85		5.7[c]	261 yr	>7 yr
Vessey *et al.* [1982]	?		6.0	4317 yr	24 mo
Jones and Forrest [1992]	1728	7.2			
Bracher and Santow [1992]	262	8.1			
Schirm *et al.* [1982]	1223	9.6[s]			
Grady *et al.* [1983]	1223	9.7[c,s]			
Vaughan *et al.* [1977]	696	10.1[s]			
Grady *et al.* [1986]	526	13.8[s]			
Westoff *et al.* [1961]	~212		13.8	10062 mo	?

a Based on Table 6 in Trussell and Kost [1987] and Table 10:2 in Hatcher *et al.* [1990], and updated to include subsequent studies.

b 24-month probability of failure; 12-month probability not published.

c Calculated by James Trussell from data in the article.

g Most of these studies incorrectly report the loss to follow-up probabilities as the number of women lost at any one time during the study divided by the total number of women entering the study. Thus, these are the percentages presented in the table. However, the correct measure of LFU would be a gross life table probability. When available, 12-month probabilities are denoted by the letter "g."

r Nonresponse rate of the entire study.

s Standardized: Vaughan *et al.* [1977] (1973 NSFG) — intention (the average of probabilities

Characteristics of the Sample[w]	*LFU*[g]	*Comments*
77% males aged 40+; all married	4.8[c]	Britain; spermicidally lubricated
All married	2.9[t]	Britain; Hull survey
All white; at recruitment aged 25-39 and married	1.0[v]	Oxford/FPA study
All white; at recruitment aged 25-39 and married	?	Oxford/FPA study
	?	Britain
All white aged 25-34; all married at recruitment	0.3[t,v]	Oxford/FPA study
Aged 15-44[t]	21.[r]	1988 NSFG; failure rate = 15.8[s]
16% aged <20, 65% aged 20-29, 19% aged 30+; 48% parity 0	25.[r]	Australian Family Survey
Aged 15-44; all married[t]	18.2[r]	NSFG 1973, 1976
Aged 15-44; all married[t]	18.2[r]	NSFG 1973, 1976
Aged 15-44; all married[t]	19.0[r]	NSFG 1973
Aged 15-44; all married[t]	20.6[r]	NSFG 1982
All married	5.7[r]	FGMA study

for preventers and delayers); Grady *et al.* [1983] (1973 and 1976 NSFG) — intention (our calculation, the average of probabilities for preventers and delayers); Schirm *et al.* [1982] (1973 and 1976 NSFG) — intention, age, and income; Grady *et al.* [1986] (1982 NSFG) — intention, age, poverty status, and parity; Jones and Forrest [1992] (1988 NSFG) — duration, age, marital status and poverty status.

t Total for all methods in the study.

v The authors report that LFU for "relevant reasons (withdrawal of cooperation or loss of contact)" was 0.3% per year in the 1982 study and "less than 1%" per year for all reasons in the 1974 study. In the 1982 study, women had on average been followed for 9.5 years; if 0.3% are LFU per year, then 2.8% would be LFU in 9.5 years. LFU including death and emigration is about twice as high as LFU for "relevant reasons."

w Unless otherwise noted, characteristics refer to females.

Table 2. Summary of studies of contraceptive failure: spermicides[a]

			Failure Rate					
				Pearl Index				
Reference	*Method Brand*	*N for Analysis*	*Life-table 12 Month*	*Index*	*Total Exposure*	*Maximum Exposure*	*Sample Characteristics*	*LFU[g]*
Edelman [1980]	S'positive	200		0.0	2682 mo	?		?
Squire *et al.* [1979]	Semicid Suppository	326	0.3				69% aged 20-34; 55% married	0.0[c]
Brehm and Haase [1975]	Patentex Oval	10017		0.9[c]	63759 cy	?	18% aged <21, 20% aged >35; 46% parity 0	?
Florence [1977]	a-gen 53	103		1.2[c]	2255 cy	61 mo	17% aged <21, 51% aged 21-30, 20% aged 31-40, 12% aged 40+	?
Dimpfl, Salomon and Schicketanz [1984]	Patentex Oval	482	1.5				22% aged <21, 25% aged >31; 44% parity 0; 60% married	?
Bushnell [1965]	Emko Vaginal Foam	130		1.8	2737 mo	57 mo	Aged 17-51; 76% aged 20-35	?
Carpenter and Martin [1970]	Emko Prefil Vaginal Foam	1778		3.4[c]	17200 cy	18 cy	69% aged 21-35	14.2[c]

Wolf *et al.* [1957]	Preceptin Vaginal Gel	112	4.2[c]	1145 mo	29 mo	All aged 13-40; all married	8.9[c]
Bernstein [1971]	Emko Prefil Vaginal Foam	2932	4.3[c]	28332 cy	20 cy	70% aged 21-35	16.1[c]
Tyler [1965]	Delfen Vaginal Foam	672	6.0	9486 cy	>16 mo	Rates for full applicator doses & half doses combined.	?
Kleppinger [1965]	Delfen Vaginal Foam	138	7.5	1116 mo	19 mo	53% aged 21-30; 27% postpartum	0.0[c,g]
Dubrow and Kuder [1958]	Delfen Vaginal Cream	338	7.6	633 mo	12 mo	Mean age = 25; 39% black; 45% Puerto Rican[t]	59.5[c]
Dubrow and Kuder [1958]	Preceptin Vaginal Gel	835	8.1	3728 mo	23 mo	Mean age = 25; 39% black; 45% Puerto Rican[t]	45.1[c]
Wolf *et al.* [1957]	Delfen Vaginal Cream	875	8.9[c]	5232 mo	30 mo	All aged 13-40; all married	13.0[c]
Rovinsky [1964]	Delfen Vaginal Cream	251	9.1	2915 mo	67 mo	70% aged 20-34; 55% Puerto Rican	28.0[c]
Vessey *et al.* [1982]		?	11.9	303 yr	?	All white; at recruitment aged 25-39 and married	0.3[t,v]

Table 2. Summary of studies of contraceptive failure: spermicides[a] (continued)

			Failure Rate					
				Pearl Index				
Reference	*Method Brand*	*N for Analysis*	*Life-table 12 Month*	*Index*	*Total Exposure*	*Maximum Exposure*	*Sample Characteristics*	*LFU*[g]
Jones and Forrest [1992]		267	13.4				Aged 15-44[t]	21.[r]
Vaughan *et al.* [1977]		596	14.9[s]				Aged 15-44; all married[t]	19.0[r]
Grady *et al.* [1983]		1106	17.5[c,s]				Aged 15-44; all married[t]	18.2[r]
Schirm *et al.* [1982]		1106	17.9[s]				Aged 15-44; all married[t]	18.2[r]
Mears [1962]	Emko Vaginal Foam	425		18.0[c]	722 cy	3 cy	Pearl index of 9.3 among consistent and 48.4 among inconsistent users	>20.[c,t]
Kasabach [1962]	Koromex A Jelly	242		21.0[c]	2058 mo	24 mo	36% aged 25-35; all married; all parous	19.3[c]

Bracher and Santow [1992]		89	21.5				27% aged <20, 56% aged 20-29, 17% aged 30+; 49% parity 0; 87% married or cohabiting	25.[r]
Grady *et al.* [1986]		284	21.8[s]				Aged 15-44; all married[t]	20.6[r]
Frank [1962]	Koromex A Jelly	824		24.8[c]	5767 mo	12 mo	72% aged 21-35	17.0[c]
Tietze and Lewit [1967]	Emko Vaginal Foam	779	28.3				86% aged <30; all married; 75% nonwhite	6.9[g]
Tietze and Lewit [1967]	Cooper Creme and Creme Jel, Koromex A, Lactikol Creme and Jelly, Lanesta Gel	806	36.8				79% aged <30; all married; 75% nonwhite	3.4[g]
Mears [1962]	Volpar Foaming Tablets	425		48.2[c]	728 cy	3 cy	Pearl index of 44.4 among consistent and 64.1 among inconsistent users	>20.[c,t]
Mears and Please [1962]	Vaginal Creme	678		49.6[c]	707 cy	3 cy	Pearl index of 31.4 among consistent and 132.0 among inconsistent users	>41.[c,t]

Table 2. Summary of studies of contraceptive failure: spermicides[a] (continued)

Reference	Method Brand	N for Analysis	Failure Rate: Life-table 12 Month	Failure Rate: Pearl Index: Index	Pearl Index: Total Exposure	Pearl Index: Maximum Exposure	Sample Characteristics	LFU[g]
Mears and Please [1962]	Volpar Foaming Tablets	678		59.0[c]	705 cy	3 cy	Pearl index of 47.8 among consistent and 106.7 among inconsistent users	>41.[c,t]

a Based on Table 2 in Trussell and Kost [1987] and Table 11:2 in Hatcher *et al.* [1990], and updated to include subsequent studies.

c Calculated by James Trussell from data in the article.

g Most of these studies incorrectly report the loss to follow-up probabilities as the number of women lost at any one time during the study divided by the total number of women entering the study. Thus, these are the percentages presented in the table. However, the correct measure of LFU would be a gross life table probability. When available, 12-month probabilities are denoted by the letter "g."

r Nonresponse rate of the entire study.

s Standardized: Vaughan *et al.* [1977] (1973 NSFG) — intention (the average of probabilities for preventers and delayers); Grady *et al.* [1983] (1973 and 1976 NSFG) — intention (our calculation, the average of probabilities for preventers and delayers); Schirm *et al.* [1982] (1973 and 1976 NSFG) — intention, age, and income; Grady *et al.* [1986] (1982 NSFG) — intention, age, poverty status, and parity; Jones and Forrest [1992] (1988 NSFG) — duration, age, marital status and poverty status.

t Total for all methods in the study.

v The authors report that LFU for "relevant reasons (withdrawal of cooperation or loss of contact)" was 0.3% per year in the 1982 study and "less than 1%" per year for all reasons in the 1974 study. In the 1982 study, women had on average been followed for 9.5 years; if 0.3% are LFU per year, then 2.8% would be LFU in 9.5 years. LFU including death and emigration is about twice as high as LFU for "relevant reasons."

TABLE 3. Summary of studies of contraceptive failure: sponge[a]

Reference	N for Analysis	Life-table 12-Month Failure Rate	Characteristics of the Sample	LFU[g]	Comments
Jones and Forrest [1992]	227	14.5	Aged 15-44[t]	21.[r]	1988 NSFG
Edelman *et al.* [1984]	722	17.0	89% aged 20-34; 28% married; 77% ≥13 years education; 94% white, 49% never-married; 38% used oral contraceptives prior to entering study	33.2[c,g]	Random assignment to diaphragm or sponge
McIntyre and Higgins [1986]	723	17.4	89% aged 20-34; 28% married; 77% ≥13 years education; 94% white, 49% never-married; 38% used oral contraceptives prior to entering study	33.2[c,g]	A reanalysis of data used by Edelman *et al.* [1984]; random assignment to diaphragm or sponge; much higher rate for parous women
Bounds and Guillebaud [1984]	126	24.5	92% aged 20-34; all married/consensual union; "most" ≥13 years education; 99% white	1.7[g]	Britain; random assignment to diaphragm or sponge

a Based on Table 4 in Trussell and Kost [1987] and Table 12.3 in Hatcher *et al.* [1990], and updated to include subsequent studies.

c Calculated by James Trussell from data in the article.

g Most of these studies incorrectly report the loss to follow-up probabilities as the number of women lost at any one time during the study divided by the total number of women entering the study. Thus, these are the percentages presented in the table. However, the correct measure of LFU would be a gross life table probability. When available, 12-month probabilities are denoted by the letter "g."

Table 4. Summary of studies of contraceptive failure: cervical cap[a,b]

		Failure Rate						
			Pearl Index					
Reference	*N for Analysis*	*Life-table 12 Month*	*Index*	*Total Exposure*	*Maximum Exposure*	*Sample Characteristics*	*LFU[g]*	*Comments*
Denniston and Putney [1981]	110	8.0[d,n]				98% aged 20-35; 70% nulliparous	20.9[c]	
Cagen [1986]	620	8.1[n]				87% aged 20-34; 80% always used spermicide and 14% never did	38.5[c]	LFU = "no response"
Koch [1982]	372	8.4				76% aged 20-29; 65% college graduates	8.0[c]	Women advised also to use condom for the first several cap uses
Richwald *et al*. [1989]	3433	11.3				Mean aged 29.0; 91% white non-Hispanic; 80% unmarried; almost 60% college graduates; 64% one or more previous pregnancies; 6.1% failure rate among perfect users, 11.9% among imperfect users	18.[g]	Women advised to use extra spermicide or use condoms during first 2 months of use; 15 sites; 14 in Los Angeles, 1 Santa Fe
Powell *et al.* [1986][b]	477	16.6				67% aged 25-34; "about half" unmarried; 97% high school graduates; 17% using pill when fitted for cap	43.8[c]	Canada; back-up methods encouraged (including the "morning after pill" in cases of cap dislodgement; 23 women used method)

Bernstein *et al.* [1986]	687[c]	17.4				95% aged ≤35, 16% married, 96% ≥high school completion	26.3[c,g]	Random assignment to the diaphragm or cervical cap
Kassell and McElroy [1981]	90		18.1[c]	731 mo	12 mo	Mean age = 23.6; mean education = 14.7 years	10.0[c]	
Boehm [1983]	47		18.1[c]	397 mo	12 mo		31.6	All women reported exclusive use of cap
Lehfeldt and Sivin [1984]	130	19.1				37% aged 16-25; 72% college graduates; 91% nulliparous	7.2[c]	All women agreed to exclusive use of cap
Smith and Lee [1984][b]	33	27.0				80% aged 20-29; clients at university student health service	1.5-4.6[c]	Regular users for whom the cap (with spermicide) was the sole method used "during the fertile portion of the cycle"

a Based on Table 4 in Trussell and Kost [1987] and Table 12.5 in Hatcher *et al.* [1990], and updated to include subsequent studies.

b All entries pertain to the Prentif Cavity-rim cap except for Smith and Lee [1984] and Powell *et al.* [1986], which included both Cavity-rim and Vimule caps.

c Calculated by James Trussell from data in the article.

d 6-month net probability of failure; 12-month probability not available.

g Most of these studies incorrectly report the loss to follow-up probabilities as the number of women lost at any one time during the study divided by the total number of women entering the study. Thus, these are the percentages presented in the table. However, the correct measure of LFU would be a gross life table probability. When available, 12-month probabilities are denoted by the letter "g."

n Only net rates available for this study.

Table 5. Summary of studies of contraceptive failure: diaphragm[a]

Reference	N for Analysis	Failure Rate: Life-table 12 Month	Failure Rate: Pearl Index: Index	Failure Rate: Pearl Index: Total Exposure	Failure Rate: Pearl Index: Maximum Exposure	Characteristics of the Sample	LFU[g]	Comments
Lane *et al*. [1976]	2168	2.1[c]				61% aged 21-34; 71% unmarried; 92% white	1.1[c,g]	Failure rate downward biased due to improper exposure allocated to women LFU
Vessey and Wiggins [1974]	4052		2.4	5909 mo	>60 mo	All white; at recruitment aged 25-39 and married; all had been using the diaphragm for at least 5 months at enrollment; no previous pill use	1.0[v]	Oxford/FPA study
Vessey *et al*. [1982]	?		5.5	2582 yr	24 mo	All white and aged 25-34; all married at recruitment; at enrollment, all women had been using the diaphragm, IUD, or pill successfully for at least 5 months	0.3[t,v]	Oxford/FPA study
Loudon *et al*. [1991]	269		8.7	2350 mo	12 mo	Mean age 28.6; 57% gravidity 0; 68% married or cohabiting; 54% already using diaphragm at start of trial	>3.7	Britain; random assignment of spermicide: either C-film or jelly
Dubrow and Kuder [1958]	873		9.3	5814 mo	48 mo	Mean age = 25; 39% black; 45% Puerto Rican; 93% ≤12 years education[t]	38.0[c]	

Jones and Forrest [1992]	472	10.4	Aged 15-44[t]	21.[r]	1988 NSFG; failure rate when standardized/corrected for estimated underreporting of abortion = 22.0[s]
Hall [1973]	347	10.6	Approximately 75% aged 20-24; 47% black; 38% Hispanic; all postpartum	16.0	
Bounds and Guillebaud [1984]	123	10.9	90% aged 20-34; all married/ consensual union; "most" ≥13 years education; 96% white	0.0	Britain; random assignment to diaphragm or sponge
Edelman *et al.* [1984]	717	12.5	88% aged 20-34; 55% never married; 76% ≥13 years education; 94% white; 39% used oral contraceptives prior to entering study	37.8[c,g]	Random assignment to diaphragm or sponge
McIntyre and Higgins [1986]	717	12.9	88% aged 20-34; 55% never married; 76% ≥13 years education; 94% white; 39% used oral contraceptives prior to entering study	37.8[c,g]	A reanalysis of data used by Edelman *et al.* [1984]; random assignment to diaphragm or sponge
Vaughan *et al.* [1977]	166	13.1[s]	Aged 15-44; all married[t]	19.0[r]	NSFG 1973
Grady *et al.* [1983]	349	14.3[c,s]	Aged 15-44; all married[t]	18.2[r]	NSFG 1973, 1976
Bernstein *et al.* [1986]	707[c]	16.7	96% aged ≤35; 17% married; 97% ≥high school completion	33.5[c,g]	Random assignment to the diaphragm or cervical cap

Table 5. Summary of studies of contraceptive failure: diaphragm[a] (continued)

Reference	N for Analysis	Failure Rate				Characteristics of the Sample	LFU[g]	Comments
		Life-table 12 Month	Pearl Index					
			Index	Total Exposure	Maximum Exposure			
Grady *et al.* [1986]	257	17.0[s]				Aged 15-44; all married[t]	20.6[r]	NSFG 1982
Tietze and Lewit [1967]	1197	17.9				86% aged <30; all married; 60% ≥high school completion; 50% white	7.2[g]	
Schirm *et al.* [1982]	349	18.6[s]				Aged 15-44; all married[t]	18.2[r]	NSFG 1973, 1976
Bracher and Santow [1992]	219	21.0				12% aged <20, 77% aged 20-29, 11% aged 30+; 56% parity 0; 87% married or cohabiting	25.[r]	Australian Family Survey; first use of method

a Based on Table 5 in Trussell and Kost [1987] and Table 12.4 in Hatcher *et al.* [1990], and updated to include subsequent studies.

c Calculated by James Trussell from data in the article.

g Most of these studies incorrectly report the loss to follow-up probabilities as the number of women lost at any one time during the study divided by the total number of women entering the study. Thus, these are the percentages presented in the table. However, the correct measure of LFU would be a gross life table probability. When available, 12-month probabilities are denoted by the letter "g."

r Nonresponse rate of the entire study.

s Standardized: Vaughan *et al.* [1977] (1973 NSFG) — intention (the average of probabilities for preventers and delayers); Grady *et al.* [1983] (1973 and 1976 NSFG) — intention (our calculation, the average of probabilities for preventers and delayers); Schirm *et al.* [1982] (1973 and 1976 NSFG) — intention, age, and income; Grady *et al.* [1986] (1982 NSFG) — intention, age, poverty status, and parity; Jones and Forrest [1992] (1988 NSFG) — duration, age, marital status and poverty status.

t Total for all methods in the study.

v The authors report that LFU for "relevant reasons (withdrawal of cooperation or loss of contact)" was 0.3% per year in the 1982 study and "less than 1%" per year for all reasons in the 1974 study. In the 1982 study, women had on average been followed for 9.5 years; if 0.3% are LFU per year, then 2.8% would be LFU in 9.5 years. LFU including death and emigration is about twice as high as LFU for "relevant reasons."

Table 6. Summary of studies of contraceptive failure: Vaginal Contraceptive Film

References	Method Brand	N for Analysis	Pearl Index Failure Rate: Index	Pearl Index Failure Rate: Total Exposure	Pearl Index Failure Rate: Maximum Exposure	Characteristics of the Sample	LFU[g]	Comments
Iizuka *et al*. [1980]	Vaginal Contraceptive Film (C-Film)	168	0.6	2161 mo	?	All women had been pregnant before; 20% aged <25, 64% aged 25-34, 17% aged 35+	?	Japan
Brigato *et al*. [1982]	Vaginal Contraceptive Film (C-Film)	37	3.9[c]	924 mo	?		?	Italy
Apothecus [1992]	Vaginal Contraceptive Film (C-Film)	824[c]	7.9[c]	6695[c] mo	?		?	Belgium, Netherlands, Britain, Germany, Switzerland, Denmark, Sweden, Israel, Egypt; results never published; quality of study unknown
Frankman, Raabe and Ingemansson [1975]	Vaginal Contraceptive Film (C-Film)	237	9.0	1866 mo	23 mo		?	Sweden; data included in Apothecus [1992]
Smith *et al*. [1974]	Vaginal Contraceptive Film (C-Film)	63[c]	55.7[c]	194[c] mo	<15[c] mo	Aged 16-35	9.5[c]	Britain; data included in Apothecus [1992]; trial terminated for ethical reasons

c Calculated by James Trussell from data in the article.

g Most of these studies incorrectly report the loss to follow-up probabilities as the number of women lost at any one time during the study divided by the total number of women entering the study. Thus, these are the percentages presented in the table. However, the correct measure of LFU would be a gross life table probability. When available, 12-month probabilities are denoted by the letter "g."

REFERENCES

Apothecus Pharmaceutical Corporation (1992): VCF®: Vaginal Contraceptive Film™. East Norwich NY: Apothecus, Inc.

Bernstein GS (1971): Clinical effectiveness of an aerosol contraceptive foam. Contraception 3:37-43.

Bernstein GS, Clark VA, Coulson AH, Kilzer L, Nakamura R, Walsh T (1986): Use effectiveness study of cervical caps. Final report. Washington DC: National Institute of Child Health and Human Development.

Boehm D (1983): The cervical cap: effectiveness as a contraceptive. J Nurse Midwifery 28:3-6.

Bounds W, Guillebaud J (1984): Randomised comparison of the use-effectiveness and patient acceptability of the Collatex (Today) contraceptive sponge and the diaphragm. Br J Fam Plann 10:69-75.

Bracher M, Santow G (1992): Premature discontinuation of contraception in Australia. Fam Plann Perspect 24:58-65.

Brehm H, Haase W (1975): Die alternative zur hormonalen kontrazeption? Med Welt 26:1610-1617.

Brigato G, Pisano G, Bergamasco A, Pasqualini M, Cutugno G, Luppari T (1982): Vaginal topical chemical contraception with C-Film. Ginecol Clinica 3:77-80. (In Italian; translation supplied by Apothecus, Inc.)

Bushnell LF (1965): Aerosol foam: a practical and effective method of contraception. Pac Med Surg 73:353-355.

Cagen R (1986): The cervical cap as a barrier contraceptive. Contraception 33:487-496.

Carpenter G, Martin JB (1970): Clinical evaluation of a vaginal contraceptive foam. Adv Plann Parent 5:170-175.

Denniston GC, Putney D (1981): The cavity rim cervical cap. Adv Plann Parent 16:77-80.

Dimpfl J, Salomon W, Schicketanz KH (1984): Die spermizide barriere. Sexualmedizin 13:95-98.

Dominik R, Trussell J, Walsh T (1993): Computing perfect-use failure rates. Research Triangle NC: Family Health International.

Dubrow H, Kuder K (1958): Combined postpartum and family-planning clinic. Obstet Gynecol 11:586-590.

Edelman DA (1980): Nonprescription vaginal contraception. Int J Gynecol Obstet 18:340-344.

Edelman DA, McIntyre SL, Harper J (1984): A comparative trial of the Today contraceptive sponge and diaphragm. Am J Obstet Gynecol 150:869-876.

FHI and CONRAD (1993): The Reality female condom: efficacy and clinical acceptability of a new barrier contraceptive. Research Triangle NC: Family Health International.

FHI (1985): Development and testing of vaginal contraceptives. Addendum to final report. Washington DC: National Institute of Child Health and Human Development.

Florence N (1977): Das kontrazeptive vaginal-suppositorium. Sexualmedizin 6:385-386.

Frank R (1962): Clinical evaluation of a simple jelly-alone method of contraception. Fertil Steril 13:458-464.

Frankman O, Raabe N, Ingemansson CA (1975): Clinical evaluation of C-Film, a vaginal contraceptive. J Int Med Res 3:292-296.

Glass R, Vessey M, Wiggins P (1974): Use-effectiveness of the condom in a selected family planning clinic population in the United Kingdom. Contraception 10:591-598.

Grady WR, Hayward MD, Yagi J (1986): Contraceptive failure in the United States: estimates from the 1982 National Survey of Family Growth. Fam Plann Perspect 18:200-209.

Grady WR, Hirsch MB, Keen N, Vaughan B (1983): Contraceptive failure and continuation among married women in the United States, 1970-75. Stud Fam Plann 14:9-19.

Hall RE (1973): Continuation and pregnancy rates with four contraceptive methods. Am J Obstet Gynecol 116:671-681.

Hatcher RA, Stewart F, Trussell J, Kowal D, Guest F, Stewart GK, Cates W (1990): "Contraceptive Technology 1990-1992." New York NY: Irvington Publishers.

Iizuka R, Kobayashi T, Kawakami S, Nakamura Y, Ikeuchi M, Chin B, Mochimaru F, Sumi K, Sato H, Yamaguchi J, Ohno T, Shiina M, Maeda N, Tokoro H, Suzuki T, Hayashi K, Takahashi T, Akatsuka M, Kasuga Y, Kurokawa H (1980): Clinical experience with the Vaginal Contraceptive Film containing the spermicide polyoxyethylene nonylphenyl ether (C-Film study group). Jpn J Fertil Steril 25:64-68. (In Japanese; translation supplied by Apothecus, Inc.)

John APK (1973): Contraception in a practice community. J R Coll Gen Pract 23:665-675.

Jones EF, Forrest JD (1992a): Contraceptive failure rates based on the 1988 NSFG. Fam Plann Perspect 24:12-19.

Jones EF, Forrest JD (1992b): Underreporting of abortion in surveys of US women, 1976-1988. Demography 29:113-126.

Kasabach HY (1962): Clinical evaluation of vaginal jelly alone in the management of fertility. Clin Med 69:894-897.

Kassell NC, McElroy MP (1981): Emma Goldman Clinic for Women study. In King L, (ed.): "The cervical cap handbook for users and fitters." Iowa City IA: Emma Goldman Clinic for Women.

Kestelman P, Trussell J (1991): Efficacy of simultaneous use of condoms and spermicides. Fam Plann Perspect 23:226-227, 232.

Kleppinger RK (1965): A vaginal contraceptive foam. Penn Med J 68:31-34.

Koch JP (1982): The Prentif contraceptive cervical cap: a contemporary study of its clinical safety and effectiveness. Contraception 25:135-159.

Lane ME, Arceo R, Sobrero AJ (1976): Successful use of the diaphragm and jelly by a young population: report of a clinical study. Fam Plann Perspect 8:81-86.

Lehfeldt H, Sivin I (1984): Use effectiveness of the Prentif cervical cap in private practice: a prospective study. Contraception 30:331-338.

Loudon NB, Barden ME, Hepburn WB, Prescott RJ (1991): A comparative study of the effectiveness and acceptability of the diaphragm used with spermicide in the form of C-film or a cream or jelly. Br J Fam Plann 17:41-44.

McIntyre SL, Higgins JE (1986): Parity and use-effectiveness with the contraceptive sponge. Am J Obstet Gynecol 155:796-801.

Mears E (1962): Chemical contraceptive trial: II. J Reprod Fertil 4:337-343.

Mears E, Please NW (1962): Chemical contraceptive trial. J Reprod Fertil 3:138-147.

Peel J (1972): The Hull family survey: II. Family planning in the first five years of marriage. J Biosoc Sci 4:333-346.

Potts M (1993): Personal communication to James Trussell.

Potts M, McDevitt J (1975): A use-effectiveness trial of spermicidally lubricated condoms. Contraception 11:701-710.

Powell MG, Mears BJ, Deber RB, Ferguson D (1986): Contraception with the cervical cap: effectiveness, safety, continuity of use, and user satisfaction. Contraception 33:215-232.

Richwald GA, Greenland S, Gerber MM, Potik R, Kersey L, Comas MA (1989): Effectiveness of the cavity-rim cervical cap: results of a large clinical study. Obstet Gynecol 74:143-148.

Rovinsky JJ (1964): Clinical effectiveness of a contraceptive cream. Obstet Gynecol 23:125-131.

Schirm AL, Trussell J, Menken J, Grady WR (1982): Contraceptive failure in the United States: the impact of social, economic, and demographic factors. Fam Plann Perspect 14:68-75.

Smith GG, Lee RJ (1984): The use of cervical caps at the University of California, Berkeley: a survey. Contraception 30:115-123.

Smith M, Vessey MP, Bounds W, Warren J (1974): C-Film as a contraceptive. Br Med J 4:291.

Squire JJ, Berger GS, Keith L (1979): A retrospective clinical study of a vaginal contraceptive suppository. J Reprod Med 22:319-323.

Tietze C, Lewit S (1974): Statistical evaluation of contraceptive methods. Clin Obstet Gynecol 17:121-138.

Tietze C, Lewit S (1973): Recommended procedures for the statistical evaluation of intrauterine contraception. Stud Fam Plann 4:35-42.

Tietze C, Lewit S (1967): Comparison of three contraceptive methods: diaphragm with jelly or cream, vaginal foam, and jelly/cream alone. J Sex Res 3:295-311.

Tietze C, Poliakoff SR, Rock J (1951): The clinical effectiveness of the rhythm method of contraception. Fertil Steril 2:444-450.

Trussell J, Grummer-Strawn L (1990): Contraceptive failure of the ovulation method of periodic abstinence. Fam Plann Perspect 22:65-75.

Trussell J, Kost K (1987): Contraceptive failure in the United States: A critical review of the literature. Stud Fam Plann 18:237-283.

Trussell J, Strickler J, Vaughan B (1993): Contraceptive efficacy of the diaphragm, sponge and cervical cap. Fam Plann Perspect 25:101-105,135.

Trussell J, Sturgen K, Strickler J, Dominik R (1993): Contraceptive efficacy of the Reality female condom: comparison with other barrier methods. Fam Plann Perspect 26.

Trussell J, Westoff CF (1980): Contraceptive practice and trends in coital frequency. Fam Plann Perspect 12:246-249.

Tyler ET (1965): Current developments in systemic contraception. Pac Med Surg 73:79-85.

Vaughan B, Trussell J, Menken J, Jones EF (1977): Contraceptive failure among married women in the United States, 1970-1973. Fam Plann Perspect 9:251-258.

Vessey M, Lawless M, Yeates D (1982): Efficacy of different contraceptive methods. Lancet 1:841-842.

Vessey MP, Villard-Mackintosh L, McPherson K, Yeates D (1988): Factors influencing use-effectiveness of the condom. Br J Fam Plann 14:40-43.

Vessey M, Wiggins P (1974): Use-effectiveness of the diaphragm in a selected family planning clinic population in the United Kingdom. Contraception 9:15-21.

Westoff CF, Potter RG, Sagi PC, Mishler EG (1961): "Family growth in metropolitan America." Princeton NJ: Princeton University Press.

Wolf L, Olson HJ, Tyler ET (1957): Observations on the clinical use of cream-alone and gel-alone methods of contraception. Obstet Gynecol 10:316-321.

DISCUSSION

Zena Stein: What was the chemical pregnancy rate in the women's condom study? That could make a tremendous difference to the reported pregnancy rate.

James Trussell: We know, for example, in some modern oral contraceptive trials in which pregnancy tests are done every month that lots of pregnancies are found that never would have been discovered otherwise. Had they been the pill failure rate would have gone up rather significantly. In the Reality trial, Rosalie, how much difference would it have made had you not done those pregnancy tests?

Rosalie Dominik: I think what made a greater difference are the pregnancy tests done at the end of and two weeks after the trial, because there were about 4 pregnancies picked up at that time. The chemically confirmed pregnancy tests that would not have been clinically confirmed had a much smaller impact, although they did have some impact on the rate. We didn't estimate what the rate would have been without those, but it's certainly a small percentage of the difference relative to the end-of-study pregnancy tests that were done.

Penny Hitchcock: From your perusal of the literature, what would you say we know about STD prevention and control from these studies?

James Trussell: Nothing from these studies, but you will hear from Felicia Stewart all of the information that has been pulled together on the effect of spermicides and condoms on STDs.

Penny Hitchcock: Given your analysis of methods, etc., that you used looking at contraceptive efficacy, do you feel that we have studies that allow us to make conclusions about STDs, if you take methods into account?

James Trussell: There aren't any clinical trials in which people were randomized to use condoms or nothing or spermicide or nothing for ethical reasons. Almost all the information that we tease out is based upon subjects who are not using what they're supposed to be using. You can figure out, for example, the efficacy of spermicides alone because there are some people who use spermicides but not the condom. One can use very simple mathematical models to try to get some feel for what protection one would get. If you assume that couples are having intercourse twice a week and that they have the typical pregnancy rate observed in the United States observed of 14 percent per month which would lead to 85 percent getting pregnant in the first year, then you can figure out what the probability of failure for the device is during mid-cycle, assuming that one of those acts would occur mid-cycle. You could figure out the contraceptive failure rate during perfect use, then take that failure rate, for example, a condom burst rate, and apply it to the probability of STD transmission per particular act and figure out the implied reduction in STDs. The surprising thing is that even with a relatively high pregnancy rate, one can get a phenomenal impact on HIV transmission. The reason is that if the probability of transmission per act is relatively low, as it would be for HIV but not, for example, for chlamydia, then even a modest failure rate like 2 percent is going to drastically drive down the probability of infection per act and therefore drive down the cumulative probability of getting infected.

David Archer: When you're looking at perfect use failure rates in clinical trials, do you take the number of volunteers who are perfect users each month from the inception of the study, or do you wait till the end of the study and try to get an average?

James Trussell: Let's suppose that we had information every cycle on whether or not the product was used correctly and consistently. We did, for example, in the WHO trial of the ovulation method periodic abstinence. Women were interviewed every month so we knew every month whether they were perfect users or imperfect users. You calculate how many months of perfect use exposure you have in month 1, how many in month 2, how many in month 3, and so on. Those are the denominators for life table failure rates. The numerators are the pregnancies that occurred during those cycles of perfect use. The other way, which you alluded to, is to ask people at the end of the trial whether they used the method perfectly throughout the trial. We found that's not a very good way to do it. Twenty five percent of the women who said at the end of the trial that they had used the method correctly and consistently throughout the trial had, in fact, admitted during one of their interval visits that they had not used it correctly and consistently. What do we have going on here? If a subject manages to make it all the way through the trial and not get pregnant, she is going to tell you at the end that she was used the method just like she was supposed to. So we're going to get an over-estimate of perfect use exposure if we ask about it at the end of the trial. It really is critically important, I think, to collect information on a month-by-month basis.

David Archer: My second question has to do with the sponge and cap having higher failure rates than the diaphragm in those data. Could you speculate why that occurred, because in the clinical studies we've carried out, they look like they should be comparable.

James Trussell: It seems that they should be comparable; they're all barriers and at least part of their efficacy should be related to the spermicide use. For the cap, I think, at least part of the reason may be that it's much harder to fit a cap and to get it to stay on the cervix during intercourse than a diaphragm. If it falls off during intercourse it's considered a perfect use failure. The sponge contains a whole gram of nonoxynol-9 and one might think that it would provide more protection rather than less. But perhaps the diaphragm holds the spermicide nearer the cervix than does the sponge. If the cap stays on, it would certainly hold spermicide very near the cervix. If the data from the Fem Cap are replicable, we are seeing a lower typical use failure rate for that cap than what we saw as a perfect use failure rate for the Prentiff Cervical Cap. So if it stays on and has a better seal it ought to work even better.

Erica Gollub: Your numbers and your calculations are based primarily on the Bernstein study which was randomized. I think one of the real problems with cervical cap studies is that not everyone can be fitted. There are non-fit rates ranging from 10-50 percent, depending on the investigator and his or her fitting standards. I know that there were potential investigators who applied for that grant who were not intending to randomize knowing that it's very difficult to randomize a woman into a cervical cap study. There are women who simply were fitted with cervical caps that should not have been, and that is why I think we see the higher failure rates compared to the diaphragm.

A much lower percentage of women cannot be fitted with a diaphragm due to anatomical considerations.

James Trussell: Let me add two comments to that, though. Don't forget that in the other big cervical cap trial conducted by Gary Richwald and his colleagues, they asked women at the end of the trial whether they used the method correctly and consistently at every act of intercourse. The failure rate using only those women was 6 percent. That's lower than the 10 percent in the other trial, but remember that in the Ritchral trial during the first month women were encouraged to use as an adjunct method, particularly the condom. That would make the perfect use failure rate go down because it reflected the failure not just of the cap but of the cap plus spermicide in the first month. Also we have the lesson from the Bernstein trial that 25 percent of the women who said they used the method correctly and consistently at the end of the trial actually hadn't done so. If you take those women out of the denominator for the Richwald the failure trial rate will go up to about 8 percent, so that the two studies are not all that dissimilar.

Lillian Yin: Where the 353 women in the female condom study the number of women who entered or completed the study?

James Trussell: 353 women entered the trial; some of them that is disappeared, that is were lost to follow-up or never came back for even the first interview. The same is true with the cap and the sponge and the diaphragm trials. All those numbers I showed were numbers who actually entered the trial.

Alexandre Psychoyos: Does the sponge failure rate apply to different sponges or to just one kind of sponge?

James Trussell: The only sponge that I referred to is the Today Sponge.

Ron Frezieres: Prior to randomization in the Bernstein trial, all individuals were assessed to be sure that they could be fit with both the cap and the diaphragm. If they couldn't be fit with both they weren't randomized into the trial.

3

ACCEPTABILITY OF BARRIER METHODS FOR PREVENTION OF UNWANTED PREGNANCY AND INFECTION

Nancy E. Williamson and Carol Joanis

Agency for International Development
R&D/Office of Health/HIV-AIDS Division
Washington, DC 20523-1817 (NEW)
Contraceptive Use and Epidemiology Division
Family Health International
P.O. Box 13950
Research Triangle Park, NC 27709 (CJ)

INTRODUCTION

Interest in promoting barrier method use is recent and can be traced to the rise in the incidence of STDs, including HIV. Though the need for increased barrier usage has been determined, only limited or dated information exists on the contraceptive efficacy of most barriers. This is particularly true of condoms. The situation is somewhat better for STD efficacy data.

Efficacy, safety and acceptability are usually considered when speaking of the desirable qualities of drugs or contraceptives. Efficacy is often viewed as the primary performance criterion in the evaluation and recommendation of a contraceptive method. Acceptability, on the other hand, is seen as having less value. This is unfortunate because the three are interrelated. If a method is determined to be unsafe, it will not be used. If it is judged less effective than other methods, it will be less acceptable; and finally, if a product is considered less acceptable, compliance, correct use and efficacy will be reduced or sacrificed. This fact is sometimes overlooked by product manufacturers, similarly, health care providers and government officials (i.e. those with less product acceptability information) sometimes favor, prescribe or mandate one method over another based on efficacy figures alone. The result is often a mismatch between user needs and expectations and the product itself.

Barrier Contraceptives: Current Status and Future Prospects, pages 53–67

While the three desirable qualities of barrier methods are closely related, this paper will focus on acceptability and, particularly, on the discrepancies between ideal product characteristics and actual methods and the influence of technology on changing product performance characteristics. The reason for this focus on method preferences and performance is that it is probably easier to improve technology or methods and how programs present them than it is to change the users' behaviors and attitudes. At the same time, the focus on methods cannot be exclusive since it is known that even if methods were absolutely perfect, some people would choose not to use them.

DEFINITIONS OF ACCEPTABILITY

Defining acceptability is a complex process. From the consumer standpoint, it involves an assessment of opinions, wants and needs. Every user must evaluate his/her barrier choice in light of such issues as cost versus benefit, risk versus benefit and a given method's impact on his/her overall quality of life. Acceptability is affected further by access to and availability of different methods or variations of those methods (eg. different sized condoms, spermicides with different delivery mechanisms, etc.). Also, acceptability is influenced by such factors as provider bias and the provider's ability to offer high quality service.

Further, acceptability is a relative concept: Something is more or less acceptable than something else (another method, getting pregnant, risking infection). The concept of acceptability is not synonymous with method "continuation." Continuation rates consider a "discontinuation" to be a failure when actually, an individual may not have wanted to use the method for a long time. Also, continuation rates look at use from an outsider's perspective whereas the concepts of acceptability or satisfaction are from the user's perspective.

In addition to defining acceptability in terms of users and providers, it is important to note that there are three stages of acceptability. These are given in a recent World Health Organization (WHO) monograph, "Introduction of New Contraceptives into Family Planning Programmes, 1990," and are as follows:

> "Hypothetical acceptability": the "verbal assessment of potential users in response to some description or representation of a new method or even an imaginary method" (p. 14);
>
> "Initial acceptability": the "willingness of individuals to try a new method in a fully informed manner." (p. 14-15); and,
>
> "Experiential acceptability": the evaluations and behaviors of individuals who have actually tried the method." (p. 15).

These three categories of acceptability could be rephrased as questions: Do you like the idea? Are you willing to try it? Will you stick with it?

IDEAL CHARACTERISTICS OF BARRIER METHODS VERSUS THE REALITY OF CURRENT METHODS

Based on several decades of research on contraceptive use around the world, one can make a long list of desirable traits of a barrier method, classifying the traits under the rubrics of "effectiveness," "safety," and "acceptability." As one can see by looking at Table 1, the ideal method would need to jump over many hurdles.

TABLE 1. Ideal Characteristics of Barrier Methods for Preventing Pregnancy and Infection: Users

<u>Effective</u>
Prevents pregnancy and/or infection
Bioavailable (in case of spermicide, virucide)
Low rate of breakage and slippage and no pinholes
A virucide which is not a spermicide

<u>Safe</u>
No major or minor side effects for either partner
No burning sensation (some warmth may be acceptable)
No irritation
No allergic reactions
Should not increase incidence of yeast infections, Pap smear abnormalities, toxic shock syndrome, urinary tract infections
Should not contribute to fetal abnormalities in case of contraceptive failure
Can be used by a person with genital irritation/infection

<u>Acceptable</u>
Initial Appeal of the Product ("Initial acceptability")
Attractive packaging
Credible appearance (i.e., it looks like a legitimate medical product)
Attractive advertising
Its name has positive connotation in local language
Available when/where needed
Good placement of product in store
Useful information on package
Credible claims made for product (effective for FP/STDs/HIV?)
If sized, multiple sizes available
Number of uses per package
Reasonable price (initial price and per-use price)
Subsidized product available for low income consumers
Endorsement by trusted health professional or salesperson

Endorsement by friends/relatives
FDA or in-country regulatory approval

Obtaining and Storing the Product
Independent of health professionals
No extra supplies needed (spermicide, lubricant, applicator, inserter, water)
Not bulky to store, ship, carry
Reusable and easy to clean
Does not need additional lubricant/spermicide if reused
Can be stored in harsh environments
Long shelf life

Using the Product ("Experiential acceptability")
Package does not leak
Easy removal from package
Requires only minimal touching of genitals
Easy to position/don
User can tell if it is placed correctly
Benefits from, but does not require an inserter
Quick acting
Rarely dislodges before/during/after sex
Comfortable to men and women
Does not reduce sensation of men or women
Does not reduce intimacy between partners
Does not require cooperation from male partner
Does not require cooperation from female partner
Soft material
Not messy
Pleasant smelling
Pleasant tasting
Right size
No rough surfaces, seams
Unobtrusive, not ugly
Can be used without partner's knowledge
Not noisy
Product does not stick to fingers
Use does not interrupt sex act
Use does not contribute to loss of erection
Has right amount of lubricant
Does not contribute to vaginal dryness
User can add any type of lubricant
Lubricant is appropriately slippery
Product is long acting (e.g., up to 24 hours)

Product does not need re-application between sex acts
Product not too slippery to handle
Product can be taken out immediately after sex
User can douche immediately after sex
Device be left in for days/weeks

Disposing of the Product
Product is easy to remove
Product does not leak/slip off during removal
Easy to dispose of (biodegradable, not bulky)

Many people (though not all) prefer unsafe/unprotected sexual intercourse over the use of a barrier method. They want the feeling of a "natural" sexual act. Thus, they would like the method to be tasteless, odorless, convenient, invisible and immaterial, in addition to being safe and effective. This is a very hard bill to fill and current methods are far from this ideal.

A review of the acceptability issues associated with six current barrier methods (latex male condoms, plastic male condoms, female condoms (Reality), diaphragms, foaming tablets, and vaginal film) is shown in Table 2. A comparison of Tables 1 and 2 will differentiate between user preferences and product realities. This comparison is helpful in identifying where the greatest deficiencies are.

TABLE 2. Acceptability Issues with Six Barrier Methods

Male Latex Condoms
Packages sometimes leak
Cannot be stored in harsh environments which reduce shelf life
Without instruction, may be donned incorrectly (backwards, pulled on rather than rolled on)
May break or slip off; requires spermicide as back-up in these instances
Some men report reduction of sensation or loss of erection
May reduce intimacy (i.e., barrier)
Requires male cooperation
Cannot use with oil-based lubricants
Can have unpleasant smell
May have too much or too little lubricant
Interrupts sex act
New condom needs to be used for each sex act

Male Plastic Condom

New concept; people are unfamiliar with it
May reduce intimacy (i.e., barrier)
Requires male cooperation
May contribute to loss of erection
Interrupts sex act
May slip off, break; requires spermicide as back-up in some instances
New condom needs to be used for each sex act
Product not biodegradable

Female Condom (Reality)

New concept; people are unfamiliar with it
High price
Bulky to ship, store and use
Potentially reusable, but may be hard to clean and requires additional lubrication
Requires touching female genitals to use
Sometimes difficult to position (insert)
"Penile misrouting" occasionally occurs
May be uncomfortable to either or both partners
May not be the right size (currently available in one size)
Some people find it obtrusive, ugly
Some people find it noisy
Can be used with any type of lubricant
More difficult to dispose of than latex male condoms
Product not biodegradable

Diaphragm with Spermicide

Prescriptive (needs to be sized by health professional)
Not easily available
Relatively expensive; subsidized price seldom available
Requires refitting if weight gain/loss or pregnancy occurs
Extra supplies needed (spermicide, soap, water)
Bulky to store, ship (including spermicide)
Requires touching female genitals to use
Can be difficult to position, slippery or messy
May dislodge during sex
Requires cooperation from female partner
Oil-based lubricants should not be added
Woman cannot douche for 6-8 hours
Sometimes difficult to remove
Must be washed with soap and water and dried before storing

Vaginal Foaming Tablets

More expensive than condoms
Requires touching female genitals (though not as much as with diaphragm or female condom)
Cannot have sex for 10-15 minutes after insertion
Effective for only one hour
Requires cooperation from female partner
May be considered messy
User cannot douche for 6-8 hours
Requires re-application spermicide between sex acts
May cause vaginal or penile irritation

Vaginal Contraceptive Film

New concept in developing countries
Appearance may be an issue (it looks like a piece of plastic; people may question whether it is a contraceptive; bogus drug problem)
Requires cooperation from female partner
Product may stick to fingers during insertion if fingers are not dry
Product must be placed high in vagina which may be difficult or unacceptable for some women
Must wait 5 minutes between insertion and intercourse (though 15 minutes is ideal)
Remains effective no more than one hour
User should not douche for 6 hours after sex
Needs re-application between sex acts

INFLUENCE OF TECHNOLOGY ON PRODUCT ACCEPTABILITY

At this juncture, it is important to note that barrier contraceptives are "personal products" and that consumers are "loyal" in their choice and use of personal products. That is, barring a life situation change, the consumer will rarely change products or brands unless that product fails repeatedly, is not available or is substantially improved along a critical performance parameter. It is through technology (changes in materials or design) that performance parameters are improved or changed.

What are critical performance parameters? Defined simply, they are product attributes or use characteristics that are judged important (by the consumer) to the continued use of the product. These attributes may be quantitative or qualitative.

In the case of male condoms, breakage and slippage are the primary quantitative or functional performance parameters. Condoms that break put people at risk of pregnancy and STDs. Typical breakage rates seen in studies range between 1 and 5 percent of devices used, irrespective of culture. Most of this breakage occurs with a very few study participants. Thus, breakage is a rare event for most people.

Slippage, on the other hand, occurs more frequently or in roughly 4 to 11 percent of devices used, again irrespective of culture. Further, 60 percent of this slippage occurs at withdrawal and can be attributed more to user failure than method failure.

Technology can impact both condom breakage and slippage. Through the use of stronger materials (in this case plastics), breakage from prolonged use or improper use may be reduced. Plastics are resistant to the elements that weaken latex condoms, notably ozone, sunlight, high temperature and moisture. Thus, condom shelf life may be prolonged and be less dependent on proper user storage. By changing the design (shape and size) of the condom retention mechanism, slippage may be reduced by increasing the force necessary to remove the condom on an erect penis.

While there are individuals who stop using condoms because of breakage/slippage problems, most consumers cease use because of inconvenience and/or because use of the condom affects on one or more of the qualitative performance parameters (i.e. those effecting aesthetics or sensory attributes).

Qualitative performance parameters for male condoms are typically those listed here.

General appearance: How does this product look on my body?
Strength of the condom: Does this product look strong enough, or conversely, does it look so thick it couldn't possibly break?
Ease of putting on: This is a critical performance parameter and one that many latex condom users complain about. In the dark, it is easy to put on a condom upside down. If the condom material is very thin, it is possible for the material to "roll back on itself" making the condom difficult or impossible to unroll. The "ease of putting on" parameter seriously affects convenience.
Odor: Smell is one of the major complaints about caecum or natural lambskin condoms and is an occasional complaint about latex.
Comfort at base of the condom: When using elastic materials, it is possible to have a "garter effect" or cutting feeling at the base or open end of the condom. While this is not a common problem, it is one that bears attention in the development of new condom products.
Noise: Noise, like odor, is another sensory attribute. It has been one element to overcome in the development of the plastic condom.
Feeling of the lubricant: In this case, viscosity is the issue. Some people prefer the dryer feeling of silicones and others prefer the very wet feel of aqueous based lubricants. The amount of lubricant desired varies from couple to couple and is dependent on several variables including the time of month, age of the female participant and user preference. Lubricant preferences are highly individual.
General fit, length and width of the condom: Again, this is subject to individual preference. Length and width acceptability measurements are not as much of an issue with the latex condom because of its elastic properties. These can be very important parameters, however, in the development of a plastic condom.
Comfort in use and sensitivity: These are two of the most important qualitative parameters. These criteria, though individual and largely perceptual, contribute significantly to user failure. In other words, the condom doesn't feel as good as wearing no protection and the user chooses not to use the condom.

As mentioned previously, critical performance parameters must be improved for users to switch from one product or brand to another. Product changes which improve upon these characteristics will impact positively on the behavior of new users as well. For approximately five years, FHI has been in the process of developing a plastic condom and we have made progress in producing a product that improves upon these important sensory attributes.

EXAMPLES OF RECENT RESEARCH ON BARRIER METHODS FOR CONTRACEPTION

To illustrate some of the points made about method acceptability, some recent research findings will be presented from developing countries. For many years, there have been controversies in Bangladesh, about whether condoms were being used for the purposes intended or whether consumption could be attributed to use of condoms as balloons, smuggling into other countries, or used during menses as a hygiene device. Table 3 gives some unexpected findings from a recent in-depth survey of 402 condoms users in Bangladesh.

TABLE 3. Selected Results from Bangladesh Condom Use/Misuse Study (402 current condom users)

Smell: More than one-third of respondents reported that condoms sometimes have a bad smell when the package is opened.
Styles of using condoms: Thirteen percent reported that they unrolled the condom before putting it on at least sometimes; 13 percent sometimes used two condoms at once; 6 percent sometimes added additional lubricants; 40 percent sometimes began sex without a condom but ended sex with a condom. Condom re-use was very rare. More than half of the respondents tried to time condom use during the risky part of the menstrual cycle.
Disposal: Twenty-seven percent reported problems in disposing of condoms. Rural people were less likely to report disposal problems than semi-urban or urban people.
Most common complaints: Condoms reduce pleasure. "Boredom."
Most common advantages: Condoms have few problems, can be used according to own convenience, have no side effects and are safer than other methods. Condoms prolong sexual activity and makes insertion/penetration easier.

Source: Folmar et al., 1992.

Similarly, in India, there has been a condom social marketing program for a long time, but only recently have individuals been seriously interviewed and asked their opinions of the products (Table 4). The government-distributed free condom, Nirodh, has been, until just a few months ago, an unlubricated device. In a recent survey of men in three Indian states, however, this lack of lubrication was seen as a problem for a quarter (27 percent) of those currently using Nirodh (free). By contrast, none of the users of a new socially marketed condom, Masti, had this particular complaint. This problem could have been identified and corrected a decade ago or more.

TABLE 4. Selected Results from Condom Research in India Problems Experienced with Brand Currently Used: Three States (N=716 current condom users)

Problem	Nirodh (free)	Nirodh	Nirodh Deluxe	Kohinoor	Masti
Breakage	**18%**	3%	**16%**	**13%**	**18%**
Poor fitting	4%	3%	1%	2%	3%
Difficulty in donning	4%	0%	1%	2%	3%
Too thick	6%	1%	0%	3%	3%
Too thin	6%	4%	6%	6%	6%
Defective	12%	4%	3%	2%	9%
Too wet	0%	1%	**11%**	**10%**	3%
Too dry	**27%**	**17%**	1%	2%	0%
Disposal	12%	8%	9%	18%	10%
Have not experienced problems	57%	73%	68%	60%	58%

Source: Thapa, 1992, based on Table 28, p. 60.

GENERAL OBSERVATIONS ON THE ACCEPTABILITY OF BARRIER METHODS

- There are two distinct problems associated with barrier usage: (1) getting people to initially try a method; and (2) getting them to continue its use over a substantial period of time. The factors affecting these two behaviors are quite different as noted in Table 1.
- The history of contraceptive development suggests that the personal preferences of people other than the end users are not very relevant in deciding what will appeal to others. Instead, early field testing for acceptability among end users is crucial.
- Although we need to improve barrier methods, we should not worry about getting perfect products since people use things even though they have complaints about them.

- We should not be overly discouraged about cultural barriers: If a product works, meets a real need, is attractively packaged and promoted, is available, and is affordable, this can overcome many cultural barriers, including Islam and the low status of women. This is a lesson we have learned from family planning.
- We can often find unexpected things if we take the trouble to go out to ask people about their experiences. It is better to do this early in the development process rather than later.

RESEARCH PRIORITIES

There are many unresolved acceptability issues regarding currently available or currently developed barrier methods.

Family Planning and STDs/AIDS:

- Dual Methods: Will people use both condoms and spermicides? This is recommended by some family planning programs.
  If they are advised to do so, will women who use methods like pills, IUDs, injectables, implants, or sterilization also use a barrier method when they are at risk of STDs/HIV?
- Is it better to suggest that people negotiate use of barriers or just make it a non-negotiable demand with no discussion?
- Which arguments are most effective in making barrier methods acceptable and used?
  - Lifestyle, positive, norm
  - Security
  - Fear
  - Habit (i.e., seatbelt analogy)
  - Concern for others: safety of spouse, children
  - Concern for self
- Should advertising for barrier methods portray people as they are or as they aspire to be? Or even beyond where they aspire to be?
- Should advertising for barrier methods use separate messages for sexual activity in marriage and outside marriage? Separate messages for family planning and for AIDS/STDs?

Male Latex Condoms

- Would condom use increase if spermicidally-lubricated condoms were provided?

Female Condom

- Given its cost, what is the role of the female condom in either family planning or HIV/STD programs? Could it be used as a backup method or a threat by the female partner to get men to use male condoms?
- Could a female condom be developed that was even more acceptable?
- What are men's views of the female condom? If given the choice of no sex, a male condom, or a female condom, which would they choose?
- Does use of a female condom increase the risk of male violence?
- Will people actually try to re-use it? How many times?

Plastic Male Condoms

- Are these more/less acceptable than latex condoms?
- How should the new concept be presented?
- Who will prefer them? Who will not? Will they attract people who do not use latex condoms or latex condom users who switch?

Diaphragms

- Would they be of interest to a significant number of women in LDCs if they were added to the current cafeteria?
- Can the recommended regimen of use be simplified without reducing efficacy (i.e., leave in longer, eliminate spermicide, self-fitting)?

Chemical barrier methods (spermicides, virucides)

- Which drug delivery systems are most acceptable (foams, gels, film, etc.)?
- Would acceptability be increased significantly if the products were bioavailable more quickly (i.e., there was no waiting time)?

CONCLUSION

It is important to feed insights about acceptability into new product development, but this is not an easy task since what people want may not be technically feasible or financially possible.

This task was somewhat challenging at this meeting since participants were divided into working groups; the people concerned about acceptability were in different groups from those concerned about development. But in the plenaries and informally, we who represent very different professional orientations were able to speak together.

Finally, the social scientists ask for patience from the "hard scientists." In the social sciences, it is difficult to make hard and fast generalizations since individuals, cultures, and societies differ so greatly. Thus, if we find that barrier use can be increased in one setting, by appealing to concern for the health of future children, or changing the specifications of a product, this does not mean it will work in another.

ACKNOWLEDGMENTS

The authors gratefully acknowledge the financial assistance and support given by Family Health International (FHI). Our research was funded by FHI under a Cooperative Agreement with the United States Agency for International Development (USAID). FHI is an international, non-profit organization that conducts research and provides technical assistance in health, family planning, sexually transmitted diseases (STD) and the Acquired Immune Deficiency Syndrome (AIDS). FHI is based in Research Triangle Park, North Carolina, USA. The views expressed in this paper are not necessarily those of USAID.

REFERENCES

Folmar S, Nurul Alam SM, Raihan Sharif, AHM (1992): Condom Use in Bangladesh: Final Report. Research Triangle Park, NC: Family Health International.

Hatcher RA, Stewart F, Trussell J, Kowal D, Guest F, Stewart GK, and Cates W (1990): Contraceptive Technology, 1990-1992, 15th Revised Edition, New York: Irvington Publishers.

Thapa S (1992): Social Marketing of Condoms in India. Study Report. Research Triangle Park, NC: Family Health International.

Williamson N (1991): Summary International Acceptability Task Force Meeting, July 16-17, 1991, Research Triangle Park, NC: Family Health International.

DISCUSSION

Zena Stein: This question is addressed to Ms. Cottingham. Are you testing the acceptability of the diaphragm with spermicide? I was thinking about the point of Dr. Trussell saying that the spermicide added something useful to the condom, but quite different issues clearly would apply to the use of the diaphragm with spermicide for protection against infections. In the Brazil studies, was the diaphragm used with a spermicide?

Debbie Rogow: The working group is going to undertake introductory trials based on information obtained from previous clinical trials. We are going to be looking at the diaphragm with jelly. Separately, there are follow-up clinical studies planned in Brazil which will look at the efficacy of the diaphragm in preventing both pregnancy and infection, with and without jelly.

Jackie Forrest: Nancy Williamson and Carol Joanis mentioned that there is not only the issue of the user acceptability but also provider acceptability. I was just trying to think of what we've heard already of factors that might relate to provider acceptability, whether it be in a commercial sense or whether it be in a family planning program. On my list I've got things that were mentioned such as cost, storage issues and the issues raised by James Trussell of how many pockets we would need for a year's supply [of some products]. Also, we need to know if a method is applicable to all or most women or only a select group, the ease of fitting and the training one needs, an estimation of efficacy and acceptability. There is also the question of contraceptive focus versus STD focus. Could you take a minute and maybe add to that list? While we may like to focus only on potential users in our discussions on developing new methods, they have to go through some kind of provider network, and that can very much influence whether they ever reach the user. Could you say something about this?

Nancy Williamson: Yes, I think your question is better than the answer; you have really answered your own question. I think providers are concerned about the amount of time and effort it takes to provide a particular method. On the other hand, some people are enamored by high tech. Sometimes providers love laproscopes and they like Norplant because it's new and so forth. But, in general, providers are trying to minimize the time requirements. Some like pills because you can just hand them out without any instruction, although pills do not work too well if you don't give any instruction. I think the ease in counseling is an issue when providers are busy. Much more work needs to be done on the role of providers. In some countries you find a focus on female sterilization, in some places condoms are fairly widespread; it doesn't seem to be solely due to the culture but due to provider biases in many settings.

Alfred Shihata: I cannot agree with you that for any barrier method to be accepted it should be invisible and not felt by either the man or the woman. I am doing an acceptability study at the moment and I am trying to find out why methods are not acceptable. I have divided intercourse into three phases: the foreplay, the intercourse, and the intimacy that follows intercourse. So, I took, for example, the condom. It interrupts spontaneity of intercourse at two important phases, the foreplay and as soon as ejaculation happens, if it is correctly used. These two interruptions, plus the reduction of sensitivity, affect the acceptability of the method. Any female barrier, if it is put in hours before intercourse, would be invisible. In my view if it cannot be seen or felt by either partner during intercourse, and if it does not reduce sensitivity of intercourse it would be acceptable.

Nancy Williamson: I think that is a very valuable distinction and the reason I mentioned the contraceptive film is that in many senses that seems to be a very unobtrusive method. If you look at the little piece of film, it is not messy, it can be put in before hand, and it doesn't interrupt the sex act; it has a lot of positive attributes. The only question is, does it work? We still don't know about that. As you suggest, a very useful way of looking at acceptability is through the different stages of sex.

Clark Taylor: In your talk, Nancy, you mentioned that experienced users of condoms in Bangladesh have a long list of reasons why they find condoms advantageous. I wonder if you would give us that list and tell us if some of them reported increased pleasure and increased enjoyment?

Nancy Williamson: They mentioned safety issues as most important, but they did mention the longer sexual intercourse, and the ease of insertion with a lubricated condom. People certainly do see a lot of advantages and one of the conclusions of the study was that we try to promote these advantages more instead of only emphasizing the safety issue. I think that it is very important to find out what people like about condoms and other barrier methods.

Elizabeth Ngugi: I would like to comment that the gentle loving couples need to be seen differently from commercial sex workers in the issue of acceptability. I suggest that in your working group you need to look at this issue.

Nancy Williamson: Yes, that is absolutely the case. There is still a lot of skepticism among people that work in the AIDS field over whether it is even going to be possible to get married couples to use condoms. It introduces the element of distrust into a relationship. We found in the AIDS field that it is very easy to get prostitutes to use condoms in many settings. Their clients sometimes resist but the prostitutes are very, very interested. The issue of using condoms in marriage, or in long-term relationships, is a very difficult issue and I think that it is an important distinction.

Sergia Galván: I would like to hear more about the influence that providers have on the use of the diaphragm, particularly physicians, including the gynecologists.

Nancy Williamson: Jane Cottingham pointed out that there are provider biases against the diaphragm. This may be due, in part, to the more intensive counseling requirements; it is an important issue.

4

FEMINIST PERSPECTIVE ON BARRIER USE

Judy Norsigian

Boston Women's Health Book Collective
240a Elm Street
Somerville, MA 02144

Of course, there is no one feminist perspective on anything, let alone contraception. My own views are those of a lay person who has talked about this topic with hundreds of women both in my own country and in other countries for almost 20 years. Today, I would like to cover a bit of history, concluding with the major reasons most women's health advocates place such great emphasis on barrier methods.

The issue of barrier methods poses substantial challenges to all who want to see more successful prevention of both unintended pregnancy and sexually transmitted diseases (STDs). As we discussed earlier, one of the challenges we face is changing the negative perceptions of most health and medical care providers towards barrier contraception. Another challenge is broadening our views of lovemaking and sex to acknowledge the pleasure and satisfaction of sexual intimacy other than sexual intercourse per se. For example, the experience of sexual pleasure and of stimulation to orgasm for many women does not come primarily from penetration by a penis. This is especially critical when we talk about interrupting the "sex act" or the "main meal," which for men and women sometimes means something quite different. Understanding what is pleasurable for both partners in a sexual relationship will help guide us towards developing more useful barrier methods.

Women have long been interested in barrier methods and promoting greater barrier method use. An early example of this in the United States can be found in the work of Mary Ware Dennett, an activist less well known than Margaret Sanger. In the 1920s Dennett authored among other publications "The Sex Side of Life," a booklet designed for teenagers which included frank discussion of barrier methods. Even then, activists and educators acknowledged the role of barriers in disease as well as pregnancy prevention.

Barrier Contraceptives: Current Status and Future Prospects, pages 69–74

Skipping up to the late 60s and early 70s, a number of women's groups emphasized barrier contraception as part of larger campaigns calling for legalized abortion and increased availability of contraceptive methods. For those of us in the *Boston Women's Health Book Collective* much of our early understanding of the importance of barrier contraception in preventing STDs resulted from a long-standing relationship with Edward Brecher, whom I consider one of the most important sex educators of the past few decades. From him and from staff at the Centers for Disease Control - for example, Dr. Paul Wiesner, then head of the Venereal Disease Control Division - we were able to gather important though not well-publicized information that was included in the 1973 edition of our book, <u>Our Bodies, Ourselves</u>. We also pursued this issue in our own discussions with women in small group settings and decided to allocate substantial space in the book to barrier methods and STD prevention.

Others, such as Dr. Robert Hatcher in Atlanta, also produced materials beginning in the 1970s with similar emphasis on the disease prevention role of barrier methods. He was particularly concerned with issues of acceptability and with making barrier method use more attractive to younger populations. Judith Bruce and Bruce Shearer produced an outstanding monograph entitled <u>Barrier Methods Reconsidered</u>, published by the Population Council. Many of you at this meeting also have been active professionals in this arena. Most recently, the barrier methods discussion in the birth control chapter in the 1992 update of our book, <u>The New Our Bodies, Ourselves</u> benefitted considerably from the careful reading and comments of Dr. Felicia Stewart and Professor James Trussell. I mention these examples in particular, because they represent positive models of collaboration among professionals and laypersons.

Women's health advocates began to call formally for greater barrier method use at the First International Women and Health Meeting in Rome in 1977. Subsequent Women and Health Meetings, for example in Costa Rica (1987) and in Manila (1990) repeated this call to greater action. Women argued for more barrier method research as well as for greater availability of existing barrier methods. Many women repeatedly indicated how difficult it was to obtain barrier contraception in developing country settings, even when there was provider interest and support. Increased interest in STD prevention, especially HIV/AIDS prevention - probably more than any other single factor - was responsible for this heightened concern.

In 1986 the Population Council hosted what I think was the first of several dialogues about contraception and fertility regulation involving scientists and women's health advocates. In 1991 in Geneva the World Health Organization (WHO) and the International Women's Health Coalition hosted a dialogue involving about two dozen participants. The report from this meeting provides an excellent overview of the areas where scientists and women's health advocates both agree and disagree. In November 1992 WHO sponsored a similar dialogue in Manila, where feminist advocates argued for more barrier method research, especially on acceptability and efficacy of the diaphragm in their own country.

More conventional conferences and meetings, like the Society for the Advancement of Contraception and the various international symposia on contraceptive research and development, such as the one held earlier this month (March 1993) in

Mexico City, have only recently incorporated into their agendas women's and lay perspectives on contraceptive research. Among the more controversial issues addressed at these meetings has been contraceptive vaccine research, which some women's health advocates would like to see halted altogether. Another has been the extent to which female-controlled barrier methods should be emphasized in countries where formal family planning programs have had relatively little experience making them available.

I'd like now to summarize why feminists and other women's health advocates have been drawn to barrier methods, especially in recent years. First, as I discussed earlier, there has been a longstanding concern with STDs such as gonorrhea and the tremendous toll that these diseases, often involving pelvic inflammatory disease (PID), take on women's lives. The HIV/AIDS crisis has only intensified this concern.

A second reason for many women had to do with avoiding the unpleasant effects and occasional health problems associated with the more effective contraceptives. Especially in the 70s and early 80s, women complained frequently about negative reactions to the high dose estrogen oral contraceptive and IUDs, notably the Dalkon Shield. Even yesterday on my way here I met a young Dominican gentleman who, after striking up a rather lively conversation, said, "My girlfriend wants to know what else you can use besides the pill, because it makes her gain weight and it makes her feel funny. What else can she use?" It turned out his knowledge of available contraceptives was quite limited, although he was a fairly well educated man. When I showed him an article with a picture of the Reality female condom, he asked me for a copy to share with a number of younger women whom he knew to be sexually active (not mutually monogamous like his partner and him). He seemed quite eager to advise these women about a new way to protect themselves from STDs.

A third reason for this keen interest in barrier methods has to do with learning more about one's body. Using a female barrier method properly often requires a better understanding of human reproduction, considerable comfort with touching ones genitals (so-called "private parts"), and a willingness to communicate verbally with one's sexual partner about using the contraceptive. Achieving this kind of knowledge and comfort with one's body is often an important first step towards better sexual relationships and more effective fertility control. It can also be a key to successful birthing experiences later on. Furthermore, the experiences of women's groups in places such as rural India and Ahman, Jordan, demonstrate that even women with strong cultural biases against touching themselves can, in fairly short time, develop a keen interest in using barrier methods to protect themselves from pregnancy and STDs. Cultural barriers do exist, but they should not be overstated.

A fourth reason - and not unrelated to the one just mentioned - has to do with achieving a greater sense of control in our sexual lives. The use of a women-controlled method is extremely important to many women. They want something that's not provider-dependent and which permits ready access to reversibility, especially if they should change their minds on short notice about the issue of preventing pregnancy. However, partner cooperation and support are often necessary for the successful use of female barrier methods. Obtaining this key ingredient may require deep societal change - for example, changes in deep-seated attitudes held by both men and women as well as changes in the power differential between men and women. In the meantime, women who want to use barrier methods often have great difficulty in their sexual negotiations with men.

To move this agenda forward women have called for the following: more policy-makers and more scientists committed to the development of barrier methods, especially those methods that offer both pregnancy prevention and protection from STDs; better training of family planning providers to help eliminate false negative assumptions about barrier methods; more study of barrier methods in developing country settings; and increased US AID funding for distribution of female-controlled barrier methods such as the vaginal contraceptive film and the female condom.

Our own organization has received numerous enquiries from other countries about the Reality condom, Lea's Shield, and the Femcap. Half of these are not coming from feminist or women's organizations, so I suspect that receptivity to barrier methods is already increasing, especially in places with heightened awareness of the STD/AIDS problem. In the U.S. we will soon have the opportunity to measure the popularity of the first female-controlled method with the potential of protecting against HIV infection. Many of us will be eagerly watching the sales figures for the Reality condom, as they will give us some indication of how successful women will be in obtaining partner cooperation. Also of interest will be the matter of sustained sales - will something like Reality, even if acceptable at first, be able to keep its users satisfied for more than a short period? Using something over and over again is different from being willing to try it once or twice. As with so many facets of our lives, the media could play a major role in shaping public opinion about barrier methods of contraception. Let's hope it is a positive role.

DISCUSSION

Penelope Hitchcock: I want to thank you for crystallizing some of the very important issues that are driving the feminist machinery here. I want to emphasize one aspect that you did not mention that I think is, frankly, an embarrassment to us. That is that we currently have no barrier methods that allow one to get pregnant but still protect against infection. I think it is extraordinary if you think about the human condition from a biological standpoint that our most important commitment is to have and raise healthy children. One of the things I think we have learned from the AIDS epidemic is that people, in the face of HIV infection, still wish to have families and I think it is a challenge for us to meet the family's and the woman's needs for protection against infection as well as contraception.

Judy Norsigian: Absolutely. Chris Elias and Lori Hisey have just finished a monograph on this very subject, and the need for microbicides that are not also spermicidal. I think it is one of the important first pieces of writing on this topic, and there is definitely a tremendous interest among women.

Jacqueline Forrest: One of the points that came up in your talk was something that is confusing me and that is that there really is a clear distinction between a method that is user-controlled and a method that is provider-dependent. They are treated as though those are the two choices, one or the other. It seems to me that that does not make sense. I am thinking of something like the female condom, for instance, and having a great deal of difficulty understanding how we can call that user-controlled when in fact the partner may exert as much control as the woman on whether it's going to be continually used, because it's something that he knows about and he can remove. Confidentiality of use may be as important in the user-control of the method as anything else. I could see under that definition something like Norplant, which is very provider-dependent being very user-controllable if in fact the provider system is completely responsive to whether the woman wants it inserted or removed. I really would appreciate it if you could give me some thinking into what you mean when you are talking about user-controlled methods. What are the characteristics that makes something user-controlled and another method not user-controlled?

Judy Norsigian: In the interest of time let me just pick on one issue and that's your ability to discontinue the method at will, not make an appointment to go get removal. In that sense the diaphragm, though it's provider-dependent in that you need to be fitted properly, is in the woman's control in the sense that she can stop using it at any point. I think there's an overlap here: Things are not necessarily uniquely provider-dependent or user-controlled. Something like the female condom is user-controlled in the sense that she can take use and she can remove it. Obviously if your partner feels this thing down there and objects, you're in trouble. But I think that there are gradations and I think for many women it's the ability to change their minds in a given setting about using or not using contraception.

Roberto Rivera: It seems very clear to me that when we speak about the factors that affect acceptability or correct use or the characteristics of an ideal barrier method that all decisions have to do with the human sexuality, with sexual behavior or with sexual practices. These are things about which we still know very little. If we want to move ahead in this area of acceptability and correct use, simultaneously there has to be a better knowledge of human sexuality. I don't know if family planning people are the ones who know how to do this. Really, there should be a closer collaboration and important input from people who come from the sexuality field.

Judy Norsigian: I think you've said it in a very beautiful way. There are many religious groups that have said we've got to get the parents comfortable with talking about sexuality issues so that these are not such taboo subjects even in the home. The ability to talk about all this has to begin somewhere. It's certainly not going to begin with the first time you're going to have sex. If you've never done any rehearsing, it's very difficult and being a little more open about sexuality is extremely important. I think that we have to move to the point where as a society -- and in the United States we have a real problem -- we can talk about these issues, not just have it blaring in the media where sex and violence are equated, but talk about it in a way that really helps people and addresses the need for human sexuality education.

Pramilla Senanayake: I very much enjoyed the points you made about the feminist point of view and I agree with most of them. But I also have some concerns about some of the points made on user-dependent classification. I think the distinction is better made not by user-dependent or provider-dependent but more on the discontinuation aspects that you mentioned, i.e., can the women discontinue the method at her own discretion. I think the oral contraceptive very much falls into that category because she can certainly stop it when she wants to, and I would like to put Norplant in that same category. I think the fact that you have to go to a provider to get your Norplant out is not necessarily the issue. The issue is can she or can she not get it removed at will. I know that some providers have been very difficult in allowing the woman to have that method out, and I think we need to solve that aspect of it. There is another point that has been made that worries me somewhat with regard to barriers. That has to do with effectiveness, and the message somehow is that barriers are fine because there's back-up abortion. There isn't back-up abortion in many, many parts of the world, either because it's not available or because the women themselves may not want to have an abortion. So I think, to some extent, effectiveness is important in looking at barrier versus other methods. I would also like to ask one question on the study in Sao Paulo. After diaphragms were made available, what was the acceptance? Was it much more widely accepted because it was available or was there some problem in the numbers there? Can I just have clarification on that, please?

Judy Norsigian: On the question of efficacy, I think we have to remember that for many women whether or not they're going to get pregnant with a particular method may weigh much less heavily than their sense of inconvenience or side effects associated with the product, whether they're real or imagined. Many women don't consider a higher rate of pregnancy as important an issue as other things. I think sometimes the providers place efficacy at the top of the list, whereas women sometimes place it lower down, and not just because they can get abortions as back-up. I think many women don't even think that far ahead when they're having sex.

With respect to the question you just raised about the diaphragm, two diaphragm studies in southern Brazil were reported on, one in which women kept the diaphragm in all the time, did not use a spermicide and just took it out daily to wash it; they had a 2.6 percent failure rate. Another one in which it was used more conventionally at the time of intercourse, I think there was a 7.8 percent failure rate. For me, the critical issue in introducing new research on the diaphragm or any barrier method is that we begin with providers who do not have a strong negative bias against these methods. That will filter down to everyone. I think we need to begin with people who have more positive attitudes and then demonstrate with good studies that maybe these things are more effectively used by women and their partners than we had imagined. I'm very worried that initially conducting trials in settings where there are strong negative provider biases is going to influence the outcome.

PART II: BARRIERS RESEARCH

5

VAGINAL ANATOMY AND PHYSIOLOGY AS THEY RELATE TO SEXUAL INTERCOURSE

Subir Roy

USC School of Medicine
Department of Obstetrics and Gynecology
Women's Hospital
1240 North Mission Road
Los Angeles, CA 90033

This chapter includes a review of the embryology and anatomy of the vagina followed by a consideration of its physiology with special reference to issues that relate to vaginal contraceptive development.

EMBRYOLOGY

The undifferentiated gonad is a junction of the urogenital sinus with the Müllerian ducts (Figure 1) [Growdon, 1986]. Following invagination, the urogenital sinus forms the distal third of the vagina and with its surrounding muscles comprises the urogenital diaphragm. Following confluence of the medial portions of the Müllerian ducts, the upper two-thirds of the vagina, cervix and uterus are formed with two distinct and separate oviducts.

ANATOMY

The muscles comprising the urogenital diaphragm (Figure 2) [Moore, 1986] adjacent to the vagina include the bulbocavernosus muscles, which connect anteriorly to the symphysis pubis, surround the most external portion of the urethra and vagina, join the transverse perineal muscles and then the anal sphincter before attaching to the coccyx inferiorly. The vestibular bulbs are behind the bulbocavernosus muscles. These structures respond during sexual arousal and intercourse. Similarly, the Bartholin's glands produce secretions in the distal most portion of the vagina during stimulatory activity.

Barrier Contraceptives: Current Status and Future Prospects, pages 77–89

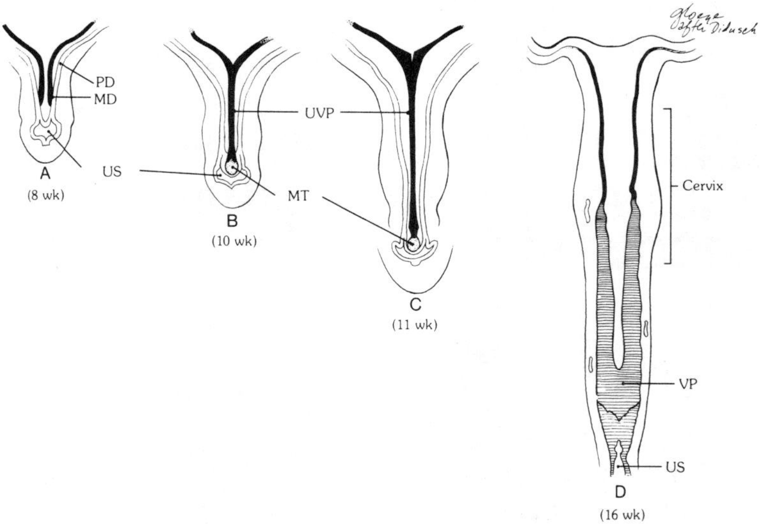

Fig. 1. Early embryologic development of the genital tract (A, B, C) and vaginal plate (D). PD, paramesonephric duct; MD, mesonephric duct; US, urogenital sinus; MT, Müllerian tubercle; UVP, uterovaginal primordium; VP, vaginal plate. [Growdon WA, 1986].

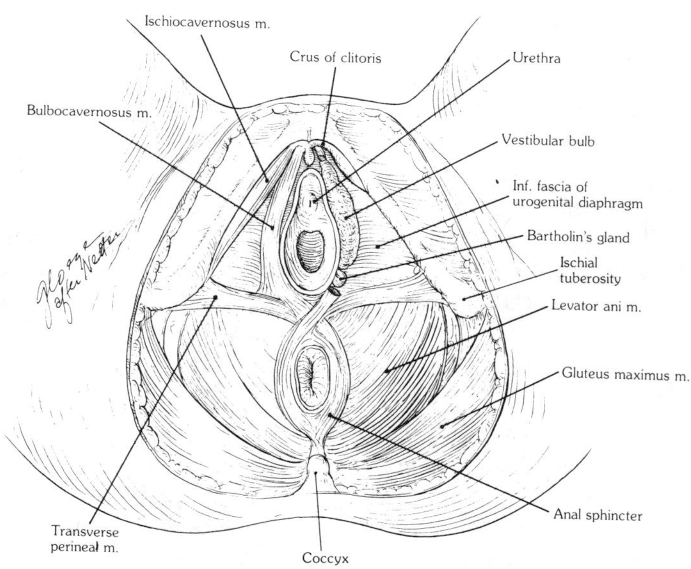

Fig. 2. The perineum, showing superficial structures on the left and deep structures on the right. [Moore JG, 1986].

The pelvic diaphragm is comprised of the levator ani muscle, which arises along the arcuate tendon bilaterally and is joined in the midline, where it is penetrated by the urethra anteriorly, the vagina in the middle and the anus posteriorly (Figure 3) [Curtis, 1946]. The pelvic diaphragm surrounds the distal vagina and is attached to the urogenital diaphragm (Figure 4) [Graney and Vontver, 1990]. Additional support for the vagina comes from the cardinal ligaments bilaterally, which surround the upper third of the vagina and which are the main suspensory ligaments of the uterus (Figure 5) [Nichols and Milley, 1972].

The arterial blood supply for the vagina comes bilaterally from descending branches of the uterine artery to the upper vagina and by direct vaginal arteries to the remainder of the vagina. The venous return is devoid of valves and forms a lacuna, essentially surrounding the entire vaginal tube in a venous plexus that connects with uterine, vesical and rectal plexuses [Williams et al. 1989].

The innervation of the vagina consists of bilateral hypogastric branches, predominantly autonomic with a few visceral afferents. Innervation of the introitus and the distal vagina consists of bilateral pudendal branches with a few somatic afferents.

Both vessels and nerves course together along connective tissue septa, entering and departing the vagina laterally between the prevesicle and vesicovaginal spaces superiorly and between the pararectal and rectovaginal spaces inferiorly (Figure 6) [De Lancey, 1992]. The "spaces" between these connective tissue septa permit the bladder, vagina and rectum to function independently of one another [Nichols and Milley, 1978]. A portion of the venous drainage of the vagina joins the superior hemorrhoidal venous return, co-mixes with the inferior mesenteric blood supply and thereby enters the portal system [Williams et al. 1989]. Thus, some vaginally administered substances may have direct portal effects analogous to oral ingestion.

Anatomists [Curtis, 1946] refer to the vagina as a distensible muscular and membranous canal extending from the cervix to the external genitalia. Typically, the anterior wall is 6.5 cm and the posterior wall is 8 cm in length. Along the midportion, it appears to form a transverse slit, "H-shaped" in appearance. It is composed of medial longitudinal columns with obliquely placed ridges, or rugae, which enable the vagina to accommodate or to stretch without laceration.

The wall of the vagina is comprised of three structures: the fibrous outer layer, the muscular middle layer and the epithelial inner layer. The fibrous layer joins with pelvic fascia and extraperitoneal tissues to form a dense sheath, which clothes and connects the vagina to surrounding tissues (e.g., the bladder, urethra, rectum and pelvic diaphragm). It serves as a substrate through which course vaginal nerves and vessels, including the venous plexus. It is comprised of bundles of connective tissue containing elastic fibers. With sexual arousal, it may function as an "erectile" tissue similar to the bulbs of the vestibule.

The muscular coat is comprised of an inner circular layer and an outer longitudinal one, both of which are irregularly arranged with considerable amounts of connective and elastic tissues having a gradual transition to the fibrous layer. In the distal portion, it intermingles with pelvic and urogenital diaphragm muscles.

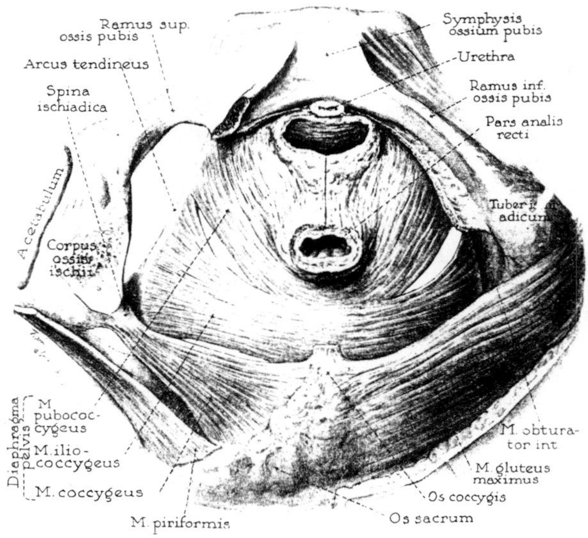

Fig. 3. Portions of the right innominate bone have been removed (cut through the inferior pubic ramus near the symphysis and the ischium superior to the level of the spine) to demonstrate the attachments, form and relationships of the pelvic diaphragm. The following features are especially important: the clefts between the adjacent muscular elements of the pelvic diaphragm; tendinous origin of the pubococcygeal and iliococcygeal divisions of the levator ani; the manner in which the pubic passes deep to the iliac division; the proximity of the urethra to the pubic symphysis, and the greater distance between vagina and anus; the complementary position which the coccygeus holds to the iliococcygeus (left), the manner in which it is covered by the gluteus maximus (right); the dorsal situation of the piriformis. [Curtis AH, 1946].

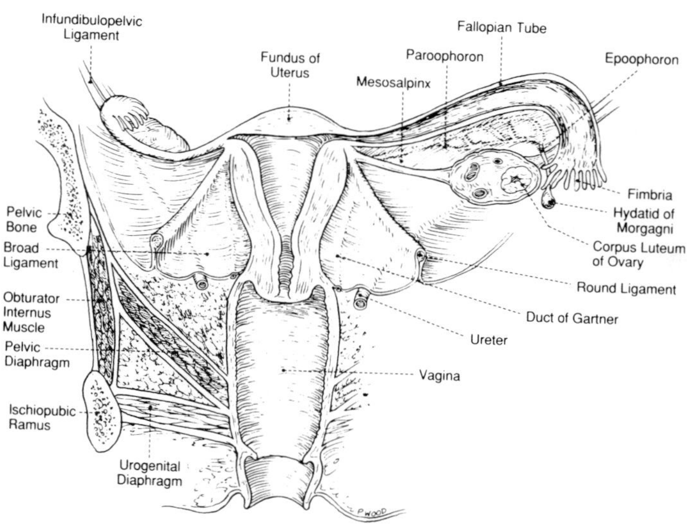

Fig. 4. Coronal section of pelvis illustrating broad ligament, endopelvic space, pelvic diaphragm, and urogenital diaphragm. [Graney and Vontver, 1990]

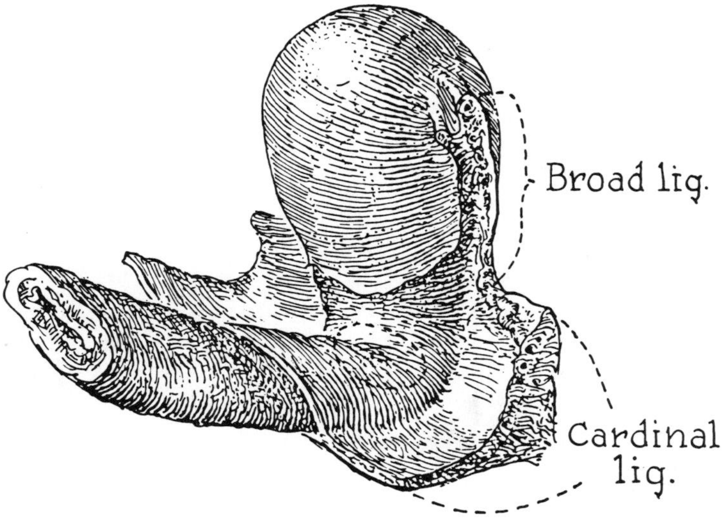

Fig. 5. The cardinal ligament is shown as it attaches to the lateral portions of both cervix and upper third of the vagina. Notice that it follows the angulation of the intersecting axes of these two organs. [Michols and Milley, 1972].

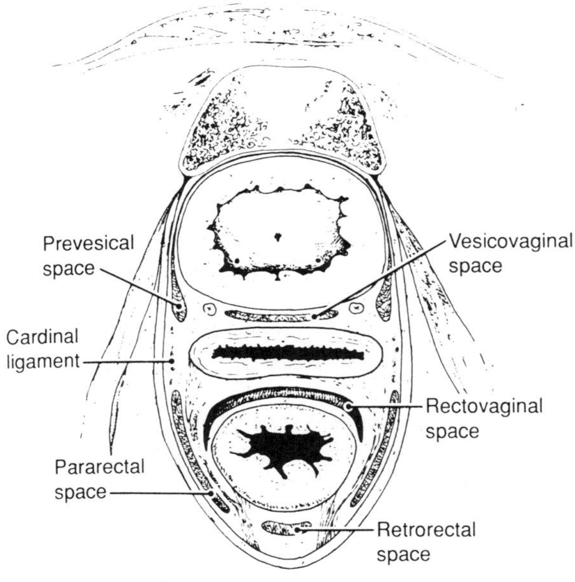

Fig. 6. Cross section of the pelvis showing cleavage planes. [De Lancey, 1992].

The epithelial layer is comprised of stratified squamous epithelium resting upon numerous vascular papillae of fibro-elastic tunica propria. This epithelium responds to estrogens with corresponding greater superficial cells and fewer intermediate and basal cells (Figure 7) [Wagner and Levin, 1978].

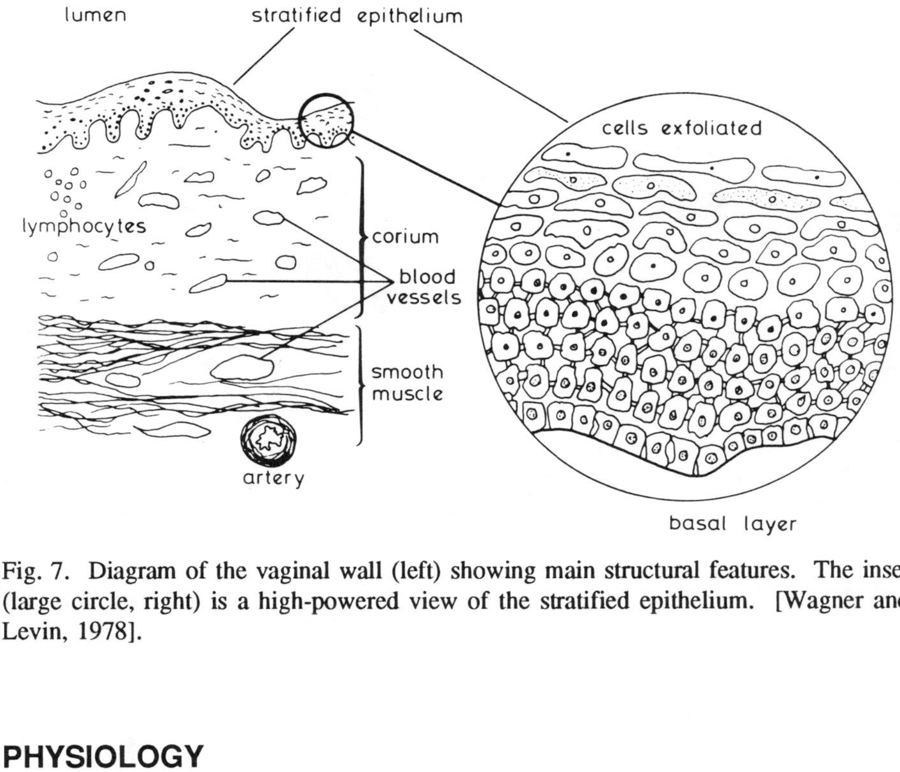

Fig. 7. Diagram of the vaginal wall (left) showing main structural features. The inset (large circle, right) is a high-powered view of the stratified epithelium. [Wagner and Levin, 1978].

PHYSIOLOGY

The normal pH from healthy women of reproductive age is between 3.4 to 6.0, from the introitus to the posterior vaginal fornix where cervical mucus pools [Voeller and Anderson, 1992]. The acidic pH is produced by the metabolic production of lactic acid by vaginal flora. Normal semen is slightly alkaline, having a pH of 7.0 to 8.4, altering vaginal pH within 8 seconds of ejaculation and maintaining this change for many hours [Fox, Meldrum and Watson, 1973]. Semen volume averages 3.3 ml (range: 0-13 ml) and has a greater buffering capacity than vaginal secretions associated with sexual arousal, which are at a more "physiologic" pH.

In normal women, bacterial flora consist of a mixture of aerobic and anaerobic flora averaging a concentration of 10 billion per cc of vaginal secretions. Under adverse conditions, these may adhere to the epithelium and may lead to vaginitis or vaginosis, whereby vaginal pH is elevated and/or local inflammatory reactions are observed.

Protection from infection is provided by cervical mucus containing IgA, cellular elements and other antimicrobial secretions [Cohen, Weber and Mardh, 1990]. Alterations of these factors may impact safety and efficacy of vaginal contraceptives.

Data regarding vaginal trauma following sexual intercourse are not consistent. Either trauma has not been observed [Soper, 1992] or it has been observed in 61 percent of patients [Norvell, Benrubi and Thompson, 1984] while it was reported in all rape victims [Slaughter, 1992]. Some vaginal trauma was observed even in some subjects who asserted abstinence and non-use of condoms [Roddy et al; 1992]. Vaginal trauma, especially if it should be facilitated by vaginal contraceptives, should be carefully studied because of the possibility of toxic shock syndrome development.

The basis for vaginal engorgement with sexual arousal and the corresponding vaginal secretion production appears to be the result of release of vasoactive intestinal peptide (VIP) [Levin, 1991]. VIP is present in nerves closely applied to blood vessels in the vaginal wall. With sexual arousal, local release of VIP increases arteriolar dilatation, venous return compression and, hence, "engorgement." This "engorged" status favors transudation of fluid into the vaginal canal, an observation first made by Masters and Johnson [1966]. A variety of mechanisms have been alleged for this phenomenon [Wagner and Levin, 1978], but the concept of VIP may be the most persuasive.

Masters and Johnson pioneered the scientific investigations regarding sexual response enabling them to elucidate four phases [1961a, 1961b, 1966]. The vaginal component of sexual response during the excitement phase, under the influence of the parasympathetic portion of the autonomic nervous system (ANS), consists of engorgement of labia and clitoris and vaginal transudation. During the plateau phase, there is marked vasocongestion throughout the body. The labia and lower third of the vagina are especially engorged--the "orgasmic platform" develops reducing the vaginal diameter by a third to a half, and the vagina lengthens 3 to 5 cm with dilation of the upper two-thirds. During the orgasmic phase, under the influence of the sympathetic portion of the ANS, myotonic contractions of the urogenital and pelvic diaphragm muscles occur, repeating at 0.8-second intervals. It is generally believed that during the resolution phase, additional excitement cycles may be stimulated at any time after orgasm.

A discussion of sexual arousal cannot be complete without at least mentioning the anterior vaginal suburethral area, which is believed to be exquisitely sensitive to direct stimulation leading to engorgement and culminating in orgasm with discharge of "prostatic-like" fluid per urethra. This process was first described by Grafenberg (1950); thereafter, this anatomic region has been referred to as the "G-spot" and is an alternate primary site for sexual satisfaction to that of the clitoris, which was championed by Masters and Johnson [1966].

Because of the very large surface area of cervical mucus (the fractal geometry is immense) it is not surprising that foaming agents, suppositories and gels may not prevent sperm transport into cervical mucus as well as mechanical barrier methods (Table I), especially in situations where ejaculation occurs opposite the cervical os with great volume (Figure 8) [Dickinson, 1938].

TABLE 1. Pct Results Subject M.A.

Mean Number Sperm Per HPF*		
Method	Base Line	Contraceptive
Foam	38.4	44.7
Suppositories	19.7	14.6
Mech. Barrier	49.6	0.3

* With progressive motility
Bernstein G (1990)

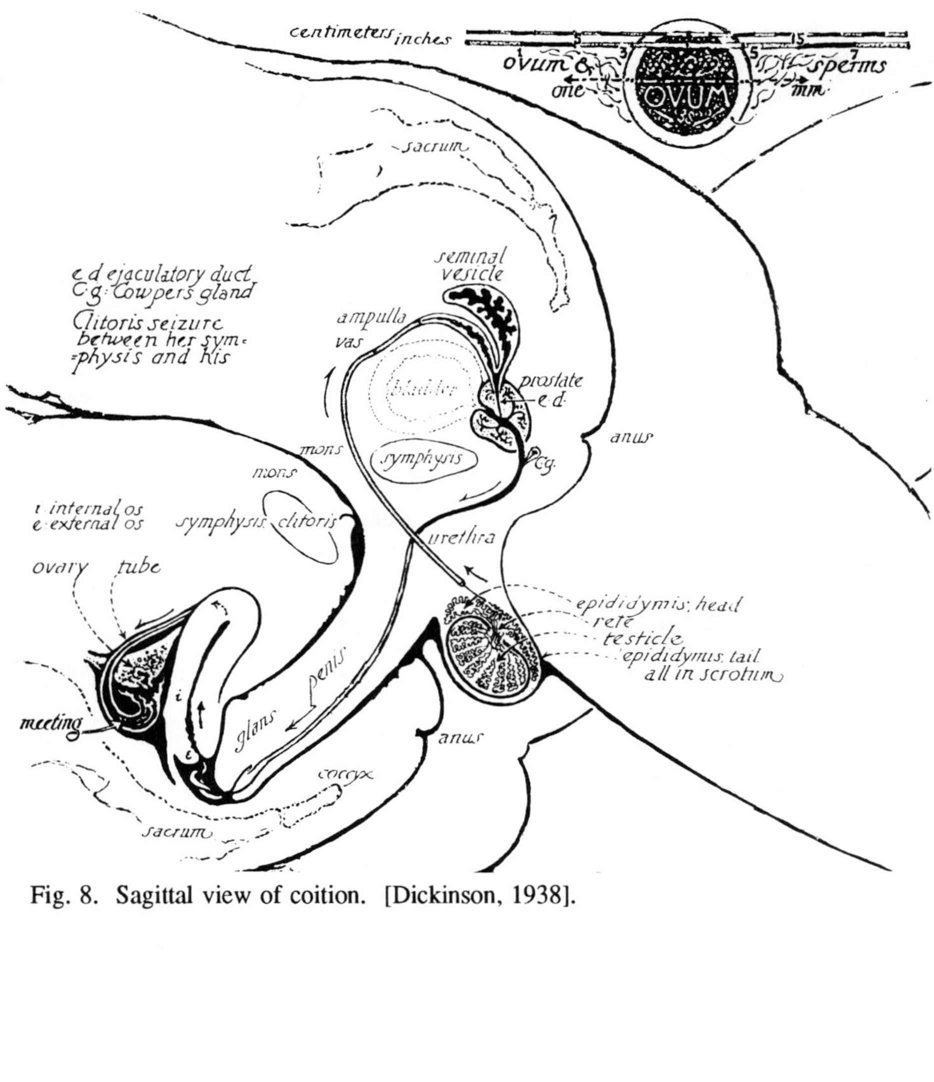

Fig. 8. Sagittal view of coition. [Dickinson, 1938].

FURTHER CONSIDERATIONS

Many women of reproductive age have anatomical changes as a consequence of parity including symptomatic cystocele, rectocele, uterine descensus or procidentia. These anatomic factors should be considered in the development of vaginal contraceptives.

Another consideration is coital position. How many vaginal contraceptives would work in non-traditional positions? In this era, should such a term as non-traditional be used?

Extremes of vaginal lubrication or semen volume should be considered in the design and testing of vaginal contraceptives. Sources of fluids during sexual intercourse are depicted in Figure 9 [Wagner and Levin, 1978]. Although the sources of fluids have been identified, the expected volumes, especially of vaginal transudation, are not well studied or understood.

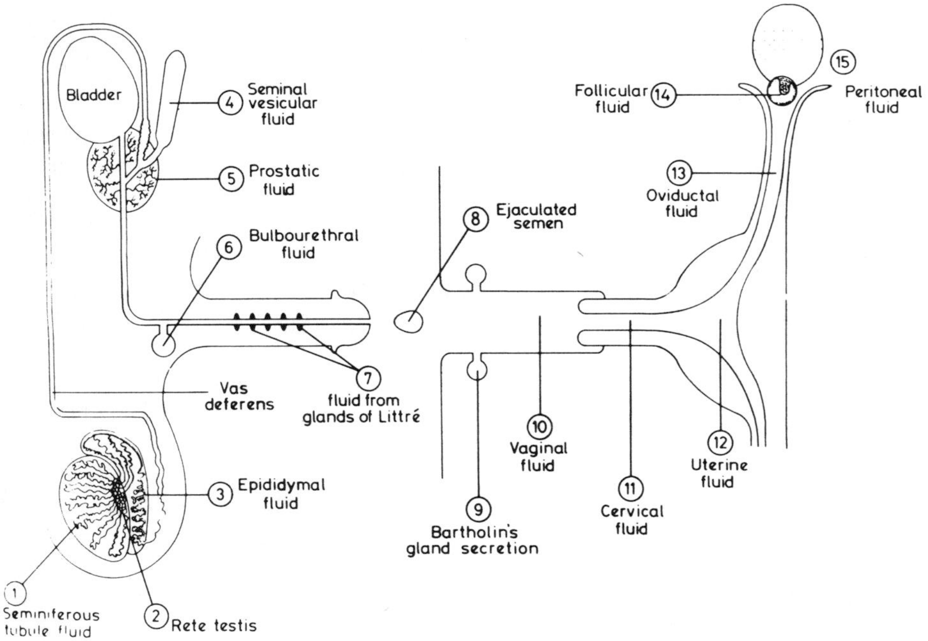

Fig. 9. Secretions manufactured by the glands and epithelia of the male and female genital tracts. [Wagner and Levin, 1978].

CONCLUSIONS

Our understanding of vaginal anatomy has not changed substantially over the past decade; however, we should remember that systemic effects may occur with vaginal administration of contraceptive compounds, especially if they return to the liver via the portal circulation. Compounds should continue to be tested with this in mind. Our understanding of sexual arousal, especially venous engorgement, has been enhanced with the identification and role of VIP. We still do not have much information regarding the amount of vaginal transudation associated with sexual arousal and how this may impact vaginal contraception. Most studies have been performed in patients with normal anatomical relationships. We should test vaginal contraceptive products in women with anatomic effects of parity engaging in coitus in positions that they normally use. Then, we will be able to provide products to patients of the real world.

REFERENCES

Anson BJ and Curtis AH (1946): Anatomy of the female pelvis and perineum. In: A textbook of gynecology (5th Ed.). W. B. Saunders Company, Philadelphia, Curtis AH (ed), p 20.

Bernstein GS (1990): Barriers: Contraceptive and Noncontraceptive effects. In: AIDS and Sex. An Integrated Biomedical and Biobehavioral Approach. Oxford University Press, New York, Voeller B, Reinisch JM, Gottlieb M (eds), p 349.

Cohen MS, Weber RD, Mardh P-A (1984): Genitourinary mucosal defenses. In: Sexually Transmitted Diseases (2nd Ed.). McGraw-Hill Inc., New York, Holmes KK, Mardh P-A, Sparling PF, Wiesner PJ (eds), pp 117-127.

De Lancey JOL (1992): Anatomy of the Female Pelvis. In: Te Linde's Operative Gynecology, (7th Ed.) Philadelphia, J.B. Lippincott Company, Thompson JD, Rock JA (eds), p 49.

Dickinson RL (1938): Control of Conception 2nd Edition. Baltimore, Williams and Wilkins Company, p 40.

Fox CA, Meldrum SJ, Watson BV (1973): Continuous measurement by radio-telemetry of vaginal pH during human coitus. J Reprod Fertil 33:69-75.

Grafenberg E (1950): The role of the urethra in female orgasm, Int J Sexol 3:145.

Graney DO, Vontver LA (1990): Anatomy and physical examination of the female genital tract. In: Sexually Transmitted Diseases 2nd Edition. McGraw-Hill Inc, Holmes KK, Mardh P-A,Sparling PF, Wiesner PJ, et al. (eds), p 110.

Growdon WA (1986): Embryology and Congenital Anomalies of the Female Genital System. In: Essentials of Obstetrics and Gynecology. W. B. Saunders, Philadelphia, Hacker NF and Moore JG, p 249.

Levin RJ (1991): VIP, vagina clitoral and periurethral glans--an update on human female genital arousal. Ex Clin Endocrinol 98(2):619.

Masters WH and Johnson VE (1961a): Orgasm, anatomy of the female. In: Encyclopedia of Sexual Behavior,, Hawthorn Books, New York, Ellis A and Abarbanel A, (eds).

Masters WH and Johnson VE (1961b): The physiology of the vaginal reproductive function. West J Surg 69:105-120.

Masters WH and Johnson VE (1961c): Intravaginal environment: I. A lethal factor. Fertil and Steril 12:560-580.

Masters WH and Johnson VE (1966): Human sexual response. Little, Brown and Co., Boston.

Moore JG (1986): Female reproductive anatomy. In: Essentials of Obstetrics and Gynecology. W. B. Saunders, Philadelphia, Hacker NF and Moore JG, p 4.

Nichols DH and Milley PS (1988): Clinical anatomy of the vulva, vagina, lower pelvis, and perineum. In: Gynecology and Obstetrics, Sciarra JJ (ed). J.B. Lippincott Company, Philadelphia, Vol. 1, Chap 1., p 9.

Nichols DH and Milley PS (1987): Functional pelvic anatomy: the soft tissue supports and spaces of the female pelvic organs In: the human vagina, Vol. 2. Elsevier/North Holland Biomedical Press, New York, Hafez ESE and Evans TN (eds),.pp 21-37.

Norvell MK, Benrubi GI, Thompson RJ (1984): Investigation of microtrauma after sexual intercourse. J Reprod Med 29:269-271.

Roddy RE, Cordero M, Cordero C, Fortney JA (July 19-24, 1992): A dosing study of nonoxynol-9 and genital irritation. International Conference on AIDS Vol 8, Issue No. 2. p B-221 (Abstract No. PoB 3783).

Slaughter (1992): Personal communication.

Soper DE (1/31/92): Evaluation of the effect of using the female condom on the female lower genital tract. Proceedings of the U.S. Food and Drug Administration Obstetrics and Gynecology Devices Panel. Rockville. MD.

Voeller B, Anderson DJ (1992): pH and related factors in the urogenital tract and rectum that effect HIV-1 transmission, Topanga, The Mariposa Foundation.

Wagner G and Levin RJ (1978): Vaginal fluid. In: the human vagina, (Vol. 2). Elsevier/North-Holland Biomedical Press, New York, Hafez ESE and Evans TN (eds), pp 121, 122.

Williams PL, Warwick R, Dyson M, Bannister Lawrence H (1989): Gray's Anatomy (37th Ed.) Churchill Livingstone, New York, pp 816.

DISCUSSION

Bob Staab: A lot of the products that we look at today are dissolved or need to dissolve prior to use. Do you, or does anyone here, have a good feel for what the volume might be or what the range of volumes might be in a normal woman in the excited state?

Subir Roy: That's an excellent question. We know about the range of values for seminal ejaculate, but do we really have any data about the amount of normal vaginal fluids that are transudated or found in the female in an excited state? I couldn't find these data and I think that's exactly the sort of information that would be very useful to all of us.

Jeffrey Spieler: James Trussell showed us that parous women had from 2-3 times as many unplanned pregnancies -- these are perfect users -- but in the diaphragm group there was no difference between nulliparous and parous women in terms of perfect use. Could you tell us why you think that might have been the case, based on anatomy and your understanding of the manner in which caps, sponges, and diaphragms are situated in the vagina?

Subir Roy: It's fair to say that a parous women has less muscular support and perhaps a greater degree of distensibility even in the pre-excited state. Therefore anatomical devices that are relatively small, like a sponge, or a cervical cap that doesn't stay on well, may become dislodged in a parous vagina. The diaphragm as used in those parous women appeared to better compensate for their anatomy than these other two devices which are relatively limited in their size.

Zena Stein: Have you got information on special properties of vaginal epithelium, including perhaps the pH and susceptibility to infection, in the young adolescent?

Subir Roy: Where they appear to differ is that many adolescents have the squamocolumnar junction on their ectocervix. The squamous cells are more resistant to transmission than are the columnar cells. If the columnar cells are out on the ectocervix, and are then exposed to vectors such as CT, GC, and HPV, they're more likely to become infected and to lead to long-term complications, including upper genital tract infection and cervical dysplasia. That's probably also affected by smoking and number of sexual partners.

Alfred Shihata: We know that menstrual blood increases the pH, making the vagina more alkaline. Absence of the cervix plug during menses also lowers the resistance of the upper tract to invasion by microorganisms. What is the consensus regarding intercourse during menstruation in the face of the presence of HIV in the semen and/or the menstrual blood. What is the risk of transmission in either case?

Subir Roy: I would prefer to let Bruce Voeller or Debbie Anderson answer the question, and neither is here.

Kamran Moghissi: When you ask whether sex during menses is safe, you must ask, "safe with respect to what?" As far as pregnancy is concerned, it probably is safe. As far as transmission of sexually transmitted diseases is concerned, it's probably less safe. There's no question that the cervical plug does prevent the transmission of certain sexually transmitted diseases. There is evidence now, and we have some new evidence in our laboratory, that cervical mucus has some bacteriostatic and bacteriocidal activities with respect to certain microorganisms. However, this is not true as far as HIV is concerned.

6

MICROBIOLOGY OF THE VAGINA AND CERVIX

Peter Mason and Lovemore Gwanzura

Department of Medical Microbiology
University of Zimbabwe Medical School
Harare, Zimbabwe

THE PHYSICAL ENVIRONMENT

The human vagina is lined by a non-secretory, stratified squamous epithelium and bathed in fluids from Bartholin's and Skene's glands, from the mucus secreting cells of the cervix and from oviductal and uterine sites [Paavonen, 1983]. This complex mixture of salts, urea, carbohydrates (including glucose and glycogen), fatty acids, proteins, iron chelators and other macromolecular substances forms a rich nutrient medium within which many organisms readily grow [Geshnizgani & Onderdonk, 1992]. The histology of the lower cervix differs considerably from that of the vagina, with large numbers of mucus secreting and ciliated epithelial cells, but the external surface is covered in fluid continuous with that of the vaginal fornix.

MICROBIAL FLORA

Despite being at a much less acid pH, the flora of the external os of the cervix is qualitatively similar to that of the vagina [Osborne et al., 1979; Larsen and Galask, 1980], and differences are only apparent with organisms that are present in low numbers [Keith et al., 1972]. A number of studies document in detail the microbial flora of the healthy vagina and cervix, and these have been reviewed by Warren [1988]. The range of organisms that is commonly present includes diverse groups (Table 1), indicating the vagina to contain many microenvironmental niches within which particular organisms may find or be able to create conditions conducive to their survival and growth [Bartlett et al., 1977]. The flora of the lower vagina, for example, may be related more closely to that of the perineum, while that of the high vagina and cervix may be considered the "true" vaginal flora [Duerden, 1980].

Barrier Contraceptives: Current Status and Future Prospects, pages 91–103

TABLE 1. Bacteriology of the healthy vagina

"NORMAL" BACTERIAL FLORA

AEROBIC		ANAEROBIC
	Gram-positive rods	
Lactobacilli		Lactobacilli
Corynebacterium		Clostridium
		Eubacterium
	Gram-positive cocci	
Streptococcus		Peptococcus
a-hemolytic		Peptostreptococcus
b-hemolytic (GBS)		
GDS		
Enterococci		
Staphylococci		
coagulase negative		
coagulase positive		
	Gram-negative rods	
Enterobacteria		Bacteroides
E.coli		fragilis
Klebsiella		melaninogenicus
Proteus		bivius
		disiens
Gardnerella		Fusobacterium
Pseudomonas		

Note: Many other organisms may occur

Quantitative studies have documented the relative importance of the various species, though with differing findings. Bartlett et al. [1977] reported that anaerobic colony counts were about 10 fold higher than aerobic colony counts, while Masfari et al. [1986], described the dominant flora as aerobic. There may be a number of reasons for these different findings [Warren, 1988], but whichever may be the case, it is clear that disturbance of the normal microflora is an important factor determining the health status of the vagina.

It is generally recognized that the dominant flora in the healthy vagina is a mixture of Lactobacillus species and anaerobes [Redondo-Lopez et al., 1990]. This is not, however, true in all populations studied, and low Lactobacillus carriage rates have been recorded, for example, in women in Africa [Ekwempu et al., 1981; Mason et al., 1983; Mason et al., 1989a]. Eschenbach et al. [1989] have correlated an absence of hydrogen peroxide producing Lactobacilli with bacterial vaginosis, and suggest these strains represent "an antimicrobial defense mechanism of the vaginal ecosystem." Other studies have shown specific antimicrobial therapy to be much more effective in the treatment of vaginitis than are non-specific procedures designed to acidify the vagina or artificially introduce Lactobacilli to stimulate this natural defense [Fredricsson et al., 1989]. Our experience in Zimbabwe is that, with the possible exception of gonococci, the absence of aerobic Lactobacilli does not correlate with infection by vaginal pathogens (Table 2) and the precise role of the dominant flora is thus still unclear. This flora is not static, but varies in response to the physical and biochemical environment and with the health status of the host. While it was thought that acid-producing bacteria were themselves responsible for the low vaginal pH, a more important mechanism is probably the utilization of glycogen by epithelial cells. Glycogen levels in the epithelium are under hormonal control, and vaginal pH is thus controlled indirectly by estrogen production.

TABLE 2. Isolation of pathogens from women in Harare

	No (%) isolates	
	Lactobacillus isolated (n = 41)	Lactobacillus not isolated (n = 148)
N. gonorrhoeae	3 (7%)	22 (15%)
T. vaginalis	8 (20%)	39 (26%)
G. vaginalis	14 (34%)	47 (32%)
Mycoplasma	13 (32%)	35 (24%)
Ureaplasma	13 (32%)	56 (38%)

DEVELOPMENT CONSIDERATIONS

Infant and Child

The vaginal tract becomes colonized, during and after birth, by exposure to organisms in the maternal genital tract and in the surroundings. There are few detailed investigations of genital microbiology in asymptomatic infants, but in neonates maternally-derived estrogen would allow the development of a "pseudomature" genital microflora, resembling that of a post pubescent female, where acid-tolerant bacteria predominate. After two to three weeks, the estrogen levels fall sharply and the vaginal epithelium exfoliates, with reduction in glycogen levels. Acid production by both epithelial cells and resident bacteria decreases dramatically, and the vaginal pH rises to neutrality. Acid-tolerant organisms no longer have an advantage, and a large variety of different species may be found, though numbers of each species may be low. Anaerobes such as Bacteroides sp., Clostridium sp. and anaerobic cocci are frequent colonizers at this age [Hammerschlag et al., 1978a,b].

Puberty

With the onset of puberty, estrogen levels increase, and glycogen is deposited in the thickening vaginal epithelium. Vaginal pH decreases to acid levels, once again selecting the acid-tolerant organisms, particularly Lactobacilli. The vagina becomes less favorable for anaerobes, and while the variety of organisms may now be more limited, total numbers of individual species may be very high, of the order of 10^8 per gram. During the menstrual cycle, changes in the structural and physiological milieu correlate with marked changes in bacterial populations. Quantitative studies have shown a 10- to 100-fold decrease in bacterial numbers in the week prior to menses followed by a dramatic increase in the number of species as menstruation commences [Larsen & Galask, 1982]. A significant decline in the prevalence of many different species occurs in the week following menstruation.

Sexual activity impacts upon vaginal microbiology, with about twice the number of species present as in non-sexually active adolescent girls [Shafer et al., 1985], but it is now recognized that the principal source of many vaginal organisms, particularly the enterobacteria, Bacteroides, peptostreptococci, Gardnerella and mycoplasmas, is the intestinal tract [Holst, 1990]. Thus while sexual transmission of components of the normal flora may occur, it is of minor epidemiological importance with a more important role being played by endogenous sources. Organisms entering the genital tract may be selected by environmental conditions and become dominant, while conditions in other sites may enable only temporary colonization. Differences in the flora between genital and intestinal tracts may be explained by these dynamic considerations.

Pregnancy

There is a progressive decrease in numbers of anaerobes during pregnancy compared with a smaller increase in aerobic Lactobacilli. After normal vaginal delivery, there is a large increase in numbers of anaerobes up to the third postpartum day, and this is followed by a gradual decrease, so that by the sixth week after delivery the flora has returned to normal [Goplerud et al., 1976; Larsen & Galask, 1980]. While genital pathogens are well characterized as having a deleterious effect on pregnancy outcome, little information is available on the effects of normal flora. In a study of fetal loss following transcervical chorionic villi sampling during early pregnancy [Hammarstrom & Marsk, 1990], a significant risk factor was the presence of high numbers and diverse species of bacteria in cervical samples. Changes in normal flora, particularly the absence of Lactobacilli, may predispose to colonization by organisms associated with the induction of preterm birth [Martius et al., 1988].

Menopause

Some studies [Osborne et al., 1979] indicate no statistically significant difference in number or type of bacterial species from the vagina or cervix of post menopausal women. Others indicate less frequent colonization by Lactobacilli and an increase in colonization by Gram negative rods and anaerobes [Larsen & Galask, 1980]. These changes are similar to those seen in prepubertal girls, and so presumably are related to changes in the glycogen content and metabolic activity of the vaginal epithelium resulting from decreased estrogen production.

Other Factors

In a recent study [Chow & Bartlett, 1989], there was no difference in the flora of women randomly assigned to the use of tampons or napkins.

Intrauterine devices or vaginal rings releasing hormonal contraceptives apparently have little effect on normal flora [Ulstein et al., 1987; Davies et al., 1992] apart from slightly increased colonization by anaerobes and streptococci [Goldacre et al., 1979]. Condoms reduce transmission of vaginal organisms to men [Holst, 1990], but resident flora are apparently little affected by use of the female condom [Soper et al., 1991]. Some spermicides may increase susceptibility to infection by killing Lactobacilli [Klebanoff, 1992].

Trauma, malignancy and the use of antibiotics may also have a significant impact on both the species present and their numbers [Warren, 1988].

CLINICAL SIGNIFICANCE

The major clinical questions that arise from studies on vaginal microflora are which are organisms most likely to be involved in infection of the genital tract and what are the factors that allow colonization and progression to disease?

With notable exceptions, the majority of bacterial organisms that are components of "normal flora" are rarely implicated in genital tract infection. Associations have been made among absolute numbers of Gardnerella vaginalis, anaerobes and vaginitis [Fredricsson et al., 1989]. In most cases, however, symptoms are associated with the presence of particular pathogens that usually are primarily sexually transmitted (Table 3). Many other organisms may cause infection in some age groups (e.g., Enterobius in young girls) or populations (e.g., Schistosoma haematobium in Africa), as well as infections that, although originating in the genital tract, exert their main pathogenic effect elsewhere (e.g., syphilis, HIV infection).

TABLE 3. Lower genital tract pathogens

Viruses	Bacteria
Herpesvirus	Neisseria gonorrhoeae
Cytomegalovirus	Mobiluncus sp.
Papillomavirus	
Mycoplasmas	Yeasts
Mycoplasma hominis	Candida albicans
Ureaplasma urealyticum	Torulopsis glabrata
Chlamydias	Protozoa
Chlamydia trachomatis	Trichomonas vaginalis

These same pathogens may, however, also be carried in the genital tract of healthy women. Lancefield group B streptococci for example may be found in the vagina of over 20 percent women, with no apparent symptoms [Mason et al., 1989a; Walsh & Hutchins, 1989]. Gonococci were recovered from 22 percent pregnant women who developed postpartum sepsis in Zimbabwe, but these women were asymptomatic before delivery when samples were collected [Mason et al., 1989b]. Clearly, the mere presence of an organism does not mean active infection, and other factors are involved.

The factors that are perhaps the most important in determining progression from colonization to infection include:

a The virulence potential of the organism
b The physical microenvironment
c The immune status of the host

Lactobacilli are frequent colonizers of the vagina, but have an extremely low virulence potential and are only rarely implicated in disease [Redondo-Lopez et al., 1990]. Their presence in the vagina, therefore, has minimal clinical implication, and indeed their presence in vaginal smears may be an indicator of vaginal health [Hillier et al., 1992].

An environment that is selective for an organism may allow it to reach sufficient numbers for expression of its virulence potential. This appears to be the case, for example, with Gardnerella vaginalis and anaerobes which may be carried in small numbers by many asymptomatic women, but when able to grow in large numbers, are associated with vaginitis [Spiegel et al., 1983]. Physical trauma, for example during childbirth, may allow organisms carried in the genital tract, such as enterobacteria, Bacteroides species or group B streptococci, to cause serious infection.

The development of an inflammatory response follows well established immune mechanisms that require recognition of the invader and adequate signaling of host defenses. Variation in immunoregulatory capability will thus play a major role in defining the degree of response that is made, and this variation may result from endogenous or exogenous factors [Larsen & Galask, 1980; Osborne et al., 1982]. The former include innate mechanisms such as the effect of vaginal mucus containing lysozymes and phagocytic cells capable of attacking and killing potential invaders.

Lactoferrin, or other chelating agents in the mucus, may limit the growth of iron requiring bacteria. Potential vaginal pathogens may induce suppression of these [Sturm, 1989] or more specific [Mason & Gwanzura, 1990] responses. The effects of general immuno-suppression on genital flora have been little studied. Ohm et al. [1978] described the microbial flora of 25 immunosuppressed transplant recipients and noted that, although there was little influence on the aerobic flora, anaerobes were isolated more frequently and in larger numbers in these, compared with normal women. Limited experience in Zimbabwe, where HIV prevalence in women attending STD clinics is over 60 percent, is that the microbial flora of HIV positive women shows large numbers of many diverse aerobic or facultatively anaerobic species. Mascellino et al., [1991] have described a 10- to 100- fold increase in numbers of aerobic and anaerobic bacteria in AIDS/ARC patients. Considering the importance of heterosexual transmission in the epidemiology of HIV infection, studies need to be undertaken urgently to define more precisely the interaction between HIV and normal vaginal microorganisms. It has been reported, for example, that the survival of HIV in the female genital tract may be affected by the activity of hydrogen peroxide producing Lactobacilli [Klebanoff & Coombs, 1991]. Preliminary unpublished studies in Zimbabwe indicate that only 10 percent of female STD clinic patients carrying hydrogen peroxide producing Lactobacilli were HIV-antibody positive, compared with over 30 percent of women with no Lactobacilli or with non-hydrogen peroxide producing strains. This finding, if confirmed by more detailed studies, has significant implications for HIV transmission in Africa because of the low prevalence of these Lactobacilli in African women.

CONCLUSION

The view was held for many years that the vaginal flora consisted almost exclusively of "Doderlein's bacilli" (Lactobacilli and other Gram-positive rods), the abundance of which was an indicator of vaginal pH and, thus, of health. Improved techniques, particularly with culture of anaerobic and fastidious organisms, have shown the flora to be a complex and dynamic mix, and the interrelationship of different species has a profound impact on the health status of the vagina. The exact role played by the "normal" flora in maintenance of this health status is, however, not well understood, and even the definition of the "normal" flora is not without controversy. It is widely recognized, however, that serious infections of the upper genital tract originate mainly from organisms that have initially become established in the vagina. Progress towards protection against these infections is, thus, dependant upon a more complete understand-ing of vaginal microbiology. With the current epidemic of heterosexually transmitted HIV infection in sub-Saharan Africa and the known relationship between HIV transmission and genital infections, it is becoming increasingly urgent to know the interactions that may occur.

REFERENCES

Bartlett G, Onderdonk AB, Drude E, Goldstein C, Anderka M, Alpert S, McCormack WM (1977): Quantitative bacteriology of the vaginal flora. J Infect Dis 1977 136:271-277.

Chow AW, Bartlett KH (1989): Sequential assessment of vaginal microflora in healthy women randomly assigned to tampon or napkin use. Rev Inf Dis 11 (Suppl 1):S68-73.

Davies GC, Feng LX, Newton JR, Dieben TO, Coelingh-Bennink HJ (1992): The effects of a combined contraceptive vaginal ring releasing ethinyloestradiol and 3-ketodegestrel on vaginal flora. Contraception 45:511-518.

Duerden BI (1980): The isolation and identification of Bacteroides spp. from the normal human vaginal flora. J Med Microbiol 13:79-97.

Ekwempu CC, Lawande RV, Egler IJ (1981): Microbial flora of the lower genital tract of women in labour in Zaria, Nigeria. J Clin Pathol 34:82-83.

Eschenbach DA, Davick PR, Williams L, Klebanoff SJ, Young-Smith KY, Critchlow CM, Holmes KK (1989): Prevalence of hydrogen peroxide-producing Lactobacillus species in normal women and women with bacterial vaginosis. J Clin Microbiol 27:251-256.

Fredricsson B, Englund K, Weintraub L, Olund A, Nord CE (1989): Bacterial vaginosis is not a simple ecological disorder. Gynec Obstet Inv 28:156-160.

Geshnizgani AM, Onderdonk AB (1992): Defined medium simulating genital tract secretions for growth of vaginal microflora. J Clin Microbiol 30:1323-1326.

Goldacre MJ, Watt B, Loudon N, Milne LJR, Loudon JDO, Vessey MP (1979): Vaginal microbial flora in normal young women. Br Med J 1:1450-1453.

Goplerud CP, Ohm MJ, Galask RP (1976): Aerobic and anaerobic flora of the cervix during pregnancy and the puerperium. Amer J Obstet Gynecol 126:858-868.

Hammarstrom M, Marsk L (1990): First trimester live pregnancy and subsequent fetal loss. Gynecol Obstet Invest 30:19-22.

Hammerschlag MR, Alpert S, Onderdonk AB, Thurston P, Drude E, McCormack WM, Bartlett JG (1978a): Anaerobic microflora of the vagina in children. Amer J Obstet Gynecol 131:853-856.

Hammerschlag MR, Alpert S, Rosner I, Thurston P, Seminc D, McComb D, McCormack MD (1978b): Normal flora of the vagina in children: normal and potentially pathogenic organisms. Pediatr 62:57-62.

Hillier SL, Krohn MA, Nugent RP, Gibbs RS (1992): Characteristics of three vaginal flora patterns assessed by gram stain among pregnant women. Amer J Obstet Gynecol 166:938-944.

Holst E (1990): Reservoir of four organisms associated with bacterial vaginosis suggests lack of sexual transmission. J Clin Microbiol 28:2035-2039.

Keith L, England D, Bartizal D, Brown E, Fields C (1972): Microbial flora of the external os of the premenopausal cervix. Br J Vener Dis 48:51-56.

Klebanoff SJ (1992): Effects of the spermicidal agent nonoxynol-9 on vaginal microbial flora. J Inf Dis 165:19-25.

Klebanoff SJ, Coombs RW (1991): Viricidal effect of Lactobacillus acidophilus on human immunodeficiency virus type 1: possible role in heterosexual transmission. J Exp Med 174:289-292.

Larsen B, Galask RP (1980): Vaginal microbial flora: practical and theoretical relevance. Obstet Gynecol 55 (Suppl): 100-113 S.

Larsen B, Galask RP (1982): Vaginal microbial flora: composition and influences of host physiology. Ann Intern Med 96:926-930.

Martius J, Krohn MA, Hillier SL, Stamm WE, Holmes KK, Eschenbach DA (1988): Relationship of vaginal Lactobacillus species, cervical Chlamydia trachomatis and bacterial vaginosis to pre-term birth. Obstet Gynecol 71:89-95.

Mascellino MT, Iona E, Iegri F, Catania S, Trinchieri V, Oliva P, Amenta L, Reverberi L, Sorice F (1991): Evaluation of vaginal microflora in patients infected with HIV. Microbiol 14:343-349.

Masfari AN, Duerden BI, Kinghorn GR (1986): Quantitative studies of vaginal bacteria. Genitourin Med 62:256-263.

Mason PR, MacCallum MJ, Patterson BA, Latif AS (1983): The vaginal flora of women in Zimbabwe. J Obstet Gynaecol E Centr Afr 2:102-104.

Mason PR, Katzenstein DA, Chimbira THK, Mtimavale L (1989a): Microbial flora of the lower genital tract of women admitted in labour to Harare Maternity Hospital. Centr Afr J Med 35:337-344.

Mason PR, Katzenstein DA, Chimbira THK, Mtimamvale L (1989b): Vaginal flora of women admitted to hospital with signs of sepsis following vaginal delivery, Caesarian section or abortion. Centr Afr J Med 35:344-351.

Mason PR, Gwanzura L (1990): Reduced lymphocyte responses to mitogens in natural and experimental trichomoniasis. Infect Immun 58:3553-3557.

Ohm, MJ, Scott JR, Galask RP (1978): Cervical-vaginal flora of immunosuppressed renal transplant patients. Amer J Obstet Gynecol 130:49-54.

Osborne NG, Wright RC, Grubin L (1979): Genital bacteriology: a comparative study of premenopausal with postmenopausal women. Amer J Obstet Gynecol 135:195-198.

Osborne NG, Grubin L, Pratson L (1982): Vaginitis in sexually active women: relationship to nine sexually transmitted organisms. Amer J Obstet Gynecol 142:962-967.

Paavonen J (1983): Physiology and ecology of the vagina. Scand J Infect Dis (suppl) 40:31-35.

Redondo-Lopez V, Cook RL, Sobel JD (1990): Emerging role of lactobacilli in the control and maintenance of the vaginal bacterial microflora. Rev Inf Dis 12:856-872.

Schafer MA, Sweet RL, Ohm-Smith MS, Shalwitz J, Beck A, Schachter J (1985): Microbiology of the lower genital tract in postmenarchal adolescent girls: differences by sexual activity, contraception and presence of nonspecific vaginitis. J Pediatr 107: 974-981.

Soper DE, Brockwell NJ, Dalton HP (1991): Evaluation of the effects of a female condom on the female lower genital tract. Contracept 44:21-29.

Spiegel CA, Davick P, Totten PA, Chen KCS, Eschenbach DA, Amsel R, Holmes KK (1983): Gardnerella vaginalis and anaerobic bacteria in the etiology of bacterial (nonspecific) vaginosis. Scand J Infect Dis Suppl 40:41-46.

Sturm AW (1989): Chemotaxis inhibition by Gardnerella vaginalis and succinate producing vaginal anaerobes: composition of vaginal discharge associated with G.vaginalis. Genitourin Med 65:109-112.

Ulstein M, Steier AJ, Hofstad T, Digranes A, Sandvei R (1987): Microflora of cervical and vaginal secretion in women using copper- and norgestrel-releasing IUCDs. Acta Obstet Gynecol Scand 66:321-322.

Walsh A, Hutchins S (1989): Group B streptococcal disease: its importance in the developing world and prospect for prevention with vaccines. Pediatr Inf Dis J 8:271-276.

Warren RE (1988): The normal flora of the vagina. In Hare MJ (ed): "Genital Tract Infection in Women." New York: Churchill-Livingstone, pp. 26-58.

DISCUSSION

David Soper: Peter, I enjoyed your review and have a comment and a question for you. One is that you speculated that Neisseria gonorrhoea might be "normal flora" and where I would agree that there are patients that have GC in their cervix and no evidence of inflammation, I don't think we can really classify that as a normal flora. That's a pathogen that may not be pathogenic at that particular time, but certainly needs treatment. What I would be most interested in as you speculate as to why the patients in your population don't have more lactobacilli in their vagina? The whole concept of normal vaginal flora being based around hydrogen peroxide producing lactobacilli and their ability to keep catalase negative microorganisms such as anaerobes kind of down in their overall quantities is a very attractive hypothesis. Were these patients that had such low incidences of lactobacilli truly normal, or did they douche six times a day?

Peter Mason: With respect to the first part, yes of course I agree with that -- that was kind of just thrown in just for controversy really, or to illustrate the problems of pathogen versus normal. We have looked for answers to why women in our part of the world do not seem to have lactobacilli. It's very difficult to get accurate information on douching by these women, but that does seem to be an important factor. Certainly in Zimbabwe women douche quite frequently and they will douche with many different compounds -- things like betadine if they can get it -- and there are many herbal compounds as well, which are used specifically for cleaning the vagina either before or after intercourse. What we're doing at the moment is trying to evaluate whether that is the reason. I've been looking at women who have say a more rural lifestyle as against the more urban lifestyle to see what that difference is. That certainly could be one of the factors, yes.

Debbie Rogow: My question is about the use of the diaphragm to control menstrual flow. One of the potentially attractive aspects for some users of the diaphragm is that it affords them with what may look like a relatively non-messy and free method of absorbing menstrual flow. I'm wondering whether given your comments about the nature of microbial growth in the vagina and near the cervix and at the cervix, if this raises concerns and, if so, what they are.

Peter Mason: The presence of hemoglobin at the time of menstruation is certainly something that stimulates the growth of bacteria which would otherwise be inhibited. The presence of blood in the genital tract for kind of a long period of time, or longer than would be expected, I would imagine would encourage the continued growth of those organisms. I would say that there were needs for studies on that. The studies that are being done on the use of diaphragm have not shown that it's particularly associated with any changes in bacterial flora, but that doesn't include the use of diaphragm for that particular purpose. I would regard the use of the diaphragm for that purpose with a great deal of caution from a microbiology point of view.

David Archer: You just mentioned that the use of spermicides has an impact on the quality of the microorganisms in vaginal flora. I wonder if the change that occurs after the introduction and continuous use of the spermicide -- leads to a new equilibrium in the vaginal flora that puts women at an increased risk of STD or HIV transmission. Is this new equilibrium dangerous from the standpoint of STD transmission, or is it just a new vaginal flora equilibrium that is different from the normal one that does not put women at risk?

Peter Mason: As I say, I'm not aware of any longitudinal studies that have been done to see at this point, if there's a change in the vaginal flora and within one week of that change there's an increased susceptibility to STD transmission. I think that that is the kind of study that needs to be done. The kind of studies that have been done have been to look at women who have used spermicides over a period of time and to see what impact that has had on the vaginal flora. It's had remarkably little impact. The only kind of major change really, unfortunately, has been in the proportion of women who have these hydrogen peroxide producing lactobacilli. And if indeed these particular lactobacilli are the important part of the flora in terms of prevention of STD, then we can expect that kind of scenario.

Michael Burnhill: I want to make two very quick points. I agree with everything you said except it comes from Harare, and back in the US, what constitutes normal is probably a poor word; we need to talk about usual flora -- the usual vaginal flora has changed over the last decade. It's moved in a direction of increasing colonization by colonic bacteria and unfortunately due to the wide-spread use of antibiotics, our normal colonic bacteria or usual colonic bacteria are more virulent. Bottom line: the usual vagina that I see in my practice is colonized with more varied bacteria. The point from all of this is that in working out vaginal contraceptives we have to be very careful that as a side product of improved spermicidal activity we don't have decreased peroxide-producing lactobacillus, because I firmly believe -- at least it's what I've seen in treating chronic vaginosis -- that you can't get back to a healthy state without being dominated by peroxide-producing lactobacilli, at least on this side of the Atlantic.

Peter Mason: I'd go along with that fully.

Elizabeth Ngugi: Just to say that the women of Africa, as far as I know, only those who practice commercial sex, douche and they douche with water, they douche with soap, they douche with herbs, but married couples don't douche as a matter of course. It would be considered irrespectable to the marriage and I really don't know of that kind of a situation. So I think when we are discussing groups of women I think we need to describe what kind of women they are so that we don't generalize -- that's one point. The other point is I'd like to know whether you studied the efficiency of transmission of gonococci from women to their sexual partners and also to their unborn children.

Peter Mason: Regarding the first point there, I would agree with you to a certain extent. The surveys that we've done have included both commercial sex workers and women who are presenting for delivery at a hospital. We found no difference in the lactobacillus prevalence rate in those two groups. If there is a reason for the lactobacillus being absent we believe it to be the same in both groups. As I say, it's only in recent years that we've been able to get information from women who are presenting for delivery. They also say they douche quite frequently, so perhaps this is a difference from further north to further south; I don't know. Regarding the transmission, we have no data on that at all.

Shakti Upadhyay: Do you think that sexual intercourse might have any affect on microbial flora in the vagina, because we know that semen first of all would increase the pH; secondly, it also has many immunosuppressive factors, so repeated sexual intercourse would have any adverse affect on such micro flora?

Peter Mason: It would certainly have a transient affect. Whether it would be a more permanent or more long-term affect I would have my doubts. The most recent evidence I think is that the majority of organisms that constitute the vaginal flora have their origin in the intestinal tract. I don't think that it would have a long-term impact on the genital flora, but a short-term impact in terms of inhibiting the growth of certain organisms, certainly--but not eliminating them, just inhibiting their growth.

7

VAGINAL BARRIER CONTRACEPTIVES AND INFECTION RISK

Felicia H. Stewart

Valley Center for Women's Health
Sacramento Sierra Medical Group
Sutter Medical Foundation
87 Scripps Drive
Sacramento, CA 95825

The relationship between genital tract infection risk and use of vaginal barrier contraceptives is not a simple one. Using these methods reduces a woman's risk for some kinds of infection, and increases the risk for others (see Table 1). Persuasive clinical studies show that serious sexually transmitted diseases -- including Chlamydial and gonococcal cervicitis and their long-term sequelae, such as pelvic inflammatory disease and ectopic pregnancy -- are less likely to occur among women using vaginal barrier methods than they are among women using other contraceptive methods or no method [Cates and Stone, 1992; Rosenberg and Gollub, 1992b]. Cervical neoplasia risk, linked to sexual transmission of human papilloma virus, is also mitigated for barrier users, as is the risk of acquiring trichomoniasis (see Table 2). On the other hand, using barrier methods or spermicide increases a woman's risk for several common genital tract problems, including urinary tract infection, candidiasis and bacterial vaginosis and for toxic shock syndrome (see Table 3). So the net impact of vaginal barrier or spermicide use cannot be explained simply as a general, beneficial microbicide effect within the vagina.

Barrier Contraceptives: Current Status and Future Prospects, pages 105–122

TABLE 1. Vaginal Barrier Method Use: Effect on Infection Risk

Decreased risk for:	Increased risk for:
Chlamydial cervicitis	Urinary tract infection
Gonococcal cervicitis	Candidiasis
Trichomoniasis	Bacterial vaginosis
Cervical neoplasia (HPV)	Toxic shock syndrome
Pelvic inflammatory Disease	
Ectopic pregnancy	

HIV risk is missing from Table 1 because clinical research results are contradictory [Bird, 1991; Cates, et al., 1992; Gollub and Stein, 1992; Stein, 1992; Stone, 1992; Peterson 1992; Voeller, 1992a]. A decreased risk for HIV transmission would be expected on the basis of protection against sexually transmitted diseases that cause genital ulcers and cervicitis, powerful cofactors in HIV risk, [Wasserheit, 1992] and on the basis of laboratory evidence that nonoxynol-9 is an effective virucide against HIV in vitro. Increased risk, however, is also biologically plausible. Nonoxynol-9 is a powerful detergent that can disrupt cell membrane integrity, including integrity of the vaginal epithelium and can cause inflammation severe enough to be visible or even symptomatic in a significant number of users. This direct effect of nonoxynol-9 on vaginal epithelium has led some researchers to suggest that the dose and frequency or duration of exposure to nonoxynol-9 may determine the outcome of vaginal barrier use impact on HIV risk. It is also possible that using vaginal barrier methods indirectly exerts an effect on HIV risk, apart from or in addition to, the direct effect of spermicide on vaginal epithelial cells. For example, an alteration in vaginal flora or vaginal pH caused by barrier use might, in turn, alter HIV infectivity or might alter a woman's susceptibility.

In order to interpret the results of clinical studies, (i.e., to identify and adjust for confounding variables) and in order to optimize existing products or develop new ones with favorable characteristics, a more sophisticated understanding of the vaginal environment is needed.

VAGINAL BARRIER METHODS

Possible Mechanisms of Effect on Infection Risk

Numerous mechanisms may be involved in determining the effect of vaginal barrier contraceptives on infection risk. Little is known about these mechanisms, and how each may contribute separately or in concert with others to infection outcome.

Direct Effects of Spermicide on Microbial Pathogens. Nonoxynol-9 and other detergent spermicides (including octoxynol, menfegol, benzalkonium chloride) and chlorhexidine are also microbicides. Laboratory studies show that nonoxynol-9 is active against *Neisseria gonorrhea, Treponema pallidum, Trichomonas vaginalis,* and *Gardnerella vaginalis,* and may also reduce infectivity of *Chlamydia trachomatis.* Antiviral activity has been demonstrated against herpes simplex virus types I and II, and against HIV free virus. Nonoxynol-9 may not have a solely protective effect in relation to pathogens, however. Common uropathogens including *Escherichia coli* and *Candida* species are much more resistant to nonoxynol-9 than are desirable, normal vaginal residents including *Lactobacillus* species [Klebanoff, 1992; McGroarty, et al., 1990a]. Spermicide use could result in selective colonization of the vagina with nonoxynol-9-resistant uropathogens (Hooten, et al., 1991b) because exposure to nonoxynol-9 enhances adherence of some uropathogens to epithelial cells, including some yeast strains [McGroarty, et al., 1990b]. An increased incidence of symptomatic candidiasis is a well-recognized consequence of vaginal barrier and spermicide use. Inflammation and local epithelial micro- and macro-ulceration and cracking are commonly observed with candidiasis, and could in turn explain an increase in HIV transmission susceptibility.

Mechanical Effects of Barrier Devices. By shielding the cervix from direct exposure to semen, a barrier device may reduce transmission risk for pathogens that require or favor columnar epithelium as a portal of entry. The presence of a barrier also may impede mixing of cervical mucus and cervical secretions with vaginal secretions, and may interfere with dispersion of semen. Shielding the cervix from exposure to prostaglandins in semen also may impact normal immune system responses to intercourse. Mast cells are predominantly located in the cervical endothelium, and their release and activation are triggered by exposure to prostaglandins.

Mechanical effects with use of barrier devices include local mechanical irritation or trauma, and transport of microbial pathogens from the external skin into the vagina during insertion.

Alteration in Endogenous Microbial Flora. Nonoxynol-9's microbicide activity is selective. For example, strains of *Lactobacillus* that produce hydrogen peroxide are more susceptible than are other *Lactobacillus* strains [McGroarty, et al., 1992]. Since production of hydrogen peroxide may play a role in maintaining optimal endogenous flora, this effect may be undesirable. Normal microbial residents of the vagina undoubtedly interact with each other in numerous ways: availability of micronutrients, metabolic by-products, and production of bacteriocins, for example, may all be important in maintaining homeostasis between members of the ecosystem. Drastic alterations in the number and type of residents, therefore, is likely to upset the optimal balance--at least temporarily--in ways that may be complex and difficult to predict.

TABLE 2. Vaginal Barrier Method Use: Decreased Infection Risk

Chlamydia & Gonorrhea		*Relative Risk*	
Rosenberg 1992	Gonorrhea	**0.35**	
	Chlamydia	**0.28**	
Magder 1988	Gonorrhea	**Protective (p less than 0.001)**	
	Chlamydia	**0.00**	
Rosenberg 1987	Gonorrhea	**0.67**	**sponge**
	Chlamydia	**0.31**	**sponge**
Quinn 1985	Gonorrhea	**0.11**	
Austin 1984	Gonorrhea	**0.45**	
Berger 1975	Gonorrhea	**0.53**	
STD Sequelae		*Relative Risk*	
Cramer 19987	Tubal infertility	**0.50**	
Kelaghan 1982	PID hospitalization	**0.40**	
Trichomoniasis		*Relative Risk*	
Rosenberg 1992	Sponge	**0.29**	
	Diaphragm	**0.24**	
Cervical Neoplasia Risk			
Parazzini 1989		*Relative Risk / Use of Barrier Method*	
Never used		**1.00**	
Used less than 2 years		**0.86**	
Used 2 years or more		**0.57**	
Celentano 1987		*Odds Ratio / Ever Used Method*	
Condom		**0.72**	
IUD		**0.60**	
Oral Contraceptives		**0.48**	
Diaphragm		**0.29**	
Vaginal spermicides		**0.28**	
Peters 1986		*Matched Risk / Duration of Use (Use less than 2 years = RR 1.0)*	
Oral contraceptives, 2-9 yrs		**1.0**	
Oral contraceptives, 10 yrs +		**1.1**	
Diaphragm, 2-9 yrs		**0.7**	
Diaphragm, 10 yrs +		**0.3**	
Spermicide alone, 2-9 yrs		**0.7**	
Spermicide alone, 10+ yrs		**0.2**	

TABLE 2. Vaginal Barrier Method Use: Decreased Infection Risk (continued)

Condom, 2-9 yrs	**0.5**
Condom, 10 yrs +	**0.4**
Vessey 1987	*Neoplasia Incidence / 1,000 Woman-Years*
All women	**0.73**
Oral Contraceptive users	**0.95**
IUD users	**0.87**
Diaphragm users	**0.17**

TABLE 3. Vaginal Barrier Method Use: Increased Infection Risk

Urinary Tract Infection		*Relative Risk*
Fihn 1985	Case control/diaphragm	**2.0**
	Cohort diaphragm vs. oc	**2.5**
Foxman 1985	Case control/diaphragm	**3.0**
Hooten 1991a	Increased *E. coli* bacteriuria after diaphragm use	
Vessey 1987	Cystitis hospital referral rate for diaphragm users (2.23) is higher than rate for non-users (1.35 / 1000 women-years of observation)	
Fihn 1986	Diaphragm users who have UTI history have increased vaginal *E. coli* in culture	
Bacterial Vaginosis		*Relative Risk*
Rosenberg 1992	Sponge	**1.15**
	Diaphragm	**1.16**
Hooten 1991b	Increased vaginal colonization with *E. coli* after diaphragm use	
Candidiasis		*Relative Risk*
Kreiss 1992 (sponge)	Vulvitis, believed to be fungal by clinical exam	**3.3**
Rosenberg 1992	Sponge	**1.20**
	Diaphragm	**1.88**
Rosenberg 1987	Sponge	**2.76**
Hooten 1991b	Increased vaginal colonization with *Candida* species after diaphragm use	
Toxic Shock Syndrome		*Relative Risk*
Faich 1986	Sponge	**7.8 -- 40.0**
Schwartz 1989	Any barrier	**14.9**
	Sponge	**10.5**
	Diaphragm	**11.7**
	Cervical Cap	**8.0**

Other Alterations in the Vaginal Environment. Insertion of any device or chemical into the vaginal alters physical characteristics of the system. Increased oxygen tension, for example, has been documented after insertion of a diaphragm or tampon [Wagner, et al., 1988]. Local oxygen tension, in turn, alters conditions for bacterial growth that might impact the relative distribution of aerobic and anaerobic species. If increased oxygen tension resulted in increased growth of *Staphylococci,* or conditions favorable for production of toxin, an increased risk for toxic shock syndrome would be plausible. Increased vaginal colonization with *Staphylococcus aureus* has been demonstrated with prolonged diaphragm wear [Baehler, et al., 1983]. Spermicide product constituents also may alter physical characteristics of vaginal fluids such as pH, surface tension, viscosity, and adhesion, as well as physio-chemical characteristics including dispersion and mixing of vaginal and cervical secretions and semen.

Direct Effects of Spermicide on Vaginal Epithelium. Detergent spermicides such as nonoxynol-9 are toxic to normal cell membranes, including the cell walls of vaginal epithelial cells. Local response to cell damage might include changes in cell permeability, and initiation of normal inflammatory responses such as increased vascularity and mobilization of lymphoid cells. With concentrated or extended exposure, macroscopic damage to epithelial integrity might be expected, and visible evidence of local tissue disruption, including ulceration, has been reported [Niruthisard, et al., 1991; Reckart, 1992]. In addition, idiosyncratic sensitivity and allergy can occur with nonoxynol-9 exposure. In this situation, an exaggerated inflammatory and immune response might be expected, not necessarily related to dose or duration of exposure [Rosenberg and Holmes, 1993].

BARRIER/INFECTION INTERACTIONS

Vaginal Ecology

In addition to effects directly related to use of a barrier and/or spermicide, the vaginal environment is also altered by innumerable biologic factors that may impact infection outcome or modulate the effect of barrier use. Important factors include normal physiologic states, changes caused by systemic disease or treatment and other health practices, such as sexual and hygiene measures [Mardh, 1991].

Individual Characteristics and Normal Physiologic States. Women are not all the same, structurally or hormonally. For example, the extent of cervical ectropion, and the character and amount of cervical secretions, vary substantially between individuals, and also with age. Some women ovulate regularly, others infrequently. Even during normal ovulatory cycles, hormone levels vary between women and in the same woman from cycle to cycle. Reproductive hormones act directly on the cervical and vaginal epithelia and their secretions, so the vaginal environment changes constantly during the cyclic, reproductive years. In addition, the endogenous hormone milieu of the prepubertal, lactating, or postmenopausal woman is fundamentally unlike that of the cycling woman. Changes in reproductive hormone levels affect immunoglobulin concentrations found in cervicovaginal secretions, with low levels found in association with ovulation [Forrest, 1991]. Women also differ in their biologic sexual response characteristics. The amount of vaginal lubrication available during intercourse is variable and may reflect individual differences, hormonal milieu and situational factors that impact readiness for arousal.

Normal endogenous flora also may change in relation to hormonal milieu, and so does the risk of vulvovaginal candidiasis. Pregnancy and administration of exogenous estrogen are associated with increased prevalence of *Lactobacillus* and yeast colonization [Larsen, 1993], and an increased incidence of clinical candidiasis; Colonization with *Bacteroides* is greater during the estrogen-dominated proliferative phase of the menstrual cycle and may be responsible for an increased risk of postoperative infection observed when hysterectomy is performed during this phase rather than the secretory phase of the cycle [Larsen, 1993]. Administration of Depo-Provera appears to reduce susceptibility to clinical candidiasis [Dennerstein, 1986].

Exogenous Drug Effects. Drugs that contain hormones -- including oral contraceptives, depot-medroxyprogesterone acetate and other injectables and progestin implants -- obviously impact the vagina's hormonal milieu. Hormones have a direct effect, and their use also results in indirect effects as the woman's normal cyclic pattern is interrupted or attenuated. Altered cervical secretions, for example, are commonly observed, and women who use oral contraceptives may have more extensive or persisting cervical ectropion than do women using other contraceptive methods. Exogenous hormones also may impact immunoglobulin levels in the cervix and vagina and local antigen absorption [Forrest, 1991]. Even prenatal exposure to exogenous hormone drugs may impact vaginal ecology as the ectropion and vaginal adenosis observed in some DES daughters shows. Non-hormonal medications also may impact the vaginal environment. For example, use of decongestants or antihistamines can alter secretions, and perhaps also vaginal immune response mechanisms. It is plausible that nonsteroidal anti-inflammatory agents or even aspirin could affect prostaglandin synthesis sufficiently to alter normal immune system responses as well. Antibiotics are another obvious example of systemic drugs that directly impact vaginal ecology. Symptomatic candidiasis is a well-recognized side effect of broad spectrum antibiotic treatment.

Systemic Diseases. Possible effects on vaginal ecology can be hypothesized for virtually any significant systemic disease. Serious illness may impact endogenous hormone production, and more direct effects are likely as well. For example, nutritional deficiency, vitamin deficiency or anemia might alter availability of micronutrients for vaginal flora [Larsen, 1993]. Liver disease might alter hormone clearance. Diabetes results in elevated blood sugar levels, and vaginal candidiasis is a common problem for women who do not have good diabetes control. Thyroid excess or insufficiency may result in abnormal reproductive hormone patterns and anovulatory cycles. Infectious diseases also may cause vaginal changes. Tuberculosis, for example, may directly involve reproductive structures, such as the ovaries, and can alter endogenous hormone production. Serious systemic infection alters immune system function, and infection treatment may have additional effects on vaginal flora.

Vaginal Hygiene Measures and Other Cultural Practices. Intravaginal products used to absorb menstrual flow and products intended for cleansing are widely used. Douching, in particular, is an exceedingly common practice in many cultures. [Aral, et al., 1992; Rosenberg, et al., 1991; Onderdonk, et al., 1992]. In addition, use of lubricants and of drying agents as part of sexual intercourse is also common [Brown, et al., 1992; Brown, et al., 1993]. Cultural norms, specific products utilized, frequency, and methods of use for all of these substances are not well documented. In addition, cultural practices in relation to general hygiene, such as bathing, are likely to impact vaginal ecology, as are male hygiene practices.

Sexual Practices and Sexual Context. Individual or cultural norms with regard to sexual practices impact vaginal biology profoundly. Cultural values, for example, are likely to impact the number and choice of sexual partners, which are the most significant determinants of infection risk. Also, the prevalence and acceptability of trauma associated with intercourse influences infection risk. Even if direct injury is not an issue, the presence or absence of adequate female arousal impacts the amount and character of vaginal secretions present during and after intercourse and also the likelihood of trauma caused by inadequate lubrication.

BARRIER METHODS

Cultural Considerations

For all women, sexual decisions are inextricably interwoven with cultural and economic forces over which the individual woman has little or no control. In this context, it is unrealistic to expect that a woman will have the power to negotiate the terms of sexual encounters, especially when the goal of such negotiation is to force her partner to do something he may not be enthusiastic about doing. For many women, the cultural context of discrepant gender power, or even threat of violence, make strategies based on women negotiating condom use ludicrous. And women themselves may object to condom use. Not all women wish to avoid pregnancy, and sexuality plays important roles in human life other than fertility and transmission of pathogens [Elias and Heise, 1993]. Sexual intimacy, for example, may be sought because of its role in maintaining and enhancing bonds between partners. Mutual commitment, in turn, has profound economic consequences for both partners but especially for women because they are economically more vulnerable than men in virtually all cultures. If using condoms is intrusive or disrupts the successfulness of a sexual liaison, or if a sense of trust is undermined, then the symbolic emotional and economic costs of condom use may be unacceptable. The power of cultural issues such as these is logically and intuitively self-evident. Their powerful role is also well documented in clinical research. In a recent monograph, Elias and Heise [Elias and Heise, 1993] provide an extensive review of relevant research, and find clear evidence that decisions about sexual intimacy for women in ordinary relationships (as well as for women engaged in commercial sex work) often involve economic necessity. Elias and Howe also document the determinative role that cultural context issues often play in the success or failure of available prevention strategies.

Reassessment of Prevention Strategy

Current prevention strategy for HIV is based on three ideas: (1) abstinence outside of marriage; (2) limiting the number of sexual partners; and (3) use of latex condoms when HIV exposure is a concern. This strategy is appropriate for some individuals; for example, U.S. senators, whose cultural values and context make these three ideas reasonable and make such a strategy feasible.

Balancing Efficacy and Acceptability. Latex condoms are widely granted "gold-standard" status as the best option for preventing HIV transmission. In theory and in clinical studies of couples who do use latex condoms consistently, their efficacy is excellent [Cates, et al., 1992]. Clinical studies that compare the impact of condom use with the impact of vaginal barrier method use on STDs in general, however, are not so clear-cut. The theoretical superiority of latex condoms must be balanced against the important impact of actual use effectiveness. For some couples at least, consistent use appears to be more feasible with vaginal barrier methods than with condoms. This translates into superior STD protection against gonococcal and Chlamydial cervicitis observed in some studies [Rosenberg, et al., 1992a; Rosenberg and Gollub, 1992b]. In addition, for some couples, cultural factors previously described make condom use an irrelevant option.

Finding More Effective Prevention Strategies. Until (if) a cure for AIDS is found or effective immunization is developed and the disease eradicated, prevention strategies that are feasible and useful in actual life will be needed. No one prevention strategy or product is likely to meet diverse needs, so multiple options must be sought. Finding ways to make prevention universal will become an increasingly important public health priority in the next decades of the epidemic. Therefore, acceptability also will become increasingly important as strategies and products are considered. Methods that can be managed autonomously by women are needed, and may prove to have a more important public health impact than condoms have been able to achieve. Some attributes of ideal approaches and products might include:

- STD prevention as a part of personal hygiene
  as important as soap,
  as normal as soap,
  as routine as soap.

- STD prevention that is separate from contraception.
- Products suitable for daily or periodic use, not directly associated with sex -- longer-acting, perhaps systemic.
- Products that are non-toxic, inexpensive, readily available, easy to distribute, suitable for local manufacture or perhaps even home concoction.
- Products that are attractive, comfortable, soothing, pleasant, convenient.
- Products and strategies that are culturally consonant.
- Products that can be implemented immediately.

Clearly, development of improved vaginal barrier methods to provide effective contraception and effective STD protection deserves a high priority.

Implications for Vaginal Barrier Method Development

Although the mechanisms are complex and knowledge is scanty, available research does suggest some interesting possibilities. Perhaps an optimal dose/exposure regimen for existing products can be clarified so they can be used more appropriately. Or perhaps local tissue toxicity of spermicides could be minimized by combining several different spermicide/microbicide chemicals, each in sub-toxic concentration. Alternatively, it may be possible to develop formulations that combine an otherwise toxic spermicidal, chemical with additives that are not themselves microbicidal or spermicidal but potentiate desirable effects of a sub-toxic dose or protect the vagina against local tissue damage. Specifically, additives intended to maintain acid pH and additives intended to prevent proliferation of candida deserve further investigation.

Maintenance of Acid pH. Exposing free HIV to acid pH ranges normally found in healthy vaginal secretions reduces its infectivity and also reduces the survival of HIV infected lymphocytes [Voeller and Anderson, 1992b]. Acid pH is also important in maintaining a normal endogenous vaginal flora balance. Acidophilic species such as Lactobacilli are favored in the normal acid environment, and pH may also have less obvious effects on solubility of nutrients and activity of bacterial enzyme systems that serve to regulate the microbial balance [Larsen, 1993]. A temporary shift in pH occurs in association with intercourse, and neutral or alkaline pH is maintained for at least several hours thereafter by the strong buffering capacity of semen. This effect may enhance HIV infectivity and increase transmission risk for both partners. Abnormally alkaline pH also occurs with bacterial vaginosis. Since spermicide use is associated with an increased incidence of bacterial vaginosis, the effect of spermicide on vaginosis risk, and hence pH, would provide a plausible explanation for an increased HIV risk. Vaginal products that help maintain optimal pH, perhaps by providing a buffer, need to be explored. Such products could be incorporated into spermicide formulation, or provided for use along with contraceptive devices and spermicide, or provided for use as a routine hygiene measure without a vaginal contraceptive [Voeller and Anderson, 1992b].

Prevention of Candida Proliferation. Candida proliferation resulting from spermicide use is undesirable for several reasons. It commonly causes vulvitis, and therefore makes spermicide use unacceptable for some women. Its overgrowth can lead to a vicious cycle of recurring vaginitis. Some women develop IgE antibodies to candida that are detectable in vaginal secretions. When candida proliferates, the IgE on mast cells recognizes it, and degranulation is triggered. This results in release of histamine and inflammatory mediators including PGE_2, which in turn suppresses the T cell lymphoid response that would otherwise act to control candida proliferation [Witkin, 1993]. Finally, severe inflammation associated with candida may disrupt epithelial integrity, causing visible ulcers. In some cases, vulvar manifestations are extremely severe, and the possibility of acquired acute hypersensitivity to candidal antigen has been suggested [Sobel, 1993]. It is plausible that these effects on the vaginal immune system or on epithelial integrity may explain the adverse effect of sponge use on HIV transmission observed in the Kreiss study [Kreiss, et al., 1992]. In that study the relative risk of vulvitis -- believed clinically to be candida -- was 3.3, matching the 3.3 relative risk of genital ulcers that is usually cited as the reason for the increased HIV susceptibility observed.

Therefore, development of products intended to mitigate candida proliferation is a reasonable idea and may be especially important for spermicide formulations used with barriers such as the sponge, diaphragm and cap which are worn for prolonged periods of time. Potassium sorbate is an effective, nontoxic, water soluble, antifungal agent that may be an appropriate product candidate. This chemical, along with sorbic acid, is used in the food and wine industry to prevent growth of molds. Efficacy of 1 percent and 3 percent potassium sorbate aqueous solutions in treatment of vaginal candidiasis has been reported in small clinical studies [McKinnon and Rodgerson, 1973, 1975], and sorbate has also been used in vaginal hygiene products.

Until 1983 potassium sorbate was the active ingredient in Summer's Eve Medicated Douche™. The FDA ruled at that time that although potassium sorbate was safe for vaginal use, it had not been proven effective. The manufacturer consequently changed the active ingredient in this product to povidone-iodine so that the term "medicated" could be included in labeling [Woodley, 1985]. Sorbate is currently included in some vaginal hygiene products, including Massengill Baby Powder Soft Cloth Towelette™, Fragrance Free Soft Cloth Towelette™, and Lady Lee Herbal Douche™. This product candidate has several important advantages: It is nontoxic and has already been approved for vaginal use; it is inexpensive and readily available; and formulation problems should not be severe. Research is needed to determine whether the antifungal activity of sorbate is maintained in the presence of nonoxynol-9.

CONCLUSION

Development of vaginal barrier contraceptives that provide STD protection and reduce HIV transmission deserves the highest scientific priority, and immediate, emergency attention. Appropriate research priorities have been eloquently described [Rosenberg and Holms, 1993] and need now the strongest possible advocacy efforts to secure funding and public support. These methods could play a crucial role in stemming the acceleration of HIV spread. Reducing the speed of transmission may even itself impact the virulence of the epidemic [Cowley, 1993]. As a first priority, the relationship between nonoxynol-9 and HIV risk must be clarified. Can an appropriate dose and duration of exposure be identified that provides effective contraception and effective microbicidal activity, yet avoids local tissue toxicity? Can formulation changes, spermicidal agent combinations, or product additives be exploited to achieve safer and more efficacious products?

Second, existing spermicide options, including products currently used in other countries but not universally available, should be made available in the U.S. and worldwide. Couples who are unable to use one product because of allergy or sensitivity may be able to use another, similar spermicide product successfully. All sexually active women and men are entitled to information about the value of spermicides and vaginal barrier methods in reducing STD risks. Withholding such information is unconscionable; failing to undertake massive public education to be sure that the information reaches people is also unconscionable.

A third priority is to investigate seriously and in a scholarly way the factors that determine successful method use. How can people best be taught and motivated to use methods effectively? What product attributes impact actual use?

Finally, research is needed to look for new products that offer advantages in efficacy, safety or acceptability.

ACKNOWLEDGMENTS

Nancy Alexander, Willard Cates, Jr., Christopher J. Elias, Erica L. Gollub, Penelope J. Hitchcock, Michael J. Rosenberg, Katherine M. Stone and Bruce Voeller provided invaluable help in gathering background materials and in formulating the concepts that are the basis for this summary.

REFERENCES

Aral SO, Mosher WD, Cates W Jr (1992): Vaginal douching among women of reproductive age in the United States: 1988. Am J Public Health 82:210-214.

Austin H, Louv W, Alexander J (1984): A case-control study of spermicides and gonorrhea. JAMA 251:2822-2824.

Baehler EA, Dillon WP, Dryja DM, Neter E (1983): The effects of prolonged retention of diaphragms on colonization by staphylococcus aureus of the lower genital tract. Fertil Steril 39:162-166.

Bird KD (1991): The use of spermicide containing nonoxynol-9 in the prevention of HIV infection. AIDS 5:791-796.

Brown RC, Brown JE, Ayowa OB (1992): Vaginal inflammation in africa. (Letter to the Editor). NEJM 327:572.

Brown RC, Brown JE, Ayowa OB (1993): The use and physical effects of intravaginal substances in Zairian women. Publication pending. Reprint requests: Richard C. Brown, 6009 Old Phillips Road, Norfolk, Virginia 23502.

Cates W Jr, Stewart FH, Trussell J (1992): Commentary: the quest for women's prophylactic methods--hopes vs. science. Am J Public Health 82:1479-1482.

Cates W Jr, Stone KM (1992): Family planning, sexually transmitted diseases and contraceptive choice: a literature update -- Part I. Fam Plann Persp 24:75-84.

Celentano DD, Klassen AC, Weisman CS, Rosenshein NB (1987): The role of contraceptive use in cervical cancer: the Maryland cervical cancer case-control study. Am J Epidemiology 126:592-604.

Cowley, G (1993): The future of AIDS. Newsweek March 22, 1993:47-52.

Cramer DW, Goldman MB, Schiff I, Belisle S, Albrecht B, Stadel B, Gibson M, Wilson E, Stillman R, Thompson I (1987): The relationship of tubal infertility to barrier method and oral contraceptive use. JAMA 257:2446-2450.

Dennerstein GJ (1986): Depo-provera in the treatment of recurrent vulvovaginal candidiasis. J Repro Med 9:801-803.

Elias CJ, Heise L (1993): The development of microbicides: a new method of HIV prevention for women. The Population Council, Working Papers, No. 6, 1993. The Population Council, One Dag Hammarskjold Plaza, New York, NY 10017.

Faich G, Pearson K, Fleming D, Sobel S, Anello C (1986): Toxic shock syndrome and the vaginal contraceptive sponge. JAMA 255:216-218.

Fihn SD, Johnson C, Pinkstaff C, Stamm WE (1986): Diaphragm use and urinary tract infections: analysis of urodynamic and microbiological factors. J Urology 136:853-856.

Fihn SD, Latham RH, Roberts P, Running K, Stamm WE (1985): Association between diaphragm use and urinary tract infection. JAMA 254:240-245.

Forrest BD (1991): Women, HIV, an mucosal immunity. Lancet 337:835-836.

Foxman B (1990): Recurring urinary tract infection: incidence and risk factors. Am J Pub Health 80:331-33.

Gollub EL, Stein Z (1992): Nonoxynol-9 and the reduction of HIV transmission in women. AIDS 6:599-601.

Hooten TM, Hillier S, Johnson C, Roberts PL, Stamm WE (1991a): *Escherichia coli* bacteriuria and contraceptive method. JAMA 265:64-69.

Hooten TM, Fennell CL, Clark AM, Stamm WE (1991b): Nonoxynol-9: Differential antibacterial activity and enhancement of bacterial adherence to vaginal epithelial cells. J Infect Dis 164:1216-1219.

Kelaghan J, Rubin GL, Ory HW, Layde PM (1982): Barrier-method contraceptives and pelvic inflammatory disease. JAMA 248:185.

Klebanoff SJ (1992): Effects of the spermicidal agent nonoxynol-9 on vaginal microbial flora. J Infect Dis 165:19-25.

Kreiss J, Ngugi E, Holmes K, Ndinya-Achola J, Waiyaki P, Roberts PL, Ruminjo I, Sajabi R, Kimata J, Fleming TR, Anzala A, Holton D, Plummer F (1992): Efficacy of nonoxynol-9 contraceptive sponge use in preventing heterosexual acquisition of HIV in Nairobi prostitutes. JAMA 268:477-482.

Larsen B (1993): Vaginal flora in health and disease. Clin Obstet Gynecol 36:107-121.

Mardh, PA (1991): The vaginal ecosystem. Obstet Gynecol 165:1163-1168.

McGroarty JA, Chong S, Reid G, Bruce AW (1990a): Influence of the spermicidal compound nonoxynol-9 on the growth and adhesion of urogenital bacteria in vitro. Current Microbiology 21:219-223.

McGroarty JA, Soboh F, Bruce AW, Reid G (1990b): The spermicidal compound nonoxynol-9 increases adhesion of *Candida* species to human epithelial cells in vitro. Infection and Immunity 58:2005-2007.

McGroarty JA, Tomeczek L, Pond DG, Reid G, Bruce AW (1992): Hydrogen peroxide production by lactobacillus species: correlation with susceptibility to the spermicidal compound nonoxynol-9. J Infect Dis 165:1142-1144.

McKinnon DA, Rodgerson EB (1973): A new treatment for yeast vaginitis. Obstet Gynecol 42:460-465.

McKinnon DA, Rodgerson EB (1975): Use of potassium sorbate for treatment of fungal infections. Obstet Gynecol 45:108-110.

Niruthisard S, Roddy RE, Chutivongse S (1991): The effects of frequent nonoxynol-9 use on the vaginal an cervical mucosa. Sex Transm Dis 18:176-179.

Onderdonk AB, Delaney ML, Hinkson PL, DuBois AM (1992): Quantitative and qualitative effects of douche preparations on vaginal microflora. Obstet Gynecol 80:333-338.

Parazzini F, Negri E, LaVecchia C, Fedele L (1989): Barrier methods of contraception and the risk of cervical neoplasia. Contraception 40:519-530.

Peters RK, Thomas D, Hagan DG, Mack TM, Henderson BE (1986): Risk factors for invasive cervical cancer among Latinas and non-Latinas in Los Angeles County. JNCI 77:1063-1077.

Rekart M (1992): The toxicity and local effects of the spermicide nonoxynol 9. J Acq Immuno Def Syn 5:425-426.

Remington KM, Buller RS, Kelly JR (1987): Effect of the today contraceptive sponge on growth and toxic shock syndrome toxin-1 production by staphylococcus aureus. Obstet Gynecol 69:563.

Rosenberg MJ, Holms KK (1993): Virucides in prevention of HIV infection: research priorities. Sex Transm Dis 20:41-44.

Rosenberg MJ, Davidson AJ, Chen J-H, Judson FN, Douglas JM (1992a): Barrier contraceptives and sexually transmitted diseases in women: a comparison of female-dependent methods and condoms. Am J Public Health 82:669-674.

Rosenberg MJ, Gollub EL (1992b): Commentary: methods women can use that may prevent sexually transmitted disease, including HIV. Am J Public Health. 82:1473-1478.

Rosenberg MJ, Phillips RS, Holmes MD (1991): Vaginal douching: who and why? J Reprod Medicine 36:753-758.

Rosenberg MJ, Rojanapithayakorn W, Feldblum PJ, Higgins JE (1987): Effect of the contraceptive sponge on chlamydial infection, gonorrhea, and candidiasis: a compar ative clinical trial. JAMA 257:2308-2312.

Schwartz B, Gaventa S, Broome CV, Reingold AL, Hightower AW, Perlman JA, Wolf PH (1989): Nonmenstrual toxic shock syndrome associated with barrier contraceptives: report of a case-control study. Reviews of Infections Diseases 2(1): S43-S49.

Sobel JD (1993): Candidal vulvovaginitis. Clin Obstet Gynecol 36:153-165.

Stein Z (1992): Editorial: the double bind in science policy and the protection of women from HIV infection. Am J Publ Health 82:1471.

Stone KM, Grimes DA, Magder LS (1986): Personal protection against sexually transmitted diseases. Am J Obstet Gynecol 155:180-188.

Stone KM, Peterson HB (1992): Spermicides, HIV, and the vaginal sponge. [editorial]. JAMA 268:521-523.

Vessey MP, Metcalfe MA, McPherson K, Yeates D (1987): Urinary tract infection in relation to diaphragm use and obesity. Intl J Epidemiology 16:441-444.

Voeller B (1992a): Spermicides for controlling the spread of HIV. AIDS 6:341-342.

Voeller B, Anderson DJ (1992b): pH and related factors in the urogenital tract and rectum that affect HIV-1 transmission. Mariposa Occasional Paper #16, The Mariposa Education and Research Foundation, 3123 Schweitzer Drive, Topanga, California 90290.

Wagner G, Levin RJ, Bohr L (1988): Diaphragm insertion increases human vaginal oxygen tension. Am J Obstet Gynecol 158:1040-1043.

Wasserheit JN (1992):Epidemiological synergy: interrelationships between human immunodeficiency virus infection and other sexually transmitted diseases. Sex Transm Dis 77:19-61.

Witkin SS (1993): Immunology of the vagina. Clinical Obstet Gynecol 36:122-128.

Woodley KS (1985): Personal communication (letter); C B Fleet Company, Inc.

DISCUSSION

James Koch: Regarding the relative risk of toxic shock syndrome associated with various methods of birth control, did the figures you showed indicate the relative risk of toxic shock syndrome or the relative risk of finding the appropriate staph organism?

Felicia Stewart: Those were the relative risks for toxic shock syndrome. They're based on an estimate of the incidence of non-menstrual toxic shock, and there certainly have been criticisms of that approach. But it's been pretty persuasively documented that there probably is at least some impact and there is *in vitro* evidence that there is an increase in S. aureus growth.

Gerald Bernstein: I want to address the topic of colonization with candida and E. coli in spermicide users. Two or three years ago at the annual meeting of the American Society of Microbiology, two interesting papers were presented. The investigators treated the organisms *in vitro* with N-9 and found that it altered the surface chemistry and increased the organisms' ability to adhere to epithelial cells, which of course is a virulence factor.

The other thing I wanted to comment on was the toxic shock data. At one time there was a committee at NICHD that was organized to monitor a CDC study that was being done on toxic shock. They were monitoring a phase of that study that had to do with contraception because it was being done with NIH money. That committee put together a report that was published as a letter in JAMA that I think puts this whole thing in perspective. Even if the relative risk is as high as 40 percent, non-menstrual toxic shock is so rare that if you look at the attributable risk, it means an increase by maybe 2 or 3 per million women/cases per year. It's almost infinitesimally small, so I think for a rare disease relative risk doesn't mean very much. I think that aspect of this has been really blown out of proportion.

Felicia Stewart: I couldn't agree with you more, Gerry. Do you think we ought to take away the restrictions on duration of use and use of these products during menstrual periods?

Gerald Bernstein: No, I don't because I think the risk would be increased, either during menstruation or after prolonged wear. I don't think you can leave a foreign body in the vagina for a really long time without getting into trouble with it. Short-term -- by that I mean 24 hours or so -- is probably OK, but long-term you'll really alter the vaginal flora when you do that, not in a good direction either.

Penelope Hitchcock: There are at least two studies that indicate that douching increases the risk of pelvic inflammatory disease if there is existing cervicitis due to gonococcal or chlamydial infection. I think we may on occasion be faced with behaviors that we cannot cope with as far as protection against infection is concerned. Circumcision or lack thereof might fall into that category, as might anal intercourse because of the CD4-like receptors on rectal epithelium. There may be some behaviors that we're going to have to identify as being of very high risk and for which we just don't have protective options.

Felicia Stewart: There was a survey the year before last on douching which found it to be staggeringly common. If you actually start asking patients in a way that enables them to say yes, it's just absolutely flabbergasting how many women douche. This would be a very severe confounding variable for sexually transmitted disease risk.

Shakti Upadhyay: I think one of the most important recommendations of this meeting should be that new methods should not induce inflammation. Anything which is toxic to the cell walls of the vaginal epithelium will also induce a local inflammation following repeated use. So whether this would be one of the criteria for the selection of these barrier contraceptives and if it's unavoidable probably one could think in terms of using an anti-inflammatory agent along with this local barrier contraceptive method.

Michael Burnhill: Regarding those two retrospective studies which showed that women with PID were more likely to have douched, I'm sure they were more likely to have douched because they had some symptoms and they thought they could treat them at home. For the last decade I have been recommending vaginal irrigation. I discovered you'd better change the name so that you can get people to do this. I suspect that I've had at least 500 women a year irrigating their vaginas with a number of substances such as boric acid solution or peroxide, etc. I have yet to see one woman whose vaginal ecosystem moved in the wrong direction, nor have I seen anybody who developed pelvic inflammatory disease.

Judy Norsigian: My question is about the role of diet on the vaginal flora. I've heard a lot of things from a lot of people, some of them practitioners, some of them nutritionists, some of them homeopaths or naturopaths, and I'm wondering if you know of anything in the literature that takes a look at this in a more formal way.

Felicia Stewart: I think there really is a very strong correlation between an increased incidence of candidiasis and diabetes. Candidiasis does appear clinically to be linked to blood sugar that's out of control, which you would think would cause a local increase in blood sugar. Half of the March issue of Clinical Obstetrics and Gynecology was devoted to vaginal ecology and all these issues. It was proposed that the venous lagoon of the vagina may be a countercurrent circulation system in the vaginal submucosa. The reason for this speculation is the observation that after giving an oral medication like penicillin, you get a peak serum effect in an hour or two but the vaginal concentration continues to increase for as long as 15 hours. If that's the case then you can imagine that we might actually be able to do something systemic, and that what you ate -- if you bought it at a fast food restaurant -- might have an untoward effect on your vagina.

Judy Norsigian: A couple of nurse practitioners in separate cities who have reported on women with chronic candidiasis have said that radically reducing their sugar intake finally had an effect, along with some other things. And I also want to support what was just said ahead of me about douches. Although the douches that have all kinds of perfumes in them may be problematic, a lot of women have reported success with things like vinegar douches if done properly.

8

A CONTRACEPTIVE AND PROPHYLACTIC BARRIER: CURRENT CONCEPTS FOR THE MALE CONDOM

Howard Charles Price

Materials Technology Division
Family Health International
Research Triangle Park, NC 27709

INTRODUCTION

The latex condom remains the most widely available and effective male barrier device for contraception or the prevention of sexually transmitted diseases (STD). Despite its obvious and unique value for the prevention of STD, the condom remains vastly underutilized throughout the world. The reasons for the low incidence of condom use include cost and availability in the developing countries, aversion to condom use as an accessory to sex in most cultures, and a *declassé* reputation of the condom among certain cultures, such as the United States.

The poor acceptability of condoms derives, in part, from the momentary interruption of the sex act and, in part, from the design and performance characteristics of condoms. When condoms are accepted and used correctly, then it is the design and performance characteristics that determine the overall success rate of this method of contraception and disease prevention. Studies on the effectiveness of condoms have been reviewed in recent publications [Liskin et.al., 1990: Cates and Stone, 1992].

Education is unquestionably the key to achieving greater acceptance of the male condom. The practical aspects of the device --- its fit, form, and function --- will always strongly influence the success of this barrier method and merit full consideration in any strategy seeking to markedly improve condom utilization and effectiveness. The application of appropriate performance requirements and testing methods provides assurance of better quality latex condoms. New condom designs, materials, and manufacturing

Barrier Contraceptives: Current Status and Future Prospects, pages 123–133

technologies offer the potential for improvements in both the perception and the performance of male condoms.

MALE CONDOM ATTRIBUTES AND FAILURE MODELS

To appreciate how the general status of condoms may be improved, it is helpful to have a full understanding of component features and performance issues. There are several readily identifiable, physical performance-related quality attributes for latex male condoms and a number of failure modes which can be related, at least in part, to those attributes. Physical performance quality attributes and failure modes are listed in Table 1.

TABLE 1. Physical Performance Quality Attributes And Failure Modes for Male Condoms

Quality Attributes	Condom Failure Modes
Strength and extensibility	Tearing
Design and dimensions	Slippage, tearing
Permeability	Leakage
Inherent stability	Reduced shelf life
Packaging attributes	Reduced shelf life, tearing
Lubricity	Slippage, tearing

Certain attributes or properties of the latex condom are interactive in effecting the performance of the device. For example, an increase in the ultimate tensile strength of the condom may not result in an improved tear resistance if there is a concomitant decrease in extensibility. Also, improvement in the physical and performance aspects of the condom must be moderated to accommodate acceptability/pleasure requirements. Because these sensory requirements are fundamental to product success, they can also be considered quality attributes. Sensory quality attributes are listed in Table 2.

TABLE 2. Sensory Quality Attributes For Male Condoms

Implementation:	Donning: Package access, gripping, unrolling Doffing: Removing, discarding
Tactile:	Constraint, sensitivity, lubricity, texture
Visual:	Transparency, color, package appearance
Other stimuli:	Odor, flavor, sound

CONDOM QUALITY ASSURANCE STANDARDS

Many countries have published standards that specify quality attributes and test methodology for the latex condom. These standards range in authority from voluntary guidelines to mandatory requirements for permission to market the product. For the United States, performance standards for condoms were developed by the volunteer standards organization, ASTM (American Society for Testing and Materials). First published in 1976, these performance standards are currently designated D3492-89, Standard Specification for Rubber Contraceptives (Condoms) (ASTM D3492-89, 1989). The International Organization for Standardization (ISO), with worldwide representation, began publishing its International Standard Series ISO 4074 on Rubber Condoms in 1980 (ISO 4074/1-9, 1980-1990). The primary properties specified in these two representative standards are listed in Table 3.

TABLE 3. Condom Properties Specified In ASTM and ISO Standards

Property	ASTM, ISO, or Both
Chemical purity/leachables	Both
Visual defects	ASTM
Color fastness	ISO
Design/dimensions	Both
Tensile properties, with and without accelerated aging	Both
Airburst volume/pressure, with and without accelerated aging	ISO
Leakage	Both
Package	Both

The two standards documents are similar but differ in some of the properties specified, acceptance limits and test methods. Both standards undergo periodic review and revision with considerable debate as to the precision and accuracy of the tests, as well as their relevance to clinical performance. The airburst test is illustrative. This test is required by ISO, but not ASTM. Both the test method variables and the test's importance as a predictor of condom field performance have been discussion topics among the standards committees for years.

LATEX MALE CONDOM STUDIES

Stability Studies

One of the major concerns for latex condoms is their stability, particularly under adverse conditions (e.g., high temperature or elevated ozone levels). Adequate estimation of the effect of environmental conditions is important to establish the shelf life of the product and to provide guidance for organizations charged with maintaining the supply of condoms in remote locations.

Two stability studies currently are underway. A controlled laboratory study -- involving collaborative participation by the U.S. Food and Drug Administration's Center for Devices and Radiological Health (FDA-CDRH), Health Industry Manufacturer's Association (HIMA), the State of Washington, and Program for Appropriate Technology in Health (PATH) -- is evaluating thermal and ozone-related degradation of condom properties under different packaging regimens. Reports from this study are expected later this year.

FHI, with support from the U.S. Agency for International Development (USAID) is conducting a prospective five-year aging study. In this study, now in its second year, supplies of condoms which have been well-characterized in performance lab tests are stored in three locations of differing climatic conditions. The temperature and humidity of the environment is monitored continuously and the condom supplies are resampled periodically for testing to assess the extent of environmental damage.

Human Use Studies

Several research organizations have reported human use studies on male condoms [Gerofi et.al., 1991; Russell-Brown et.al., 1992] and human use studies have been the subject of recent reviews [Liston et.al., 1990; Finger, 1991]. Most recent studies have focused on how condom failures relate to either condom quality testing or cultural variables.

Emory University and FHI collaborated on a short-term practical stability study, which examined the damage condoms incurred when carried in wallets. In a preliminary account [Glasser, 1992] airburst test performance was reported to be slightly reduced after three months use, but tensile test performance was not affected.

FHI, in cooperation with PATH and supported by USAID, conducted a study evaluating the correlation between condom laboratory tests and human use breakage rates [Steiner et.al, 1992]. In this study, correlation was sought between human use breakage rates and various laboratory test parameters. The better correlated test values (linear correlation coefficients of 0.7-0.8) were ultimate tensile elongation, airburst (percent rejected) and Condom Quality Index (CQI), a value calculated from several airburst parameters. In this study, the characteristic that best correlated with breakage rate was the age of the condom (correlation coefficient of 0.92).

CONDOM MATERIALS AND MANUFACTURING PROCESSES

Materials

Lamb cecum condoms are still marketed, but natural latex rubber is the only commercially important elastomer currently used for fabricating condoms. Of the many new thermoplastics that have become available during the last half century, only a relative few have been considered for the fabrication of condoms. But for condom designs that which require a combination of relatively high extensibility, tensile strength, and relatively low permeability and cost, natural rubber has remained unsurpassed.

In recent years, the search for an alternative to the latex condom has been more rigorously pursued. This renewed interest is due, in part, to the life-and-death consequence that attend condom failures in the era of pandemic HIV, recognizing that rubber condoms undergo some loss in performance properties with storage, particularly under oxidizing conditions.

Reports of "latex" allergic reactions also add incentive to the pursuit. Kraton®[1] type block copolymers and polyurethanes are two of the more-promising synthetic elastomers being evaluated in experimental condoms (see Table 4). Both of these polymers are considered to be less susceptible to oxidative damage than latex rubber. Since they are synthetic they are free of protein impurities, which are the putative allergens in some latex products. Also, these polymers may permit the manufacture of thinner condoms, which offer an important advantage in sensitivity and acceptability.

[1] Registered trademark of Shell Chemical Co.

TABLE 4. Polymers for New and Experimental Condoms

Natural (Latex) Rubber	A thermoset elastomer obtained by vulcanization of formulated raw latex; mostly cis-1,3-polyisoprene
Kratons, SEBS polymers	Thermoplastic elastomers derived from styrene, ethylene, butylene
Polyurethanes	Thermoplastic elastomers derived from glycols (polyether or polyester) and diisocyanates

Methods

Latex rubber condoms are manufactured using a dip-coating (casting) process. Mandrels presenting the profile of the condom are coated with the latex, which is formulated to undergo vulcanization or curing to latex rubber in subsequent thermal processing steps. Natural rubber formation involves chemical reactions.
Thus, the rubber material itself is being synthesized while the condom is being formed. Natural rubber curing is analogous to setting of epoxy cement, and natural rubber is likewise classified as a thermoset polymer.

In contrast, thermoplastic polymers, such as the SEBS or polyurethanes being evaluated for new condoms, can be converted to product shape without involving a change in their chemical composition. A wide variety of processes are used to produce thermoplastic products. Among processes used to produce film or thin-walled devices are extrusion, blow-molding, solvent casting, and thermoforming.

NEWLY-MARKETED AND EXPERIMENTAL MALE CONDOMS

Newly-Marketed Condoms

The major condom manufacturers have expanded their offerings of latex condoms to include a greater dimension range (large, small, and regular). Other offerings promote greater strength, more sensitivity or other aesthetic options.
A condom presenting a novel pouched contour, purportedly enhancing pleasure, is Pleasure-Plus™ (Reddy Health Care Inc.). In a human use study, this somewhat thicker latex condom compared favorably with an established latex brand in acceptability and breakage rates [Trussell et.al., 1992a]. Polyurethane fertility condoms, for use in semen collection (not as a barrier device), are marketed by Apex Medical Technologies, Inc., under their Male-FactorPak™ and HyGene™ labels.

Experimental Condoms

The interest in development of condoms made from polymers other than natural latex rubber is reflected in patents issued during recent years. Representative condom-related U.S. patents are listed in Table 5.

Table 5. Condom Related U.S. Patents

US Patent No./Year Issued/Assignee	Patent Title
#4,576,156/1986/Ortho Pharmaceutical, Division of Johnson and Johnson	Prophylactic Device and Method
#4,684,490/1987/Deseret Medical	Process for Preparation of Polyurethane Condoms
#4,855,169/1989/Apex Medical Technologies	Prophylactic Sheath with Augmented Border
#4,964,416/1990/and 5036863/1991 Family Health International	Condom Articles & Apparatus & Method for Making & Using the Same
#5,112,900/1992/Tactyl Technologies	Elastomeric Triblock Copolymer Compositions and Articles Made Therewith

Three of the patent assignees listed in Table 5 have active male condom development programs: Apex Medical Technologies, Tactyl Technologies and Family Health International. Both Tactyl Technologies and Family Health International have conducted some human use evaluations and have 510(k) applications pending with the FDA. In a human use study conducted by Emory University, the Tactyl Technologies condom compared favorably with an established latex brand in acceptability and performance [Trussell et.al., 1992b].

CONCLUSIONS

We have surveyed the status of male condoms by reviewing the properties or quality attributes of male condoms and how some of these properties are evaluated for latex (natural rubber) condoms. We also reviewed some of the stability and human use studies on latex condoms. Finally we examined the state of development of condoms made from thermoplastics. These experimental condoms offer potential for greater storage life, reduced allergenicity, and greater sensitivity.

However good the prospect for improved condoms, the low condom usage rate in the face of the ever-increasing world burden of over-population and sexually transmitted diseases is not heartening. Until there is a profound change in the acceptance of this barrier method (or its female counterpart), through intensive education and promotion programs, the world will continue to endure higher-than-necessary human suffering.

ACKNOWLEDGEMENTS

The author wishes to thank at Family Health International; Howard Miller, Eli Carter and Dale Blevins for helpful suggestions on this paper; Joyce Edwards for manuscript preparation. The support of The United States Agency for International Development (USAID) and National Institutes of Health is acknowledged for some of the projects cited. The author is appreciative of assistance from the following in providing information on the topics presented: Mark McGlothlin (Apex Medical Technologies, Inc.); Douglas Buddenhagen, Curt Hamann (Tactyl Technologies); Keith Heintz (Reddy Health Care, Inc.). The opinions expressed in this paper are those of the author and not USAID or any other organization cited.

REFERENCES

Cates W, Stone K (1992): Family planning, sexually transmitted diseases and contraceptive choice: a literature update-part 1. Family Planning Perspectives 24:75-84.

Finger W (1991): How human use affects condom breakage. Network (3) pp 10-13.

Gerofi J, Shelley G, Donovan B (1991): A study of the relationship between tensile testing of condoms and breakage in Use. Contraception Volume 43:177-185.

Glasser G (1992): Can condoms be carried in wallets? presented at the Eleventh Annual Conference Contraception In The Nineties, Atlanta, GA.(1990).

Liskin L, Wharton C, Blackburn R (1990): Condoms - now more than ever. Population Reports Series H Number 8:8-12.

Russell-Brown P, Piedrahita C, Foldesy R, Steiner M, Townsend J (1992): Comparison of condom breakage during human use with performance in laboratory testing. Contraception 45:429-437.

Steiner M, Foldesy R, Cole D, Carter E (1992): Study to determine the correlation between condom breakage in human use and laboratory test results. Contraception 46:279-288.

Trussell J, Warner D, Hatcher R (1992a): Condom slippage and breakage rates. Family Planning Perspectives 24:20-23.

Trussell J, Warner D, Hatcher R (1992b): Condom performance during vaginal intercourse: comparison of Trojan-Enz® and Tactylon™ condoms. Contraception Volume 45:11-19.

ASTM D3492-89 (1989): Standard practice for rubber contraceptives (condoms). In Storer RA et.al., (eds): "Annual Book of ASTM Standards" Easton MD: ASTM, Section 9, Vol 09.02 pp 311-313.

ISO 4074/1 (1990): Rubber condoms - Part 1: Requirements - Condoms in consumer packages. In "International Organization for Standardization ISO 4074/1." (1st ed.) Geneva:International Organization for Standardization. 1990-08-15, pp 1-4.

ISO 4074/2 (1980): Rubber condoms - Part 2: Determination of length. In "International Organization for Standardization ISO 4074/2" (1st ed.) Geneva:International Organiza tion for Standardization. 1980-06-01, pp 1-6.

ISO 4074/3 (1980): Rubber condoms - Part 3: Determination of width. In "International Organization for Standardization ISO 4074/3." (1st ed.) Geneva:International Organiza tion for Standardization. 1980-07-01, pp 1-3.

ISO 4074/4 (1980): Rubber condoms - Part 4: Determination of colour fastness. In "International Organization for Standardization ISO 4074/4." (1st ed.) Gene va:International Organization for Standardization. 1980-15-06, pp 1-4.

ISO 4074/5 (1984): Rubber condoms - Part 5: Testing for Holes. In "International Organization for Standardization ISO 4074/5." (2nd ed.) Geneva:International Organization for Standardization. 1984-12-15, pp 1-4.

ISO 4074/6 (1981): Rubber condoms - Part 6: Determination of bursting volume and pressure. In "International Organization for Standardization ISO 4074/6." (1st ed.) Geneva:International Organization for Standardization. 1981-01-15, pp 1-4.

ISO 4074/7 (1986): Rubber condoms - Part 7: Determination of resistance to deterioration during storage. In "International Organization for Standardization ISO 4074/7". (1st ed.) Geneva:International Organization for Standardization. 1986-05-15, pp 1-4.

ISO 4074/8 (1984): Rubber condoms - Part 8: Determination of mass. In "International Organization for Standardization ISO 4074/8." (1st ed.) Geneva:International Organiza tion for Standardization. 1984-11-01, pp 1-2.

ISO 4074/9 (1980): Rubber condoms - Part 9: Determination of tensile properties. In "International Organization for Standardization ISO 4074/9". (1st ed.) Gene va:Interna tional Organization for Standardization. 1980-10-01, pp 1-4.

DISCUSSION

Elizabeth Ngugi: At the meeting three weeks ago of Southern and East African countries, we were wondering whether we can have cheaper, or cheap, mechanisms of testing condoms at random when we get them because most of our countries do not manufacture condoms. I wonder whether you can devise a cheap but effective system that those countries can have for better quality control of condoms?

Howard Price: I think you have a good question but I don't think I have the answer at this point. These tests, such as air burst tests, are fairly sophisticated; the simple tensile test doesn't seem to be as specific or as sensitive to degradation of condoms. This is certainly an area for research and development: to come up with a field-type test that is reliable and sensitive enough to give you confidence that you are controlling the quality of the condoms in storage.

Clark Taylor: How do you test for the pleasurability factor in these condoms? For example, in the limited experience we have had with the polyurethane versus latex, what you get with the polyurethane is a lack of memory which makes it much more like the natural condoms than like latex and that is great for some people, but not for others, because it is that very stretching and squeeze of latex that other people really love. Also with the lubrication, some men have a lot of preseminal fluid, some men do not have any, so the amount of lubricant in a condom is really defined a lot by the needs of the individual. The last question is do you know anything about Teflon condoms as they certainly offer a lot more possibilities for texturing?

Howard Price: Teflon; I never heard of a teflon condom.

Clark Taylor: I ran across them in a Texas company that was just experimenting with them and I never heard anymore.

Howard Price: It would be interesting. Teflon has a relatively low extensibility and somewhat higher modulus than either of these other materials, so I don't know how it would perform. Also, it would be naturally lubricous, but maybe too lubricous. On your first question: I'm certainly not the one to ask how you establish what is the most acceptable in terms of performance on these new designs. Do bear in mind, though, that it is not simply the material but also the design that will be important here, because a loose-fitting condom that has a ring retention will be, again, perceived differently than one which is a clinging type condom.

Jacqueline Forrest: I have two questions: One is can you give us any estimate on when some of these new condoms might actually be on the market, and the second is do you have any sense of whether they are going to be as expensive, less expensive or more expensive than currently available condoms?

Howard Price: First question: I can give you only a ballpark estimate because we cannot really predict the FDA and all the other hurdles, including marketing and scaling up from manufacturing. I would think in the next three years we should be able to begin to see one of these at least out there. That may be slightly optimistic but again it depends on the approval process or the determination of the equivalency process. Now that does not mean any of them are going to be widely available. The capability for scaling up is there, I think Tactyl Technology is already moving in that direction.

Your other question was cost. I think initially they have all got to cost more than a latex condom. That is again because you are doing small quantities and you are trying to recoup development costs; that is why drugs cost a lot when they first come out and when generic drugs come out they can cut that cost down. I think for all of them that are mass-produced, particularly like those in our process which is not a dipping process, we can probably get that cost down if we can get the volume up. The price may be within a factor of 2 or 3 of latex but that is just a guess.

PART III: RESEARCH METHODOLOGIES

9

VAGINAL CONTRACEPTIVE DRUGS AND DEVICES: REGULATORY REQUIREMENTS

Lisa D. Rarick and Lillian Yin

Center for Drug Evaluation and Research and
Center for Devices and Radiological Health
U.S. Food and Drug Administration
Rockville, MD 20857

INTRODUCTION

The purpose of this chapter is to review the regulatory requirements for vaginal contraceptive drugs and devices in the Division of Metabolism and Endocrine Drug Products (DMEDP) of the Center for Drug Evaluation and Research (CDER), and the Division of Obstetric and Gynecologic Devices, of the Center for Devices and Radiological Health (CDRH) at the U.S. Food and Drug Administration (FDA). The FDA is encouraged to see so much interest in this long overlooked area of drug and device development, and applauds the efforts of CONRAD, FHI, Profamilia, WHO, and all agencies and individual participants attending this workshop, and appreciative of this opportunity to discuss the current regulatory requirements for contraceptive drugs and devices used vaginally. Two other divisions at CDER have recently become involved in discussions regarding these products: the Division of Anti-viral Drug Products and the Division of Anti-infective Drug Products. Both of these Divisions are currently discussing and formulating requirements for vaginal products requesting sexually transmitted disease (STD) prevention claims. The first section of this chapter addresses the regulatory requirements for vaginal drugs and the second section for vaginal devices.

DRUGS

Those involved in reproductive endocrinology and the development of contraceptive products have seen many positive developments during the last 30 years that have influenced the nature of the FDA requirements for contraceptive product development.

Barrier Contraceptives: Current Status and Future Prospects, pages 137–146
Published 1994 Wiley-Liss, Inc.

The ability to diagnose pregnancy, for example, has progressed significantly from the diagnosis made by physical signs and symptoms (such as amenorrhea with enlarged uterus, auscultation of a fetal heart, etc.) to the current use of a drop of urine in a very sensitive yet simple test that can confirm pregnancy even before a missed menses, and the use of ultrasound imaging to determine more accurately the approximate date of conception.

Clearly much has changed. As we develop more sophisticated methods, we must also accept more complicated results. Usually, improved technology will reveal higher pregnancy rates than traditionally expected. It is with this in mind that we must critically look at the DMEDP's requirements for contraceptive studies and approvals.

In this section, the following will be reviewed: the regulatory history of vaginal contraceptive drug products, the current requirements for chemistry, pharmacology, and clinical testing including some additional issues especially as they relate to HIV and STD prevention.

Regulatory History of Vaginal Contraceptive Drug Products

Although vaginal contraceptives existed prior to regulatory requirements, let us start our discussion with a look at recent significant regulatory measures in this area.

In 1972 there was a notice in the U.S. Federal Register (FR) proposing the review of safety, effectiveness and labeling by an independent advisory panel of all "over the counter" (OTC) drugs. Towards this end, a May, 1973 FR notice requested data and information on all active ingredients used in OTC products. The panel members (including Elizabeth Connell, MD--chair, Evelyn Benson, RPh, Cynthia Cooke, MD, Myron Gordon, MD, William MacColl, MD, William Pearlman, MD, and Louise Tyrer, MD) met often, at times with several outside consultants, and held two symposia over the next several years (1973 to 1978).

The recommendations of the panel, along with comments from the FDA, were published in the FR in December 1980 as a notice of the "Establishment of a Monograph; Proposed Rulemaking."

This report identified nonoxynol-9 (N-9) and octoxynol-9 (O-9) as generally considered safe and effective. It was at this time that the products containing mercury--for example, phenylmercuric acetate and phenylmercuric nitrate--were found to be <u>not</u> generally recognized as safe and effective.

This notice also described guidelines for new product development (other than N-9 and O-9) and outlined the provisions for marketing of N-9 and O-9 products under the monograph (required testing procedures, labeling, etc.).

Although this 1980 notice was a proposed rule, no further official action has since been published. Thus, the US market today contains many products which, as long as they remain within the monograph conditions (active ingredients, testing and labeling), have not been required to submit safety or efficacy data and are not regulated to the extent demanded of "new drugs." For example, good manufacturing procedures (GMP) inspections are not performed and there is no verification of product purity, stability, consistency, etc.

The DMEDP has consistently recommended that the proposed rule be modified to require that efficacy information and more chemistry information be required. In other words, New Drug Applications (NDAs) would be required for the products already on the market. The FDA, of course, recognizes that some areas, such as pharmacology requirements, could be abbreviated for those drugs with long standing histories of safe use.

With this background, we turn to a review of the usual requirements for a "new" vaginal contraceptive drug. Because the goal of this workshop is also to consider the prevention of STDs, we will try to point out where differences in requirements might exist.

Chemistry Requirements

Presentation of identity, purity, strength, consistency of production (from batch to batch), stability and meeting of GMP standards provide the majority of requirements for an adequate chemistry review. While much of this information must be submitted with the initial Investigational New Drug (IND) application (such as identity, purity and strength), details regarding consistency of the final product formulation and GMP inspections are usually required later in the process.

Pharmacology Requirements

In terms of pharmacology requirements, it is important to note that many of these requirements are subject to modification. Discussion with the DMEDP regarding unique or unusual circumstances is always welcome.

Prior to human exposure (clinical Phase I trials), one- to two-week vaginal irritation studies in rabbits and one- to two-week vaginal toxicity studies in two species (rats and one non-rodent) are necessary.

Before progressing to clinical Phase II trials, vaginal toxicity studies should be performed (again in two species) for at least as long as the proposed clinical plan -- up to a maximum of six months in rats and 12 months in a non-rodent species (monkeys, dogs or rabbits). It is at this time that a penile irritation study should also be performed (usually in the rabbit).

Normally, all the pharmacology studies are performed with the drug given at three dose levels. The usual recommendation is that the highest dose be toxic. We understand that with vaginal spermicides there may be limits to the ability to achieve a toxic dose.

In the unusual circumstance in which a drug is suspected or found to be absorbed vaginally in the humans, and has little or no absorption in animals when given vaginally, single dose and short (one month) studies may be required. In these studies, the drug is given to the animals by a route that allows for significant systemic absorption in order to determine potential toxicities.

By the same token, if a drug is shown to be absorbed by humans, pharmacokinetic studies should be performed in each species used in toxicity and carcinogenicity testing. In evaluating pharmacokinetic data half-life, C_{max}, AUC, tissue distribution and metabolite profiles should be presented and compared to human data.

Once it is planned to test a drug in women at risk of pregnancy, in other words prior to Phase II, some reproductive testing should be completed. Segments 1 and 2, which include studies of fertility and reproductive performance in rats and a teratologic evaluation conducted in two species (rat and rabbit), must be completed before Phase II. Segment 3 studies -- a perinatal and postnatal study in rats -- are usually submitted at the time of the NDA. The FDA recognizes that segment 3 may be modified in the case of contraceptives (which would not be anticipated to be used during the last trimester of pregnancy). A product intended for use throughout pregnancy (such as to prevent STDs) may require further investigation.

Genotoxicology studies are usually performed prior to Phase III clinical testing and include the Ames test and in vitro and in vivo testing using mammalian systems. Carcinogenicity study requirements involve two years of studies in rats, and/or sometimes mice. These are submitted at the time of the New Drug Application (NDA).

Circumstances that may make some of this testing difficult (such as oral ingestion during vaginal administration due to grooming habits) are certainly recognized and many of the pharmacology/toxicology requirements summarized here are open to discussion and appropriate revision.

Clinical Study Requirements and Related Issues

Clinical studies begin at Phase I. The purpose of a Phase I study is to assess safety. In the case of vaginal products, this usually involves a single-dose study to assess toxicity/ irritation. Women enrolled in a Phase I study should not be at risk for pregnancy. Usually this requires tubally sterilized subjects. Commonly, the Phase I study is a pharmacokinetic study (analysis of absorption) in a small number of women -- either with single dose administration of drug or sometimes with multiple doses. Frequently, a Phase I pharmacokinetic study is combined with a regimen including post-coital testing at mid-cycle, in order to get some preliminary information concerning potential efficacy of the product.

Although it might appear easy to conduct a Phase I study, investigators involved in these early studies recognize that to perform adequate post-coital testing, the circumstances are very restrictive and time consuming, and certainly require extremely reliable, compliant volunteers.

Once obvious safety questions are answered in the Phase I trial, efficacy is evaluated in both Phase II and Phase III studies. Requirements for Phase II/III are probably the most controversial and debatable. For a "new" active ingredient product not previously marketed in the United States, the FDA would expect more than one study (including one comparative study) with at least 200 women (total) who complete 12 months of use. Safety information including colposcopic findings and bacteriologic assessments are also expected in about 10 to 15 percent of subjects.

If, in the future, NDAs are to be required for N-9, O-9 and additional dosage forms of an approved "new" active ingredient product, the FDA would expect efficacy information only, as the safety of the active ingredients would not be in question. We envision as simple a protocol as possible in these situations. For example, no clinical laboratory testing would be expected other than pregnancy testing (i.e. no requirement for pap smears, etc.). Our current thinking is to require at least 600 women who complete six months (or alternatively, 200 women who complete 12 months), but this is open for discussion.

Although the OTC monograph contains a list of inclusion/exclusion criteria for study subjects, the DMEDP has no specific list of required criteria. The usual expectation is to enroll women aged 18 to 35 who are not pregnant, with normal cycles and normal pap smears. We would encourage that the subjects have no history of infertility or conditions that lead to infertility and that they have a history of regular coitus. Certainly subjects should have no contraindications to the product or to possible pregnancy.

In the past, for any type of contraceptive study, the FDA required that subjects use only the study contraceptive drug during a trial (i.e. no other methods were allowed). In the present climate of HIV infection and other STD-related issues, the FDA now recognizes that it must allow for potential concommitment condom use. We would hope that a clinical trial population would be made up of subjects who plan *not* to use condoms, but we accept that rare or occasional use might be warranted.

Historically, the FDA has not stated specific criteria regarding parity issues, previous methods used, coital frequency and sexual habits. Although there are not actual requirements in these areas, we do expect collection and discussion of the information.

In terms of requirements for a product for which the developer desires to make claims regarding efficacy in STD prevention, some clinical studies most likely will be necessary to confirm the claims.

In the discussion of STD prevention issues at this workshop, the FDA is anxious to consider suggestions regarding the relevant and clinically significant level of efficacy (decrease in infection rates) that should be required in order to claim an indication. Further, the idea of STD prevention by a vaginal drug product also brings up issues of potential fetal exposure if used during pregnancy. Thus, study design and strategies for gathering information concerning chronic use during pregnancy would need to be considered and presented for review.

This chapter has been a brief review of some current FDA regulatory issues. As mentioned previously, many of the requirements presented are subject to discussion and modification due to individual circumstances. The DMEDP welcomes suggestions for resolving problems and is available to discuss protocol issues and other difficulties prior to submissions.

DEVICES

Each year, roughly 5,000 new devices enter the market through one of the two regulatory routes: Premarket Notification Application (510(k)) or Premarket Approval Application (PMA), the latter being similar to New Drug Application (NDA). Ninety-eight percent of these new devices enter the market through the 510(k) route.

The 510(k)

The 510(k) route is followed for new devices that rely on a claim of substantially equivalent (SE) to a legally marketed device. The term "substantially equivalent" means either of the following:

The new device has the same intended use and has the same technological characteristics as a legally marketed device.

Example: A new latex condom claimed to be substantially equivalent to a legally marketed latex condom.

The new device has different technological characteristics than an existing device, but it can be demonstrated that the new device is as safe and effective as a legally marketed device and does not raise different questions regarding safety and effectiveness than the legally marketed device.

Example: A new polyurethane condom claiming to be substantially equivalent to a legally marketed latex condom.

The PMA

The PMA route is followed for new devices that cannot rely on a claim of being substantially equivalent to a legally marketed device. The term "not substantial equivalent" (NSE) means either of the following:

The new device does not have the same intended use as a legally marketed device.

Example: A menstrual cup to be used as a contraceptive device.

The new device has different technological characteristics, such as a significant change in material, design, energy source or other features, compared with a legally marketed device. It cannot be demonstrated that the new device is as safe and effective as a legally marketed device and does raise different questions regarding safety and effectiveness than the legally marketed device.

Example: Reality (a female condom) was unable to claim SE to the GB Ring (vaginal pouch) due to differences in material and in design. Reality female condom has gone through the PMA process.

What type of data are required to be submitted in a 510(k) application for a new latex condom and a new non-latex condom?

The FDA has developed general guidance for the submission of 510(k) applications for male latex condoms and non-latex condoms.

Some of the data requirements for latex condoms include:

- chemical composition for all additives, including color
- chemical composition of dusting agent and the lubricant
- meeting ASTM standards on latex condoms
- biocompatibility data
- packaging data
- label and labeling, such as intended use, instructions for use, and shelf-life if claimed
- quality control procedures
- inspection of the manufacturing site and manufacturing practices.

For non-latex condoms, the requirements to characterize the material are very extensive, such as physical and chemical characteristics, biocompatibility, permeability, detailed manufacturing procedures, and quality assurance data. Data are required for the finished product, such as physical specification, mechanical tests, biocompatibility tests, shelf-life, labels and labeling.

The FDA also requires statistically clinical data on significant breakage rate and slippage rates of the condom during use.

What type of data are required for a brand new barrier contraceptive?

The basic FDA requirements are as follows:

- pre-clinical data
- well controlled and statistically valid clinical studies
- labeling, both for professionals and the users.

During the investigational period of the Prentif™ Cervical Cap, the most noteworthy event was that the FDA had the opportunity to work with the Women's Health Groups. The New Hampshire Women's Health Group (now called the Concord Women's Health Group) and Irene Svair developed a clinical study protocol. This study protocol was used by over 70 women's health groups to study the cervical cap under the FDA Investigational Device Exemptions (IDE), similar to IND for drugs.

The FDA also developed "Premarket Testing Guidelines for Female Barrier Contraceptive Devices Also Intended to Prevent Sexually Transmitted Diseases," issued on April 4, 1990. In 1988, after reviewing the article in the *British Journal of Family Planning* titled "Preliminary Report -- A Female Condom (Femshield) A study of its User Acceptability," the FDA took keen interest in the potential advantage of this female condom over the male condom. The female condom's use is under the woman's control and it may provide protection against AIDS and STDs. It became very apparent to the FDA that this type of device should be made available to women in the United States as soon as possible. On August 25, 1989 at a public meeting of the Obstetrics and

Gynecology Device Panel, the guidelines were developed as a collaborative effort of experts from the FDA, the National Institution of Child Health and Human Development, the Centers for Disease Control, and the panel. Substantial interaction and dialogue with the audience was involved.

The FDA prepared these pre-clinical and clinical guidelines to expedite device study and evaluation, because of the profound detrimental effect of HIV and AIDS on public health. The guidelines tend to be general because of the diversity of devices of this generic type. A manufacturer should develop study protocols specific to its device with the help of these guidelines and should also discuss the study protocol with the FDA. During the IDE review process, the FDA will evaluate the study protocols for individual contraceptive devices on a case-by-case basis. Such devices must meet certain design criteria, and labeling for these devices must highlight the limitations of the safety and effectiveness data. Overall, the FDA believes that this expedited approach will allow products which can prevent AIDS and other STDs to get on the market rapidly. This will, in turn, lead to improved public health through the prevention of AIDS and STDs.

DISCUSSION

Erica Gollub: I think many of us are very happy about the new premarket guidelines for barriers, although I have to note with some dismay, that the female condom is not yet approved and it is nearly four years after the development of those guidelines. [Editor's Note: The FDA approved the female condom in April 1993.] I am very interested in how those new guidelines are going to impact on the Femcap and the Lea's Shield. I would also like to know if there are any sort of inter-divisional -- between drugs and devices -- problems with these devices going through on an expedited basis because they use spermicide and, furthermore, I would like to understand whether there is any movement towards expedited guidelines for drugs that women can use, most notably a virucide. What progress is being made in that area? If the spermicide being used is a known one that is already on the market, then the whole product is considered a combination product if the lead will be in our center.

Lillian Yin: You asked about Lea Shield and the Femcap. If they do not have the ability to protect women from sexually transmitted diseases like Reality would, then we will not let it go through the expedited review. We would treat it like a regular contraceptive device.

Erica Gollub: I'm asking about the Lea Shield and the Femcap because while they may or may not offer protection for HIV, they certainly could be assumed to offer protection against cervical pathogens such as the diaphragm has been observed to provide. That's why it's germane.

Lillian Yin: At this moment, even spermicides are still being discussed in-house. In fact, on the commissioner's level, we have initiated a group of people from the Public

Health Service, National Institutes of Health, Centers for Disease Control, and the Food and Drug Administration. We are discussing that issue and we are also involving the drugs and anti-infectious divisions and the antiviral division. I think in the very near future we should be able to get some reading as to which way the FDA is looking at spermicides as an antiviral/ antibacterial type of agent.

Lisa Rarick: You asked about spermicides in expedited review for antiviral indications or any antiviral indication whether it was a spermicide or not. In my opinion it is unfortunate, but a lot of divisions got involved -- antivirals, anti-infectives, our division, devices, etc. It slows things down but we don't have any specific guidelines for expedited review of those issues. At our last meeting in-house about the issue of spermicides and whether they protect, it was decided there are not enough data to say one way or the other.

Patrick Rowe: I was aghast to the see the FDA reintroduce the dog. We thought we got rid of the dog as an appropriate animal model. What animal model are you proposing for penile irritation?

Lisa Rarick: Very good question. You're right, the dogs are not a hard and fast requirement, nor is the monkey. Rabbits are acceptable. This was just my pharmacologists "wish list." In terms of penile irritation, he suggested the rabbit. I think the issue is that if there is a real potential -- if you find in your studies that the males are having some kind of difficulty -- then you would possibly need to look at pharmacokinetics of penile absorption and things like that.

Irv Sivin: You indicated that the FDA will now accept data for contraceptives when adjunctive contraception is used. Are there any guidelines that you have developed for the extent of adjunctive contraceptive use and for the analysis of the effects of such use?

Lisa Rarick: We have to accept that, ethically, condoms have to be recommended for use under certain situations and we have to accept that women are going to use them. We hope that the population as chosen is such that women aren't going to be in a position to be needing to use them every time they have intercourse and, thus, it would only be an occasional use in an unusual circumstance for those women. If a woman is using a condom all the time, something is wrong with your selection of that woman into the study and she needs to be dropped or the data possibly modified with leaving that woman out. There are no specific guidelines.

Laneta Dorflinger: I just want a clarification. In your requirements for the number of women in the Phase III studies for a new drug, I thought I heard you say that you expected to see at least 200 women who complete 12 months. What I didn't hear was, 200 women cumulatively or 200 women in each study?

Lisa Rarick: Cumulatively on your product. So if you have a randomized study, you would have to have 200 women on your study product cumulatively. One center can have 10 and one can have 190 for that matter.

Laneta Dorflinger: But if you, in fact, have two separate studies contributing to your final two separate pivotal studies are you expecting to see 200 women complete 12 months in each of those studies or accumulatively?

Lisa Rarick: No, just total.

Zena Stein: Just three comments on testing for STD infection. It seems to me you don't have to use pregnancy. A women could be postpregnancy age or with a uterus out and you could still be testing for STDs. The population in which you want to test for STDs must be a population that is at high risk. It seems to me that the regulations you put forward therefore regarding testing for STDs have really inappropriate criteria.

Lisa Rarick: I can answer the question about how we would set up a study for looking at STD risks.

Roberto Rivera: Separately from pregnancy?

Lisa Rarick: Exactly. You would have to do it separately from pregnancy because your population is incredibly different. It may be naive but the only thing that we've come up with so far is to choose a population that is at an incredibly high-risk for STD, whatever pathogen you're going to claim, HIV or others; counsel everybody in your group to use condoms appropriately. Give one arm of them the spermicide or whatever the product is or the virucide, and one arm of them with condoms alone. You can project how many people aren't going to use condoms. You can project what your rate should be in those populations with condom use and you try to find a difference between the two and show the spermicide was better than condoms alone.

James Trussell: I have two brief questions. The first is that I'm interested in the logic or the science underlying the requirement that trials for barrier methods be 12 months long when the typical use by a women who uses a barrier is less than 12 months. Secondly, I'm interested in how the FDA justifies the need to have a textbook answer on comparative efficacy, by now having a requirement for a comparative arm in a barrier trial when there hasn't been one before. How do you balance that textbook need against the reality that it will be so costly that we may have no new devices on the market? Are we talking about those barriers that will not protect women from STDs or are we talking about just barrier contraceptives? A brand new one such as Lea Shield?

Lillian Yin: Yes.

10

CHEMICAL VAGINAL CONTRACEPTIVES: PRECLINICAL EVALUATION

Gustavo F. Doncel

Contraceptive Research and Development Program
Department of Obstetrics and Gynecology
Eastern Virginia Medical School
Norfolk, VA 23507

Sperm-immobilizing agents, also called spermicides, have been almost the only constituents of chemical vaginal contraceptives (CVCs). They are effective if properly used, readily available, and simple to use. They are also among the few contraceptive methods that do not require physician intervention.

Although they have been used since ancient times, it was not until this century that laboratory evaluation was employed prior to *in vivo* trials [Baker, 1932 and 1935]. Testing consisted mainly of determining the effect of the new agent on sperm motility. Thus, one of the most effective types of spermicides discovered was the *surfactants*, such as nonoxynol-9, that so overwhelmingly reigned on the market, little room for innovation was left. Both spermicides and their testing methodology remained almost unchanged for decades [Hahn, 1980; Sobrero, 1980].

Today, however, concerns about the side effects of other contraceptive methods, as well as the urgent need to diminish the spreading of acquired immune deficiency syndrome (AIDS) and other sexually transmitted diseases (STDs), have produced a renaissance of vaginal contraception. Furthermore, a better knowledge of the fertilization process generates the possibility for designing of new CVCs with more specific and less harmful components. In order to develop these novel contraceptives, new and existing assays on sperm fertilizing capacity have to be set up for contraceptive evaluation. The purpose of this chapter is to outline a tentative CVC testing algorithm and briefly describe the methodology involved.

Barrier Contraceptives: Current Status and Future Prospects, pages 147–162

CHEMICAL VAGINAL CONTRACEPTIVES

Conceptually, chemical vaginal contraceptives are formulated compounds that, when placed into the vagina prior to intercourse, block sperm fertilizing ability, ultimately impeding conception.

Mechanistically, it is theoretically possible to differentiate three types of CVCs:

- **Spermicidal Agents.** They completely immobilize sperm, usually by disrupting sperm membranes.
- **Sperm Function Inhibitors.** They alter sperm functions such as motion, capacitation, acrosome reaction, binding to and penetration of zona pellucida/oolemma, and even post-fusion events like nuclear decondensation.
- **Cervical Mucus Blockers.** They modify cervical mucus (CM) rheology, impeding sperm penetration.

While still intriguing in many aspects, fertilization is considerably better understood nowadays than three decades ago. Therefore, several crucial steps in that process can be targeted to impair conception. Nonetheless, spermicidal agents (spermicides) should not be disregarded for they represent a short-term developmental strategy with potential activity against human immunodeficiency virus (HIV) and other sexually transmitted pathogens.

Ideally, candidates for new CVC should be:

- **Highly Active,** with low minimum effective concentration.
- **"Biodiffusible",** being able to considerably diffuse in CM while still displaying their biological activity.
- **Mucoadhesive,** coating and strongly adhering to mucosal surfaces such as vaginal epithelium.
- **Non-irritating,** especially for vaginal and cervical linings, thus diminishing adverse effects and the risk of increased AIDS/STD transmission by epithelial disruption.
- **Non-absorbable,** to prevent systemic side-effects.
- **Active Against STD Pathogens,** particularly HIV. This property is not intrinsically pertinent to CVCs, but quite relevant in today's health context. It can be displayed by the same active ingredient used for contraception or by a different one added to the contraceptive.

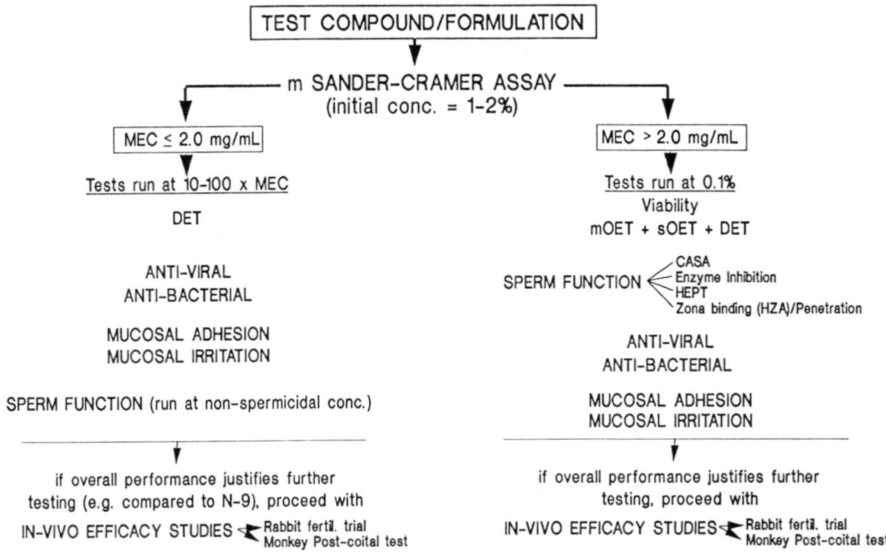

Figure 1. *Chemical Vaginal Contraceptives: Tentative preclinical screening algorithm.* m: modified. MEC: minimum effective concentration. DET: double-end test. sOET: simultaneous one-end test. mOET: modified one-end test. CASA: computer-assisted semen analysis. HEPT: hamster-egg penetration test. HZA: hemizona assay.

EVALUATION OF POTENTIAL CVCs

In 1988, the Contraceptive Research and Development (CONRAD) Program established a special project, the Spermicidal Testing Program (STP), to evaluate new putative spermicides, as well as to create and validate screening methodology. Since then, almost 400 compounds/formulations from national and international sources have been screened.

There appears to be no explicit testing algorithm for CVC candidates in the contraceptive literature; therefore, a tentative flow chart for preclinical evaluation is presented (Fig. 1). This strategic chart is based upon the theoretical nature of these agents, the ideal CVC characteristics, the experience collected with the STP, and the need to include compounds that, without being spermicidal, could still display anti-fertility effects.

Sperm Immobilizing Agents

Given an unknown compound and after its solubility has been determined, the International Planned Parenthood Federation (IPPF) Agreed Test for total spermicidal activity was first used [Kleinman, 1964]. We now prefer a slightly modified protocol of the assay originally described by Sander-Cramer [1941] which, using the same methodology as the IPPF test, does not require a high and fixed testing concentration (1g/11 mL), gives additional information about potency of compounds, and saves time from an operational standpoint. The assay uses serial two-fold dilutions of the compound in saline and normal semen samples. Its end point is total sperm immobilization in a 20-second incubation. Six to 10 different semen samples are used to achieve more reliable results, usually expressed as compound minimum effective concentration (MEC). The MEC for a given agent is obtained using the highest dilution, from the serial dilution set, which displays spermicidal activity.

The one or two percent initial concentration has been selected based on usual low availability of newly synthesized compounds, solvent-induced ill effects on sperm motility, and potency of marketed spermicides.

Vehicles used to prepare compound stock solutions depend on the intrinsic solubility of the test agents. We normally use saline solutions such as 0.9 percent NaCl and 0.01M phosphate buffer (PBS), distilled water (dH_20), dimethyl sulfoxide (DMSO), methanol, or ethyl acetate. Some of these solvents are toxic for sperm at the initial dilutions of the Sander-Cramer assay. For instance, DMSO's highest spermicidal dilution can sometimes reach 1:4 (3.0 ± 1.0 n=33).

A compound that presents a MEC ≤2.0 mg/ml in the modified Sander-Cramer assay has a significant sperm immobilizing activity and can be considered as a spermicidal agent. The MEC limit stated above is arbitrary and represents around 10 to 20 times the MEC of our standard positive control nonoxynol-9 (N-9) preparation (Ortho-OW48). With an initial concentration of 10 mg/mL in dH_20, our N-9 MEC is 0.109 ± 0.030 mg/mL (n=47); this does not seem to vary considerably when other solvents are used.

If a test compound proves to be effective at 1mg/mL, although its potency is almost 10 times weaker than N-9, it is still a low MEC and further testing may reveal other advantageous characteristics.

During 1991-1992, the CONRAD STP has evaluated 220 compounds (putative actives and bases) and 48.4 percent displayed a MEC ≤2.0 mg/mL. From these, 67 percent were new formulations containing N-9.

Biodiffusion in Cervical Mucus. In natural reproduction, sperm have to pass through the female cervical canal, a narrow and mesh-like pathway, to reach the oocytes. This makes any CVC that can achieve high bioactive concentrations in CM, or mechanically block sperm penetration theoretically better than those that do not.

Cervical mucus penetration (CMP) assays have been extensively used in infertility diagnosis [Alexander, 1981a; Schats et al., 1984; Overstreet, 1986], and can be applied for contraceptive evaluation [Alexander, 1981b; Sharman et al., 1986]. Based upon the assays described in the literature, we set up some variants to help define sperm-CM-compound interactions in a contraceptive context.

As mentioned before, an important characteristic for the test agent is the capacity to "biodiffuse" in CM, that is to diffuse in CM while keeping its biological activity. For assessing this property, we use a CMP assay variant called Double-End Test (DET), which is the next step in our testing flow once a compound has been determined to have a MEC ≤2.0 mg/mL. Briefly, the DET consists of an incubation in a test agent solution of bovine CM tubes (Penetrak ®, Serono-Baker) opened at one end, followed by another incubation of the same tube opened at the other end in a semen sample. In this way, spermatozoa migrate in an opposite direction to that of the compound, eventually ceasing their penetration where they find a compound bioactive concentration. Thus, penetration of vanguard motile sperm is recorded and compared to that of the control, i.e., the solvent. Results are expressed as percent of control sperm penetration and are inversely correlated with compound biodiffusion. Despite the controversy about the significance of the distance traveled by vanguard motile sperm (Alexander, 1981a; Katz et al., 1980), we chose this endpoint for it is operationally convenient in the initial CVC screening.

During the last two years, in the CONRAD STP, less than 20 percent of the agents with ≤2.0 mg/mL MEC showed DET sperm penetration values <60 percent, demonstrating that high biodiffusion rates are not common, especially for multi-ingredient products.

Factors potentially affecting compound-mucus interaction in the DET include electrical charge, molecular size, testing concentration and solvent used [Chantler et al., 1989; Marriot, 1989]. The latter two are experimental conditions that may considerably influence compound diffusion in CM. For instance, although it has been claimed that N-9 has a negligible CM diffusion [Chantler et al., 1992], we found that under the DET conditions, N-9 biodiffusion appears to vary significantly, depending on testing concentration and solvent used (Fig. 2). This could highlight the influence of formulation components on the CM bioactive concentration achieved by a contraceptive agent. Results obtained with two experimental N-9 containing formulations support this speculation (Table 1). It is noteworthy, however, that whether *in vitro* higher biodiffusion rates will translate into improved *in vivo* efficacy remains to be determined.

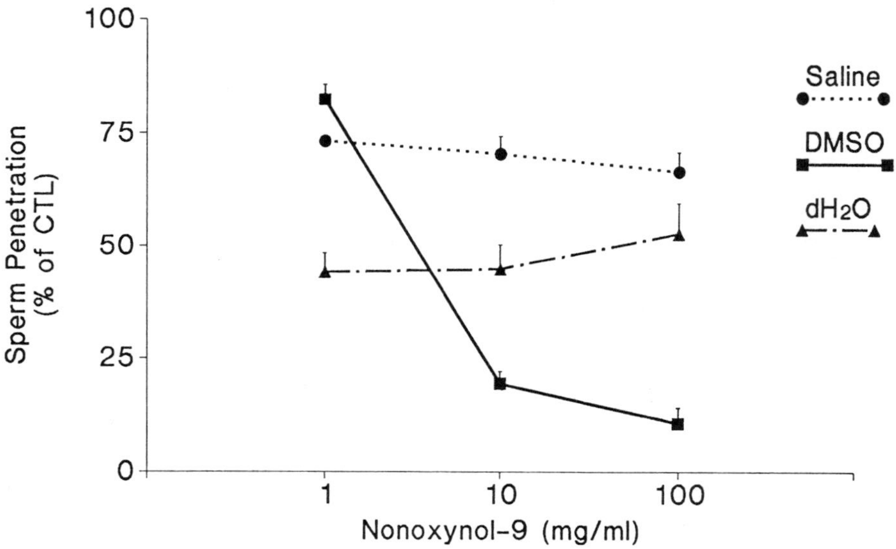

Figure 2. *Influence of solvent and concentration on nonoxynol-9 cervical mucus biodiffusion.* Data points were obtained with a double-end test (DET) and represent $\bar{x} \pm$ S.E. (n=20). The controls (CTL) were the corresponding solvents without N-9. Saline: 0.9% NaCl. DMSO: dimethyl sulfoxide. dH_2O: distilled water. Comparing concentrations, only N-9 in DMSO presented significant differences (ANOVA, $p < 0.0001$). Comparing solvents, all concentrations were significantly different (ANOVA, $p < 0.0001$).

TABLE 1. Influence of Formulation on Nonoxynol-9 Cervical Mucus Biodiffusion

Product	N-9 Content (%)	Sander-Cramer MEC (mg/mL)	Double-End Test (% CTL)	N-9 Testing Concentration (mg/mL)
Polymer-1231	4	0.119±0.14	53.6±3.9[a]	3.64
Conceptrol(R)	4	0.113±0.0	86.5±3.8[b]	3.64
Nonoxynol-9	4	0.113±0.0	61.9±0.9[c]	3.64

Results are expressed as mean±S.E. Materials were dissolved in 0.9% NaCl
% CTL: percent penetration of test sperm samples when compared to controls (incubated with 0.9% NaCl)
MEC: minimum effective concentration
Polymer-1231: experimental formulation
Conceptrol®: marketed spermicide
Statistics: a vs b (p=0.03), b vs c (p=0.0001)

More sophisticated technology, such as fourier transform infrared spectroscopy (FTIR) can be employed to measure compound diffusion in CM [Katz, 1993]. The CONRAD Program is currently supporting work in this area.

Antimicrobial Activity. CVCs are primarily intended to prevent conception. However, the AIDS pandemic and other STDs make evaluation of anti-viral and anti-bacterial properties of contraceptive compounds highly relevant.

In this regard, and as the next step in the testing flow chart, CVC candidates are sent to a CONRAD-funded retrovirology laboratory which performs cell-free and cell-associated HIV inactivation assays [Resnick et al., 1990 and 1993]. In the last two years, 84 percent of the compounds displaying a ≤2.0 mg/mL MEC in the Sander-Cramer assay significantly reduced HIV infectivity *in vitro*. Approximately 10 series of new compounds were involved. This suggests, at least in part, common mechanisms for both spermicidal and virucidal activities and highlights the existence of "bifunctional" agents.

Activity against STD pathogens, such as *Chlamydia trachomatis* and *Neisseria gonorrhoea*, needs to be evaluated. Preparations containing N-9 have been found to have *in vitro* antimicrobial activity against these pathogens [Bolch and Warren, 1973; Kelly et al., 1985]. They also proved to be effective in reducing gonococcal and chlamydial infections in clinical trials [Louv et al., 1988; North, 1990]. It is also important to screen the new agents for bactericidal activity against the normal constituents of the human vaginal flora, e.g., *Lactobacilli*, since severe disturbance of their equilibrium could have unwanted consequences [Redondo-Lopez et al., 1989; McGroarty et al., 1992].

Interactions with Mucosal Tissues. Uninterrupted coating of vaginal epithelium by a CVC is thought to be a relevant feature to prevent sexual transmission of HIV and other pathogens. Therefore, whenever possible, mucosal adhesiveness should be evaluated. It is uncommon to look for this property in a contraceptive formulation, however it is not rare when, for example, dermatologic bases are considered. Different assays have been described [Peppas and Buri, 1985; Smart et al., 1984] and can be applied to CVC candidates.

Although the real incidence and consequences of N-9-induced vaginal and cervical irritation remain elusive thus far, potential disruption of these epithelia may occur, especially when N-9 is used frequently or in high doses [Niruthisard et al., 1991]. This is of particular concern from the standpoint of STD transmission [Kreiss et al., 1992].

By and large, the rabbit vaginal irritation model has been widely used [Hahn, 1980]. Briefly, the test CVC is instilled into the vagina of approximately five rabbits per group, once daily for 10 days. Doses may vary, but should be in relation to the proposed human dose. Additional rabbits are included as sham and untreated controls. The animals are sacrificed on the day following the last treatment (11th day) and the vaginas removed, macroscopically examined and processed for histopathological evaluation.

Local toxicological studies should be the next step in the testing; however, since the rabbit assay is usually expensive, it requires a first level of decision about the relative significance of the test compound. The development of new *in vitro* assays is possible and needs to be further explored.

Sperm Function Inhibition. Chemical vaginal contraceptives can be moderately spermicidal but highly inhibitory of sperm functions. Zaneveld et al. [1988] have described potent acrosin inhibitors that display sperm-immobilizing activity but only at higher doses. Inasmuch as this dual contraceptive mechanism represents a theoretical advantage for the agent, running sperm-functional tests at non-spermicidal concentrations is an advisable step within the screening program (see below).

In vivo Preclinical Efficacy. Having collected information on spermicidal potency, CM biodiffusion, anti-microbial activity (particularly against HIV), mucosal adhesion and irritation, and sperm functional inhibitory properties, a strategic decision has to be made. If the compound's overall performance justifies further testing, *in vivo* efficacy studies (e.g., rabbit fertility trials and monkey post-coital tests) should be pursued (see Zaneveld's chapter). In this category of CVCs, i.e. spermicidal agents, a performance that warrants proceeding with the next evaluation step practically means that the test compound has to possess advantageous qualities when compared to the widely used N-9 or its formulations.

Sperm Functional Inhibitors and Cervical Mucus Blockers

Compounds that irreversibly inhibit sperm function, blocking its fertilizing capacity, may well represent the future in chemical vaginal contraception. No such compounds are on the market today. Theoretically, they would be more specific and have fewer side-effects than current surface-active spermicides.

On the right side of the CVC screening algorithm, after the test compound proves not to have a high spermicidal activity in the modified Sander-Cramer assay, i.e., MEC >2.0 mg/mL, several assays can be performed to assess its contraceptive potential (Fig. 1).

Sperm viability. A sperm viability test after a short incubation with the test compound could be the next step in the evaluation. Dead spermatozoa are no longer able to maintain a selective permeability at their plasma membrane. Therefore, electrochemically charged stains like Eosin Y can penetrate and distinctly label such cells. This type of test helps to determine whether the agent has any spermicidal activity at concentrations lower than 0.2 percent and under different experimental conditions.

Cervical mucus penetration blockage. Subsequently, a set of CMP assays enable the investigation of sperm migration blocking effects. The first assay, called "modified One-End Test" (mOET) consists of a 30 minute incubation of bovine CM tubes in the compound solution, followed by another incubation (60 minutes) of the same tubes after mixing a semen sample with the compound. The end point is penetration length of motile vanguard sperm. If sperm penetration is similar to that of the control, the agent does not have blocking activity. Conversely, if sperm migration is impeded, the compound could be active on either the CM or some sperm motion parameters. In this latter case, a "simultaneous One-End Test" (sOET), consisting of a very short compound-sperm pre-incubation and the subsequent immersion of CM tubes in that mixture, could be helpful. If positive, i.e., abnormal penetration with the test compound, emphasis may be placed on a sperm alteration.

Computer-assisted semen analysis (CASA) is a relatively new and invaluable aid to verify this hypothesis. Determination of percent of motile cells, velocity (VEL), linearity (LIN), amplitude of lateral head displacement (ALH), flagellar beat/cross frequency (B/C freq.), and other sperm motion parameters can pinpoint compound actions that otherwise would pass inadvertently. Correlation between several of these parameters and sperm penetration in CM has been clearly demonstrated [Mortimer et al., 1986; Keel and Webster, 1988]. As shown in Table 2, alteration of sperm motion parameters, particularly VEL and LIN by an experimental base polymer is associated with complete impediment of sperm penetration in CM.

If changes in motility are detected, a DET to determine compound biodiffusion could be indicated.

TABLE 2. Motion parameters and cervical mucus penetration of sperm samples co-incubated with base materials

Parameters Studied	Semen Samples (original)	Materials		
		Saline	CMC/POV	POL 1231
Sperm Motion				
Concentration	107.3±29.0	29.8±.7	34.0±2.2	28.4±7.6
Motility (%)	71.9±1.5	49.0±.9	60.7±7.3	31.3±7.8
Conc. Mot. Sp.	47.4±1.1	14.0±.9	21.1±4.0	7.0±2.3*
Velocity (um/sec)	76.0±6.5	72.2±.8	86.5±7.5	44.4±1.9*
Linearity	7.1±0.3	6.1±1.0	6.7±0.5	2.3±0.2*
ALH mean (um)	2.6±0.2	2.4±1.0	2.8±0.6	3.0±0.7*
B/C freq. (H_2)	17.1±0.6	18.5±1.0	18.3±1.4	15.9±1.4
CM Penetration				
sOET		100±0.0	100±0.0	0±0
mOET		100±0.0	100±3.2	1.1±0.9

Results are expressed as mean ± S.E.
Sperm motion parameters were obtained with a computerized system (CASA), and CMP results with two variants of the One-End Test (sOET and mOET)
Materials were tested at 0.125% in 0.9% NaCl
Sperm concentration was adjusted with Baker's buffer supplemented with BSA.
CMC/POV: base material containing carboxymethyl cellulose and povidone
POL 1231: experimental base polymer
*statistically significant when compared to CMC/POV values

Concerned about the possible influence of the pH of test compound solutions on the outcome of our CMP assays, we ran some experiments incubating equal volumes of acidified saline solutions (0.9 percent NaCl + HCl) at various pHs, and semen samples diluted with Baker's buffer to obtain $60x10^6$ motile sperm/mL. We found the mOET and sOET sperm penetrations are affected by initial saline pHs below 2.5 and 1.5 respectively. Interestingly, both pHs also affected sperm migration in the One-End Test (OET). This assay consists of a first incubation of bovine CM tubes with the test solution, followed by a second incubation after wiping clean the open end with a semen sample contained in a separate vial. Therefore, the latter finding indicates that alteration of CM pH can effectively impair CM sperm penetration. This is in agreement with the described influence of CM pH on the post-coital test outcome (Peek and Matthews, 1986).

The effect of pH on sOET sperm penetration when intact (undiluted) semen was used to better resemble *in vivo* conditions is shown in Table 3. The minor penetration observed at initial pH 1 was due to semen samples that achieved ≥3.00 after mixing pHs or possessed sperm densities $>250x10^6$/mL.

Although very preliminary, these experiments support that formulations with enough buffer capacity to keep low pHs after *in vivo* mixing with semen might be effective as contraceptives. Furthermore, HIV sensitivity to low pH has been reported (Martin et al., 1985) and even postulated as a feature to be exploited in prevention of its sexual transmission (Voeller, 1985).

TABLE 3. Influence of pH on sperm penetration in cervical mucus

Saline pH		sOET	
Initial	After mixing with semen	%CTL	n
0.51±0.01*	1.88±0.54*	0±0**	11
1.01±.00	2.67±0.56	19.4±9.2	11
1.51±.01	5.16±0.72	100±0	11
5.85±0.10[a]	6.53±0.38	100±0	11

Intact undiluted semen was used. Saline (0.9% NaCl) pH was adjusted with HCl
Acidified saline and semen were mixed in equal volumes
sOET: variant of one-end test (see text for details). %CTL: percent penetration of test sperm
samples when compared to control ones (non-acidified saline)
*mean ± SD
**mean ± SE
[a] ph of regular saline (0.9% NaCl)

Sperm function Inhibition. The next step in the screening algorithm is sperm function assays. The spermatozoon has a myriad of active enzymes that allow it to undertake its cellular functions [Baccetti, 1991]. Some of them, such as lactate dehydrogenase and creatine phosphokinase, are involved in energy production and thus, related to motility; others, like hyaluronidase and acrosin, enable the sperm to transverse the oocyte vestments.

Since many of these enzymes are crucial to ensure sperm fertilizing ability, numerous attempts have been made to block their activity for contraceptive purposes [Zaneveld, 1982]. In the rabbit fertility model, some of the inhibitors produced even better results than marketed spermicides. *in vitro* enzyme-specific assays to demonstrate compound inactivating properties may be run, especially when a related mechanism of action is suspected [Zaneveld and Jeyendran, 1990; Lee et al., 1982]

Sperm capacitation and acrosome reaction, as well as binding and penetration of zona pellucida and oolemma, are key events in human fertilization. Blockage of any of these processes by a CVC candidate would result in contraceptive activity.

Assessment of such compound properties, however relatively expensive and time-consuming, is currently possible. Protocols for identification of acrosomal status [Cross et al., 1986], hamster-egg penetration test [Yanagimachi, et al. 1976], and zona binding/penetration assays [Burkman et al., 1988; Overstreet and Hembree, 1976] can be modified to assess anti-fertility effects.

Antimicrobial activity and mucosal interactions. Upon detection of contraceptive activity in the test compound, anti-STD as well as mucosal adhesion and irritation studies should be carried out.

In vivo preclinical studies. These studies would complete the initial screen for compounds that potentially fit into the CVC categories of sperm functional inhibitors and CM blockers. If the overall compound performance justifies further testing, animal efficacy studies are indicated.

FUTURE PROSPECTS

As our understanding of the biology of sperm and fertilization increases, the methodology to assess potential CVCs improves.

It can be anticipated that more specific and less harmful CVCs will appear in the near future since knowledge about key fertilization events suitable for contraceptive targeting has already been acquired. Improved spermicidal products, sperm functional inhibitors, and cervical mucus blockers will find their way into the CVC armamentarium.

Combinations of various agents acting through different mechanisms may well represent the wave of the future. Probably, these products will have properties such as muco-adhesiveness, long-action, and quick and homogeneous dispersion.

As part of these formulations, there may also be compounds that lower the vaginal pH or impede the formation of the seminal coagulum, examples of potential additive effects to the main contraceptive action(s).

Due to increased incidence of AIDS and other STDs, bifunctional compounds or combinations of agents with both antispermatic and microbicidal activities will probably play an important role in women's reproductive health.

Finally, it is worth remembering that none of these possibilities will become a reality without the appropriate screening methodology.

ACKNOWLEDGEMENTS

I am very grateful to Ms. Rachel D. Vohsberg for her excellent technical assistance, Ms. Pauline M. Clynes for her editorial assistance, and to Ms. Adrienne R. Boothe, Ms. Stacy Moses and Ms. Martha Wilson for their help in the preparation of this manuscript.

This work was supported by the Contraceptive Research and Development (CONRAD) Program under a cooperative agreement with the U.S. Agency for International Development (USAID). The views expressed by the author do not necessarily reflect the views of CONRAD or USAID.

REFERENCES

Alexander NJ (1981): Evaluation of male infertility with an in vitro cervical mucus penetration test. Fertil Steril 36(2):201-208

Alexander NJ (1981): Cervical mucus penetration test for in vitro assay of vaginal contraceptive agents. Fertil Steril 36(4):516-520

Baccetti B (1991): "Comparative Spermatology 20 Years After". New York: Raven Press

Baker JR (1935): "The Chemical Control of Contraceptive". London: Chapman Hall.

Bolch OH Jr, Warren JC (1973): In vitro effects of Emko in Neisseria gonorrhoeae and Trichomonas vaginalis. Am J Obstet Gynecol 115:1145-1148

Burkman LJ, Coddington CC, Franken DR, Kruger TF, Rosenwaks Z, Hodgen GD (1988): The hemizona assay (HZA): development of a diagnostic test for the binding of human spermatozoa to the human hemizona pellucida to predict fertilization potential. Fertil Steril 49:688-697

Chantler E, Hemsworth M, Roach VP, Bruce JM (1992): Compatibility between the spermicide Nonoxynol-9 and mid-cycle human cervical mucus. Contraception 46:289-295

Chantler E, Sharma R, Sharman D (1989): Changes in cervical mucus that prevent penetration by spermatozoa. Symp Soc Exp Biol 43:325-336

Cross NL, Morales P, Overstreet JW, Hanson FW (1986): Two simple methods for detecting acrosome-reacted human sperm. Gamete Res 15:213-226

Hahn DW (1980): Preclinical evaluation of new vaginal contraceptives. In Zatuchni GI, Sobrero AJ, Speidel JJ (eds): "Vaginal Contraception: New Developments". Hagerstown: Harper & Row, pp 234-246

Katz DF (1993): Personal Communication

Katz DF, Overstreet JW, Hanson FW (1980): A new quantitative test for sperm penetration into cervical mucus. Fertil Steril 33:179

Keel BA, Webster BW (1988): Correlation of human sperm motility characteristics with an in vitro cervical mucus penetration test. Fertil Steril 49(1):138-143

Kelly JP, Reynolds RB, Stagno S, Louv WC, Alexander WJ (1985): In vitro activity of the spermicide nonoxynol-9 against Chlamydia trachomatis. Antimicrob Agents Chemother 27:760-762

Kleinman RL (1964): "IPPF Medical Handbook" (2nd ed.). London: International Planned Parenthood Federation.

Kreiss J, Ngugi E, Holmes K, Ndinya-Achola J, Waiyaki P, Roberts PL, Ruminjo I, Sajabi R, Kimata J, Fleming TR, Anzala A, Holton D, Plummer F (1992): Efficacy of Nonoxynol-9 contraceptive sponge use in preventing heterosexual acquisition of HIV in Nairobi prostitutes. JAMA 268(4):477-482.

Lee CYG, Moon YS, Yuan JH, Chen AF (1982): Enzyme inactivation and inhibition by Gossypol. Mol Cell Biochem 47:65-70

Louv WC, Austin H, Alexander WJ, Stagno S, Cheeks J (1988): A clinical trial of Nonoxynol-9 for preventing Gonococcal and Chlamydial infections. Infect Dis 158(3):518-523.

Marriot C (1989): Drug-mucus actions and interactions. Symp Soc Exp Biol 43:163-177

Martin LS, McDougal JS, Loskoski SL (1985): Disinfection and inactivation of the human T-lymphotropic virus type III/lymphadenopathy-associated virus. J Infect Dis 152:400-403

McGroarty JA, Tomeczek L, Pond DG, Reid G, Bruce AW (1992): Hydrogen peroxide production by Lactobacillus species: correlation with susceptibility to the spermicidal compound nonoxynol-9. J Infect Dis 165:1142-1144

Mortimer D, Pandya IJ, Sawer RS (1986): Relationship between human sperm motility characteristics and sperm penetration into human cervical mucus in vitro. Reprod Fert 78:93-102

Niruthisard S, Roddi RE, Chutivongse S (1991): The effects of frequent Nonoxynol-9 use on the vaginal cervical mucosa. Sex Transm Dis 18:176-179

North BB (1990): Effectiveness of vaginal contraceptives in prevention of sexually transmitted diseases. In Alexander NJ, Gabelnick HL, Spieler JM (ed): "Heterosexual Transmission of AIDS". New York Wiley-Liss, Inc., pp 273-290

Overstreet JW (1986): Evaluation of sperm-cervical mucus interaction. Fertil Steril 45:324

Overstreet JW, Hembree WC (1976): Penetration of the zona pellucida of non-living human oocytes by human spermatozoa in vitro. Fertil Steril 27:815-831

Peek JC, Matthews CD (1986): The pH of cervical mucus, quality of semen, and outcome of the post-coital test. Clin Reprod and Fertil 4:217-225

Peppas NA, Buri PA (1985): Surface, interfacial and molecular aspects of polymer bioadhesion on soft tissue. J Control Rel 2:257-275

Redondo-Lopez V, Cook RL, Sobel JD (1989): Emerging role of lactobacilli in the control and maintenance of the vaginal bacterial microflora. Rev Inf Dis 12:856-872

Resnick L, Busso ME, Duncan RC (1990): Anti-HIV screening technology. In Alexander NJ. Gabelnick HL, Spieler JM (eds): "Heterosexual Transmission of AIDS". New York Wiley-Liss, Inc., pp 311-326

Resnick L, Busso ME, Duncan RC, Penas JG, Doncel GF, Alexander NJ (1993): Comparative evaluation of spermicidal agents with virucidal activity against HIV in-vitro. Contraception (submitted)

Sander FV, Cramer SD (1941): A practical method for testing the spermicidal action of chemical contraceptives. Hum Fertil 6:134-153

Schats R, Aitken RJ, Templeton AA, Djahanbakhch O (1984): The role of cervical mucus-semen interaction in infertility of unknown aetiology. Br J Obstet Gynaecol 91:371-376

Sharman D, Chantler E, Dukes M, Gowland Hutchinson F, Elstein M (1986): Comparison of the action of nonoxynol-9 and chlorhexidine on sperm. Fertil Steril 45(2):259-299

Smart JD, Kellyaway ID, Worthington HEC (1984): An in vitro investigation of mucosa-adhesive materials for use in controlled drug delivery. J Pharm Pharmacol 61:295-299

Sobrero AJ (1980): Spermicidal Agents: Effectiveness, Use and Testing. In Zatuchni GI, Sobrero AJ, Speidel JJ (eds): "Vaginal Contraception: New Developments". Hagerstown: Harper & Row pp 48-65

Voeller B (1985): Spermicides: an additional potential barrier to the sexual spread of the AIDS virus LAV/HTLV-III. Mariposia Foundation Occasional Papers 4(B):1-6

Yanagimachi R, Yanagimachi H, Rogers BJ (1976): The use of zona-free animal ova as a test-system for the assessment of the fertilizing capacity of human spermatozoa. Biol Reprod 15:471-476

Zaneveld LJD, Jeyendran RS (1990): Biochemical analysis of seminal plasma and spermatozoa. In keel BA, Webster BW (eds): "CRC Handbook of the Laboratory Diagnosis and Treatment of Infertility". Boca Raton: CRC Press, pp 80-94

Zaneveld LJD, Kamiski JM, Waller DP, Quigg J, Anderson RA, Mack SR, Bauer L (1988): Aryl 4-Guanidinobenzoates: potential vaginal contraceptives. In Runnebaum B, Rabe T, Keisel L (eds): "Female Contraception". Berlin: Springer-Verlag, pp 286-292

Zaneveld LJD (1982): Sperm Enzyme inhibitors for vaginal and other contraception. In Zatuchni GI, Osborn K (eds): "Research Frontiers in Fertility Regulation". Chicago: Northwestern Univ., pp 1-14

DISCUSSION

James Koch: I am wondering if the extractability of nonoxynol-9 from either marketed gels or creams is superior and, if so, in what way? My second question relates to the concentration of nonoxynol-9 in various products. There are 2 percent nonoxynol-9 products and there are 5 percent nonoxynol products. Is there a significant functional difference between these two types of products?

Gustavo Doncel: Concerning the first question, we have not extensively studied all the different formulations existing in the market. All the ones that I have studied have very poor biodifusability. Although I have tested several experimental compounds and formulations that have significantly improved non-biodifusion, I don't have any data on commercial products concerning cervical mucus penetration, or other functional tests, or how the formulations relate to their activity. Concerning the question on concentration, in all of these experiments the concentrations are adjusted. I think that the only way to determine the efficacy of 2 percent, 4 percent, and 8 percent N-9 is through *in vivo* animal studies.

Alfred Shihata: Sperm can be killed very easy by many compounds. A pH of 3 or 4 alone can immobilize the sperm easily. Even distilled water or a placebo can immobilize the sperm. To find a spermicide is a very easy matter. I think our biggest task is to find a virucidal or bacteriocidal product.

Gustavo Doncel: I have a slightly different opinion about the resilience of sperm. I think that they are very resistant and resilient to everything. To completely immobilize sperm within a few seconds you really need strong compounds. Of course distilled water will really damage them because of high osmotic solution, but KY-jelly, personal Ortho lubricants, and some of the other lubricants existing in the market may severely affect motility and sperm penetration but only when you use them at 100 percent concentration.

11

VAGINAL CONTRACEPTIVE EFFICACY: ANIMAL MODELS

Lourens J.D. Zaneveld

Department of Obstetrics and Gynecology,
and Department of Biochemistry
Rush University
Rush-Presbyterian-St. Luke's Medical Center
1653 West Congress Parkway
Chicago, IL 60612-3864

INTRODUCTION

Agents that immobilize spermatozoa or inactivate these cells by other mechanisms can vary greatly in their in vitro and in vivo potencies. A typical example is nonoxynol-9, the most frequently used spermicide among currently marketed vaginal contraceptives. N-9 has an ED_{50} of about 40-50 ug/ml [Quigg et al., 1988], yet 60 to 100 mg need to be placed vaginally in women for contraceptive efficacy [Population Reports, 1984]. Animal models confirm these results. Concentrations of N-9 that are 1000-fold its ED_{50} or higher, only partially suppress pregnancy in the rabbit [e.g., Kaminski et al., 1985] and in the primate [Zaneveld et al., 1977]. Another example is K-Y jelly, which was reported to be spermicidal in vitro (Tagatz et al., 1972; Frishman et al., 1992) but has no such properties in vivo [Zaneveld et al., 1977; Joyce et al., 1985].

Before any claims are made regarding the contraceptive potency of agents, it is essential that their in vivo activity be evaluated. Since clinical trials are too cumbersome and expensive for screening purposes, animal models need to be used. Although spermatozoa from different animal species vary to a smaller or larger extent in their susceptibility to spermicidal agents [e.g., Alexander et al., 1981; Arambasic et al., 1992], these differences become less noticeable under in vivo conditions [Alexander et al., 1981] and high doses are normally evaluated in vivo which overshadow the differences found in vitro. Even so, it is worthwhile to compare the in vitro spermicidal activity of the agent under study between the human and the animal model. In general, spermatozoa from the primate react more like human spermatozoa than those from the rabbit.

Barrier Contraceptives: Current Status and Future Prospects, pages 163–171

Only two animal models lend themselves readily to the evaluation of vaginal contraceptives: the rabbit and the primate (stump-tailed macaque and baboon). Mice, rats and guinea pigs have a very small vagina, produce a vaginal plug after ejaculation, ejaculate into the cervix and/or mount and ejaculate a number of time before fertilization can take place. Larger animals such as the dog or sheep can be used but are not readily accessible under laboratory conditions.

In vivo studies can have two general endpoints: collect spermatozoa from the vagina and/or cervix after insemination (postcoital tests); or fertility/pregnancy (fertility tests). Postcoital tests can only be performed in the primate due to the long length of the rabbit vagina. Fertility tests can be performed in both the rabbit and the primate, although cost considerations limit the studies that can be performed in the latter.

RABBIT

The rabbit is the most widely used animal model for evaluation of the in vivo antifertility properties of vaginally placed compounds. Although it is a fairly simple model, a number of factors need to be considered before starting the studies. The following is taken from several articles [Homm et al., 1976; Williams et al., 1980; Joyce et al., 1985; Kaminski et al., 1985] as well as from personal experience.

The agent needs to be dissolved or well dispersed in a base (carrier). In vitro studies have to assure that the agent and base are compatible, that the base does not inactivate the agent and that the agent is easily released. The base should dissolve and spread readily after vaginal placement, and remain in place for a reasonable period of time. Furthermore, it should not have any effect on sperm function. Care has to taken in the selection of the base because a wrong formulation may greatly decrease the efficacy of a compound.

Test and control experiments should be performed simultaneously. Each group should contain at least four animals. Two control groups should be utilized if it is not known whether the base has antifertility activity: no formulation (sham control); and base only (vehicle control). Ideally, different concentrations of the agent can be tested so that an ED_{50} can be calculated. This allows comparisons with other compounds when experiments cannot be performed simultaneously. Otherwise, the agents need to be compared at the identical doses during the same series of experiments.

Ovulation

New Zealand White rabbits are most frequently employed for these experiments, but other types of rabbits can be used also. Rabbits are induced ovulators. This means that they do not have regular cycles (like primates) but ovulate during the coital act if the follicles are ready. Ovulation can also occur spontaneously during the handling or shipping of the animals. The rabbit becomes refractive for two to three weeks thereafter, i.e., cannot ovulate, making it important to isolate it for at least three weeks before starting the experiments.

In order to assure ovulation and to obtain many eggs, the rabbits can be superovulated. To do so, pregnant mare's serum (typically 150 IU) is injected intraperitoneally or intramuscularly approximately 96 hours (four days) before the vaginal contraceptive experiments are to be performed. Immediately after the mating or artificial insemination procedure (see further), human chorionic gonadotropin (typically 200 IU) is injected intramuscularly or intravenously.

Insertion of the Formulation and Insemination

The rabbit has a long vagina allowing the insertion of fairly large volumes of formulation (the usual volumes vary from 1.0 to 1.5 ml). A long cannula needs to be used to deposit the formulation since it should be placed about 8 cm deep in the vagina. Such a cannula can be a 12.5 cm-long polyethylene tubing with an inside diameter of about 2 mm. The tubing should be somewhat stiff with a slight bend at about 5 cm because some difficulty may be encountered in entering the cannula deep into the vagina. If blockage occurs, the cannula should be turned 90° and pushed gently forward or it should be removed and the procedure repeated. The rabbit should be restrained (not anesthetized) in the supine position while the cannula is inserted.

The formulation is entered into a syringe that can be attached to the cannula. The volume of the cannula needs to be considered when depositing the formulation in order to accurately deliver the desired amount. To do so, the cannula can first be filled with the formulation, inserted and the correct amount expressed from the syringe. After deposition of the formulation, the rabbit should be kept in the supine position for several minutes before allowing it to go back to its normal position. Mating or artificial insemination should not take place until five to 10 minutes after insertion of the formulation to allow adequate distribution of the formulation.

After insertion of the formulation, the rabbit can either be bred by a male of known fertility or can be artificially inseminated. Although the former is more natural, different males need to be used for each female if the experiments are done simultaneously so that the quality of the ejaculates may vary between test and control animals. In addition, the male penis is short so that ejaculation occurs far from the location of the formulation. Artificial insemination is generally preferred. An artificial vagina is used to collected ejaculates from two to four male animals, the semen is pooled (removing any samples that are contaminated with urine), evaluated (a high percentage of the spermatozoa should be motile) and divided into equal aliquots (0.3 to 0.5 ml). An adequate number of aliquots should be available to treat all control and test animals. The semen may be diluted with an appropriate buffer before making the aliquots. The seminal aliquot is deposited by a fresh cannula at the same site as the formulation or just before it (see above). In this fashion, differences in conception will not be due to different ejaculate characteristics.

Endpoints

Two endpoints can be used for the experiments: fertilization or implantation. Fertilization as an endpoint has the advantage that the direct effect of the compound on the fertilizing capacity of the spermatozoa can be evaluated. However, it is possible for a compound to permit fertilization but to prevent implantation, which still makes it of interest as a contraceptive. Optimally, both endpoints are used. However, fertilization as the endpoint is preferred if this is not possible. The rabbit has a fairly open cervix, unlike the primate and human, so that compounds can be transported into the uterus from the vagina, possibly preventing implantation even if the compounds have no vaginal contraceptive effect.

If fertilization is used as the endpoint, the females are sacrificed 28 to 36 hours after insemination or mating. At this point, the embryos should be in the two- to eight-cell stage if the oocytes were fertilized. To collect the embryos or unfertilized oocytes, an abdominal incision is made, the uterus and oviducts are removed in their entirety, and the embryos/oocytes flushed from the oviducts into a rounded watchglass. For flushing, a blunted 19-gauge needle attached to a syringe is entered through the uterotubal junction and 1 ml physiological saline or other medium is passed through the oviduct while holding it above the watchglass. Embryo cleavage can either be determined directly by phase contrast microscope or, more accurately, after fixation in 2.5 percent glutaraldehyde or other fixative and staining with acetolacmoid.

If implantation is used as the endpoint, the females are sacrificed 14 to 15 days after insemination or mating, the uteri exposed and incised, and the number of implanted fetuses recorded.

STUMP-TAILED MACAQUE

The nonhuman primate has several advantages over the rabbit. First of all, its reproductive tract is similar to that of the woman. Secondly, the animal goes through regular menstrual cycles and shows typical hormonal and genital tract changes (including those involving cervical mucus) as a woman, although its menstrual cycles tend to be somewhat longer (28 to 36 days; averaging 32 days for the primates under discussion). Sperm penetration into and through the cervix can only occur during the midcycle period.

The most commonly used primate for vaginal contraceptive studies is the stump-tailed macaque (*Macaca arctoides*). The advantage of this primate over any other is that the females are friendly, requiring minimal training, if any, to participate in the studies. No anesthesia is required for postcoital tests as long as only the vaginal contents are collected. Furthermore, the males and females breed at any time during the cycle and do so willingly and rapidly under supervised conditions. The females conceive readily under caged conditions.

Both postcoital and fertility studies are possible in the primate. Two types of postcoital studies can be performed: those in which only the vaginal contents are collected; or those in which both the vaginal and cervical contents are collected. The former can be done at any time of the menstrual cycle, whereas the latter can only be done during the midcycle period and requires anesthesia.

The same comments regarding the formulation and controls made above for the rabbit also apply to the primate. Test and control groups should contain at least four but preferably six animals each. The studies described below are summaries of the procedures that need to be performed and are taken from Zaneveld, et al. [1977, 1984], Zatuchni, et al. [1981] and Alexander, et al. [1981], as well as unpublished experience with the baboon.

Collection of Vaginal Contents

To perform postcoital tests in the stump-tailed macaque during which only the vaginal contents are collected, an unanesthetized female is taken from her cage by the arm, seated on a table and bent forward. A 1 ml, plastic syringe from which the tip has been removed is entered as deep as possible into the vagina. To distribute the formulation throughout the vagina, the syringe is slowly removed from the vagina while ejecting the contents.

Mating of the female can either occur immediately or can be postponed for a period of time if the longevity of the formulation is being tested. The female is placed in the cage of a proven fertile male. Mating normally takes only a very short period of time; from mounting to ejaculation, the coital act lasts approximately 90 seconds to two minutes.

The male and female are separated as soon as possible after mating, either manually or by using a transfer cage (usually requires about one minute). The female is immediately placed on the table, bent over and a tuberculin syringe containing 0.2 ml of physiological saline is inserted as deep as possible into the vagina. The saline is expelled, followed by suction with the syringe while slowly withdrawing it so that a representative sample of the vaginal contents is collected. The 0.2 ml of saline is used because the fluid volume of the macaque ejaculate is small (0.1 to 0.5 ml). The additional amount of saline guarantees adequate recovery of the vaginal spermatozoa.

Immediately after aspiration, the vaginal contents are analyzed for sperm number, motility and, if desired, morphology. In addition, the volume of the recovered contents is recorded. After collection of the vaginal contents, the female is douched with warm water or saline to remove all excess vaginal spermatozoa. Such douching allows repeat use of the animal after 24 hours.

Collection of Cervical Mucus

Postcoital tests may also involve the collection of cervical mucus. This can only be done during the midcycle period which ranges from day nine to 14 of the cycle (LH peak occurs on about day 12). To do so, the same procedures are performed as described above. However, after collection of the vaginal contents and douching with 50 ml physiological saline, the animal is anesthetized with 0.5 to 1.0 ml ketamine. Douching is done to prevent contamination of cervical mucus with vaginal spermatozoa during the collection of the mucus or while waiting for the animal to become anesthetized. The cervix is exposed with a speculum and the mucus can be collected with a 14-gauge, 2 inch angiocath intravenous Teflon catheter or similar device. The catheter is entered for approximately 0.5 cm into the cervical os and mucus is withdrawn by applying suction with a 3 ml syringe attached to the catheter. Different technologies can be applied to collect the mucus (similar to the human). The mucus is expressed on a slide and the number and motility of the spermatozoa is assessed.

Mating Studies

Mating studies can also be performed with the stump-tailed macaque. Basically, the same procedures are followed except that after mating, the female is returned to her cage without further manipulation. Pregnancy is determined by the absence of menstruation, by rectal palpation (for the presence of a fetus), and finally by parturition. Depending on the study design, the animal can only be mated once during her midcycle period or a number of times, each preceded by vaginal placement of the formulation. The pregnancy rate is high when the animals are mated repeatedly during the cycle; almost all animals become pregnant in one cycle if K-Y jelly is used.

Often, only a relatively small number of animals are available for these procedures so that a randomized assignment of the animals to the different test and control groups may bias a particular group. If the animals have been in the colony for a fairly long period of time, been bred continuously and adequate records are kept, a fertility quotient (FQ) can be assigned to each female where:

$$FQ = \frac{\text{Total number of pregnancies}}{\text{Total number of breeding cycles required to obtain these pregnancies}}$$

The animals can then be grouped so that the average FQ is approximately the same for each group [Zaneveld, et al., 1977].

BABOON

The stump-tailed macaque cannot be used for the evaluation of vaginal devices since either the male or female usually removes the device manually before mating. Device studies need to be performed in the baboon (*Papio* sp.) which has a long vagina making it impossible to remove a deeply placed device. A disadvantage of the baboon is that it needs to be anesthetized intramuscularly with ketamine (usually 10 mg/kg) before it can be handled. In addition, the animal will only mate during the midcycle period. This period lasts for about one to six days and is readily identified by the perineal turgescence (tumescence). As soon as ovulation occurs, the turgescence will start to shrink.

After the primate is anesthetized, the vagina is opened with a speculum and the device to be tested is placed as deep as possible into the vagina. Cervical coverage can be assessed if important. The baboon vagina is long but fairly narrow. Human-size devices will generally not fit the vagina or cervix, often necessitating construction of smaller devices. The animal is allowed to recover from the anesthesia (usually one to two hours) and mated with a male of proven fertility. The mating procedure can either be performed immediately after she awakens or can be postponed for a period of time to determine the contraceptive longevity of the device. However, it should be kept in mind that this must take place during the midcycle period. Mating is usually completed within two to three minutes. After mating, the baboon can be re-anesthetized, and the vaginal and cervical contents collected as described for the macaque (except that 1 ml saline is used to flush the vagina). The baboon may also be returned to her cage immediately after mating and observed for the occurrence of pregnancy (absence of menstruation, rectal palpation, parturition).

CONCLUSIONS

Adequate animal models are available for the evaluation of the postcoital spermicidal and/or contraceptive potency of vaginally placed agents. However, species differences with the human need to be taken into consideration, e.g., the susceptibility of spermatozoa to agents and the physiology of the female reproductive tract. In addition, the formulation in which the agent is applied should be carefully considered and tested.

REFERENCES

Alexander NJ, Fulgham DL, Zatuchni B, Zaneveld LJD (1981): Cervical mucus penetration test for the in vitro assay of vaginal contraceptive agents. Fertil Steril 36: 516-20.

Arambasic BM, Lazar SM, Mirsanda S (1992): Comparative in vitro study of nonoxynol-9: effects on human, bull and boar spermatozoa. Contraception 45:229-37.

Frishman GN, Luciano AA, Mayer DB (1992): Evaluation of Astroglide, a new vaginal lubricant: effects of length of exposure and concentration on sperm motility. Fertil Steril 58:630-2.

Joyce CL, Zaneveld LJD (1985): Vaginal contraceptive activity of hyaluronidase and cyclooxygenase (prostaglandin synthetase) inhibitors in the rabbit. Fertil Steril 44: 426-8.

Kaminski JM, Nuzzo NA, Bauer L, Waller DP, Zaneveld LJD (1985): Vaginal contraceptive activity of aryl 4-guanidinobenzoates (acrosin inhibitors) in rabbits. Contraception 32:183-9.

Homm RE, Doscher GE, Hummel EJ, Greenslade FC (1976): Relative antifertility activity of three vaginal contraceptive products in the rabbit: relationship to in vitro data.

Contraception 13:479-88.

Population Reports (1984): Barrier methods. New developments in vaginal contraception. Series H, No 7:157-91.

Quigg JM, Miller IF, Mack SR, Saxena SJ, Kaminski JM, Zaneveld LJD (1988): Development of the polyurethane sponge as a delivery system for aryl 4-guanidinobenzoates. Contraception 38:487-97.

Tagatz GE, Okagaki T, Sciarra JJ (1972): The effect of vaginal lubricants on sperm motility and viability in vitro. Am J Obstet Gynecol 113:88-90.

Williams WL (1980): New antifertility agents active in the rabbit vaginal contraception (RVC) method. Contraception 22:659-72.

Zaneveld LJD, Bhattacharyya AK, Kim DS, Schumacher GFB, Beluhan Z (1977): Primate model for the evaluation of vaginal contraceptives. Am J Obstet Gynecol 129: 368-73.

Zaneveld LJD, Burns JW, Goodpasture JC, Vickery BH (1984): Evaluation of the in vivo efficacy of a new vaginal contraceptive agent in stump-tailed macaques. Fertil Steril 41: 455-9.

Zatuchni B, Hahn DW, Zaneveld LJD (1981): Postcoital, vaginal, spermicidal potency of formulations: the Macaca arctoides (stump-tailed macaque) as animal model. Fertil Steril 35:683-90.

DISCUSSION

Laneta Dorflinger: All of your presentation focused on animal models for evaluating contraceptive efficacy. Can you say anything about whether these animal models would also be effective or useful for evaluating potential prevention of STDs or bactericidal or virucidal activity?

Lourens Zaneveld: No, I really can't. I think that will be the subject of a discussion tomorrow.

Penelope Hitchcock: I stumbled upon two papers in the radiology literature that have to do with movement of particles up into the human female reproductive tract. Although this was in parous women, and there were not a lot of them, basically what they did was they took a syringe full of latex particles that had some sort of metal tag on them and just simply, gently dripped this onto the cervix and kept the women on the table and then 20 minutes later rolled her into the x-ray room and took a picture of her abdomen. Those particles were, in fact, up in her tubes in 20 minutes, which suggests that our concept of the cervical plug and the sperm-dependent movement of particles and sperm, in particular, up into the upper tract may not be, in all cases or perhaps in any case, reality. You discounted one of the animal models because of rapid movement into the upper tract. Could you give us a feel for how confident you are that the human female tract doesn't mobilize things rather quickly into the upper tract?

Lourens Zaneveld: First of all, the expert on the subject will be talking right after me, Dr. Moghissi. He will hopefully address this question better than I can. Secondly, I did not discount an animal model because of sperm migration. I just gave a warning about looking at implantation studies rather than looking at fertilization studies. With regard to what you mentioned, the transport of particles or fluids from the vagina to the uterus in the human female has been the subject of much argument, starting more than 20 or 30 years ago. It is generally accepted that particles, dead sperm, and fluids do not pass through cervical mucus, yet there is an old study I think from the 1960s where some carbon black particles did transfer into cervical mucus and even through the cervical mucus. The studies you mention are also plausible. I am always intrigued by that because if you look at cervical mucus, it is 97 percent water. So anything that is water soluble ought to get into cervical mucus. Water soluble compounds ought to migrate into cervical mucus. Yet, when you do the cervical mucus penetration test, that doesn't always happen. It is clear that we need a lot more data about cervical mucus before we go much further. I don't think we can totally discount that anything you put in a vagina does not get through the cervix. But I think it is unlikely under most conditions.

Henry Gabelnick: I just wanted to add, and I can't remember exactly who did the work, but about 20 years ago people were looking at microcapsules containing steroids and delivering them through the cervix. Studies were done showing that microcapsules would indeed go into the uterine cavity rather rapidly but it would only be a small fraction of them that actually succeeded. So it was abandoned as a route of drug delivery.

Lourens Zaneveld: Isn't it sad, though, that we have to go back 20 or 30 years ago to talk about research that we need the results of today. Just a clear sign that we have ignored this field for far too long and I hope that this meeting will really give emphasis to do more research in this area.

Gustavo Doncel: I don't know if immotile sperm will be able to reach the tubes, but even if they reach the tube, for sure, they will not be able to fertilize the egg. If you block motility without killing the sperm, they are not able to bind to the zona pellucida and penetrate the egg.

Lourens Zaneveld: This is correct but I think the point that Dr. Hitchcock made is that chemicals you place in the vagina may potentially pass into the uterus or may even have an effect on implantation.

12

POSTCOITAL TESTS

Kamran S. Moghissi

Department of Obstetrics & Gynecology
Wayne State University School of Medicine
Detroit, MI 48201

INTRODUCTION

The Postcoital Test, also known as Sims-Huhner Test, is a procedure that evaluates sperm-cervical mucus interaction. It is commonly performed in early stages of the infertility investigation and is also used to assess efficacy of barrier contraceptives and hormonal formulations.

Sperm migration through the cervix involves three distinct but interrelated factors: the ability of spermatozoa to penetrate the mucus by their intrinsic motility; the fibrillar structure of cervical mucin, promoting sperm transport; and the morphological configuration of cervical crypts that contribute to the storage and preservation of spermatozoa in the cervical canal and their sustained and prolonged release to the upper tract. Human sperm is endowed with intrinsic motility, a property essential for penetration of cervical mucus and subsequent fertilization.

CERVICAL SECRETION

Cervical mucus is a complex secretion produced constantly by the secretory cells of the endocervix. Cervical mucus is a heterogeneous secretion that reveals a number of rheological properties such as consistency, flow elasticity, spinnbarkeit, thixotropy, and tack or stickiness.

The most important constituent of cervical mucus is a hydrogel rich in carbohydrates and consisting of high molecular weight glycoproteins of the mucin type [Yurewicz and Moghissi, 1981]. Most of the physical properties of cervical mucus are due to these mucins.

Ovarian hormones regulate the secretion of cervical mucus; estrogen stimulates the production of copious amounts of watery mucus; and progesterone (or progestogens) inhibits the secretory activity of cervical epithelial cells. The physical properties and certain chemical constituents of cervical mucus show cyclic variations. Cyclic alterations in the constituents of cervical mucus may influence sperm penetrability, nutrition and

Barrier Contraceptives: Current Status and Future Prospects, pages 173–184

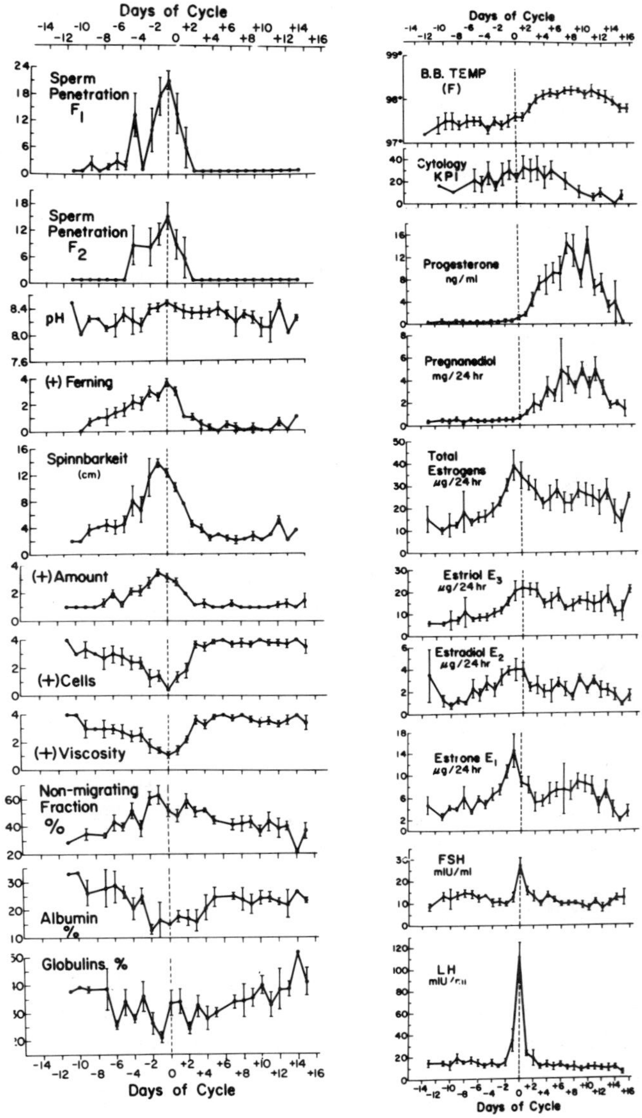

Fig. 1. Composite profile of serum gonadotropin and progesterone; urinary estrogens and pregnanediol; basal body temperature (BBT) karyopyknotic index (KPI) of vaginal cells; and cervical mucus properties throughout the menstrual cycle in 10 normal women. Day 0: day of LH peak (dotted line). Vertical bars represent one standard error of the mean. F_1 and F_2 indicate the number of sperm in the first and second microscopic fields (200X) from interface, 15 min after the start of the in vitro sperm-cervical mucus penetration test [Moghissi et al., 1972].

survival [Moghissi, 1973a,b]. Figure 1 shows serial determination of some important properties of human cervical mucus related to pituitary and ovarian hormones and sperm penetration in 10 women during a normal menstrual cycle. These data clearly demonstrate that optimal changes of cervical mucus properties, such as greatest increase in quantity, spinnbarkeit, ferning, pH and decrease in viscosity and cell content occur immediately prior to ovulation and are reversed after ovulation. Preovulatory mucus is most receptive to sperm penetration [Moghissi 1984].

Sperm penetrability of human cervical mucus begins approximately on the ninth day of a normal cycle and increases gradually to a peak at ovulation. It is usually inhibited within 1-2 days after ovulation but may persist to a lesser degree for a longer period [Moghissi, 1966].

Two phases of sperm migration through the cervix have been recognized: a rapid phase during which the leading spermatozoa penetrate the central portion of cervical canal and advance in a line parallel to mucin fibrils originating in the vicinity of the internal os; and a delayed phase. During the delayed phase sperm enter the cervical mucus and are oriented by mucin fibrils originating from the crypts and are stored within the crypts. The latter process is responsible for the storage of sperm in the cervical crypts and their gradual release over an extended period into the uterus and oviducts.

POSTCOITAL TEST

Evaluation of Cervical Mucus

The condition of cervical mucus greatly influences sperm receptivity; therefore, it should be evaluated accurately before a postcoital test is performed. Preovulatory mucus receptive to sperm penetration is profuse, thin, clear, acellular and alkaline. It exhibits 4+ ferning (crystallization) and high spinnbarkeit. Figure 2 shows a scoring system that we have devised, which has been endorsed by World Health Organization (WHO) to evaluate the quality and adequacy of cervical mucus. This system takes into account five important properties of cervical mucus that are known to affect sperm cervical mucus penetration: amount, spinnbarkeit, ferning, viscosity and cellularity. Each item receives a score from 0 to 3. A score of 3 represents optimal changes. A maximum score of 15 indicates preovulatory cervical mucus that is receptive to sperm penetration. A score less than 10 is associated with relatively unfavorable cervical secretion, and a score less than 5 represents hostile cervical mucus usually impenetrable by the sperm [Moghissi, 1977].

Several techniques for collection of cervical mucus have been described. Methods most commonly used consist of aspiration with a tuberculin syringe (without needle), pipette, polyethylene tube, or sampling with a mucus forceps. Cervical mucus should be collected and studied as close to the time of ovulation as possible. Clinical examination of cervical mucus includes determination of amount, viscosity, cellularity, pH, ferning, spinnbarkeit, and cultural studies if infection is suspected. Ferning (Fig. 3) is performed by spreading cervical mucus on a glass slide and allowing it to dry. It is customarily graded from 0 to 4+, depending on the extent of crystal formation.

CERVICAL MUCUS

Amount	0 = 0 1 = 0.1 ml 2 = 0.2 ml 3 = 0.3 ml or more
Viscosity	0 = Thick, highly viscous, premenstrual mucus 1 = Intermediate type (viscous) 2 = Mildly viscous 3 = Normal mid-cycle mucus (pre-ovulatory)
Ferning (or crystal-lization)	0 = No crystallization 1 = Atypical fern formation 2 = Primary and secondary stems 3 = Tertiary and quarternary stems
Spinnbarkeit (Place the drop of cervical mucus immediately after collection on a glass slide. Touch the cervical mucus with a cover slip and lift it gently. The length of the cervical mucus thread stretched in between is measured in cm and is scored as shown).	0 = $<$ 1 cm 1 = 1–4 cm 2 = 5–8 cm 3 = $\geqslant$ 9 cm
Cellularity (An estimate of the number of leucocytes and other cells in the cervical mucus is made at the time of post-coital test as shown).	0 = $\geqslant$ 11 cells/HPF 1 = 6–10 cells/HPF 2 = 1–5 cells/HPF 3 = 0 cells/HPF

pH of cervical mucus (do not include in total score)

TOTAL SCORE*

Fig. 2. Cervical score. A composite of the amount, spinnbarkeit, ferning, viscosity and cellularity of cervical mucus. Maximum score is 15; scores less than 10 represent unfavorable mucus and less than 5 show hostile cervical secretion.

Spinnbarkeit is measured by placing an adequate amount of cervical mucus on a microscope slide, covering it with a coverslip, and drawing the mucus between them. An estimate of spinnbarkeit (in centimeters) is made by measuring the length of the thread before it breaks. An alternative method is to pull a sample of cervical mucus from the cervix under direct observation, using a mucus forceps. The length of the mucus thread is estimated before it breaks. This method is less accurate since it does not allow for the precise measurement of the mucus thread with a ruler. The pH of cervical mucus should be measured in situ or immediately following collection. Care should be taken to assay the pH of endocervical mucus correctly since the pH of the exocervical pool is always lower than that of the endocervical canal.

TIMING

Postcoital tests should be performed as closely as possible to the time of ovulation, as determined by usual clinical means such as basal body temperature (BBT), cervical mucus changes, urinary LH test and sonography. Each couple is instructed to

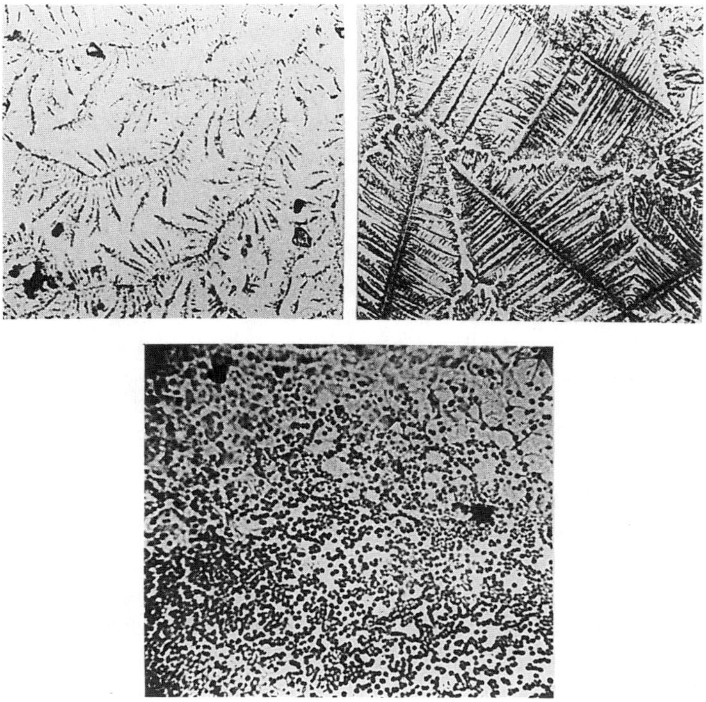

Fig. 3. Crystallization or ferning of cervical mucus. a 1+. b 4+. c Absence of ferning.

abstain from sexual intercourse for two days prior to the test, which is performed approximately six to ten hours after intercourse (standard test). A satisfactory test may be followed by a "delayed test" performed at a longer interval (18 to 24 hours) when infertility persists. The presence in the endocervix of at least 10 motile spermatozoa per high power field (400X) at this stage suggests favorable mucus and adequate survival of sperm in the cervix and would exclude cervical factor as a cause of infertility. When the initial standard test yields poor results, a second test is planned one to three post coitus (early test). A complete evaluation of cervical factor should be undertaken when both the standard and early tests are abnormal. It is also recommended that both cervical and male factors be re-evaluated when the delayed PCT is abnormal or there is persistent unexplained infertility.

TECHNIQUES OF POSTCOITAL TESTS

A nonlubricated speculum is inserted into the vagina and a sample of the posterior vaginal fornix pool is aspirated with a tuberculin syringe (without needle), a mucus syringe, pipette or polyethylene tube. With a different syringe, samples of cervical mucus are obtained from the exocervix and endocervical canal. These are placed on a separate glass slide, covered with a coverslip, and examined under a microscope at 200X and 400X. If exocervical mucus is covered with cells, debris and vaginal content, the area should be wiped dry with cotton prior to obtaining the endocervical specimen. Whenever possible, the quality of the mucus should be evaluated immediately upon collection.

INTERPRETATION

Interpreting the postcoital test requires an understanding of cervical mucus function and sperm transport. The function of the cervix as a sperm reservoir is of considerable importance in fertility. Only rarely does coitus occur at the time of ovulation. In most instances, union of gametes depends on a constant supply of sperm at the site of fertilization for some hours before and after ovulation.

Vaginal Pool Sample

Spermatozoa are usually destroyed in the vagina within two to four hours and lose their fertilizing ability within a few hours. The purpose of examining the vaginal pool sample, therefore, is to ensure that semen have actually been deposited in the vagina.

Exocervical Sample

The number of sperm in the lower part of the cervical canal varies with time elapsed after coitus. Within two to three hours after intercourse there is a large accumulation of sperm in the lower part of the cervical canal [Davajan and Kunitake, 1969; Moghissi, 1973a; Tredway et al., 1975]. In a normal woman following coitus with a fertile man, more than 25 motile sperm (2-3+ motility) per high-power field (HPF) (400X) are commonly observed in the exocervical specimen; 10 or more sperm per HPF with scores of 2-3+ motility are considered satisfactory. Less than 5 sperm/HPF, particularly when associated with sluggish or circular motion, is an indication of oligoasthenospermia or abnormality of cervical mucus [Elstein et al., 1973].

Sperm motility in cervical mucus is graded from 0 to 3 as follows: 0 = immotile; 1 = in situ motility; 2 = sluggish motility; and 3 = vigorous forward motility.

Endocervical Sample

Sperm reach the level of the internal os rapidly after ejaculation. Their number [Tredway et al., 1975] increases gradually and reaches a peak approximately two to three hours later. Thereafter their number remains relatively constant for up to 24 hours. Within two hours after coitus, the entire length of the cervix is colonized by sperm, but at 24 hours the number of sperm begin to decrease [Insler et al., 1981]. At six to 10 hours after intercourse, more than 10 sperm with adequate motility (score 3+) should normally be found per HPF. A similar number of sperm is usually detected in a delayed test (18 to 24 hours after coitus).

Postcoital Test For The Evaluation Of Barrier Contraceptives

Postcoital tests may be performed to test the efficacy of a barrier contraceptive. The fertility of the male partner should be assessed by semen analysis. Alternatively, one or more postcoital tests may be performed in no-treatment (control) cycles prior to tests in study cycles. In the treatment cycle, PCTs should be scheduled two to three hours after coitus during the periovulatory period while the subject is using the test device or formulation. When collecting cervical mucus, care should be taken to clear the surface of the cervix and avoid contamination with vaginal content. Ideally, there should be no sperm in cervical mucus. The presence of motile sperm in cervical mucus is indicative of compromised efficacy. More difficult to interpret is the presence of one or few dead sperm/HPF in the mucus. In the exocervical sample, these may represent contamination from the vaginal content. In the endocervical sample, however, dead sperm may indicate the exposure of sperm to spermicides before or after penetration into the cervical mucus. In such cases the escape of live spermatozoa from the cervical mucus to the upper reproductive tract can not be excluded with certainty.

In-Vitro Sperm-Cervical Mucus Interaction

Negative or abnormal postcoital tests are indications for in vitro cervical mucus-sperm penetration tests. These tests are also useful to test sperm-cervical mucus interaction when mucus has been altered by hormonal or contraceptive agents. Two different techniques have been used for in vitro investigation of sperm penetration of cervical mucus: the slide method and the capillary tube system. A good correlation has been found between the results obtained by these techniques [Kremer, 1968]. The results of in vitro sperm-cervical mucus penetration tests also compare fairly well with sperm concentration, motility and percentage of normal morphology in semen samples [Insler et al., 1979; Keel and Webster, 1988]. Mortimer et al. [1986] found that both the concentration of progressively motile spermatozoa and their movement characteristics are significant factors determining the outcome of homologous tests of human sperm-cervical mucus interaction. Finally, there is a significant correlation between the results of postcoital tests, in vitro tests, and semen analysis findings [Hayes et al., 1983]. The slide method as originally described has been altered to provide quantitative results [Moghissi et al., 1964] (Fig. 4).

In studies in vitro by the slide method, a sharp boundary is observed separating human cervical mucus placed in juxtaposition to semen on a microscopic slide [Moghissi et al., 1964]. At the interface, finger-like projections or phalanges of seminal fluid develop within a few minutes and penetrate the mucus. Sperm usually fill these canals before entering the mucus. Most spermatozoa penetrate the apex of the phalangeal canal and enter the mucus. In most instances a single spermatozoon appears to lead a column

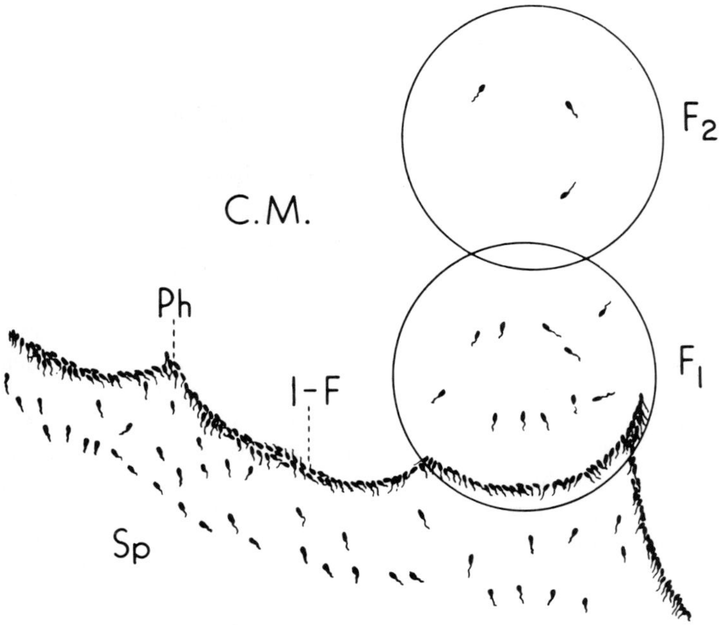

Fig. 4. Schematic representation of in vitro sperm-cervical mucus penetration test by quantitative slide method. C.M., cervical mucus; Sp, sperm; I-F, interface; Ph, phalanges; F_1, first microscopic field; F_2, second field adjacent to F_1 [Moghissi 1966].

of sperm into the mucus. After the initial resistance has been overcome by the leading sperm, others follow without difficulty. Once in the cervical mucus, sperm fan out and move at random. Some return to the seminal plasma layer, while most migrate deep into the cervical mucus until they meet with resistance from cellular debris or leukocytes. They then either stop or change direction. Both phalanx and interface formation appear to be physical phenomena resulting from the contact of two biological fluids of differing viscosities and surface tensions.

To quantitate the test, the first microscopic field from the interface, called F_1, is counted at various time intervals. The suggested times are five and 15 minutes after initiation of test. This is scored at 200X and 400X magnification. In order to study the depth of penetration, the second microscopic field adjacent to the first one, (F_2), and the third (F_3) can also be evaluated and the number of spermatozoa in these fields can be counted and recorded (Fig. 4). In good tests, at least 15 motile sperm/HPF in F_1 and at least 10 sperm/HPF in F_2 are recorded.

The capillary tube system measures the ability of spermatozoa to penetrate a column of cervical mucus in a flat capillary tube (Fig. 5). Sperm penetration in the

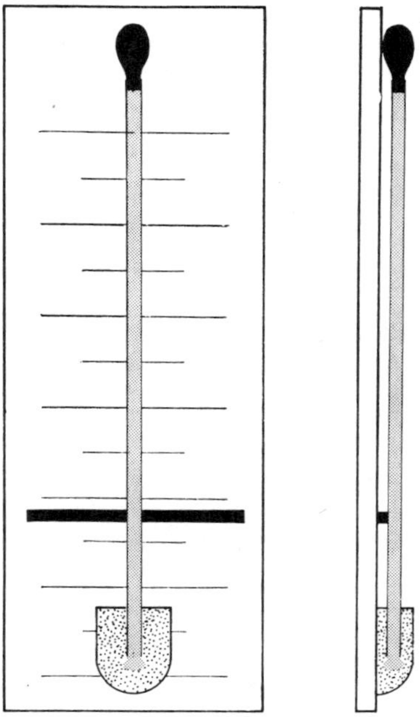

Fig. 5. A flat capillary tube for in vitro sperm-cervical mucus penetration test is filled with cervical mucus, sealed at the top and placed in a sperm reservoir (bottom). The entire system is mounted on a graduated slide.

capillary tube is assessed at 10 minute, 30 minute, 3 hour or other intervals depending on the type of study. Both linear penetration and density of penetration should be recorded. A scoring system (Fig. 2) for the evaluation of sperm penetrability has been devised [Belsey et al., 1990]. The capillary tube system is particularly useful for evaluation of midcycle cervical mucus and to study depth of penetration. The slide method is applicable to a variety of cervical mucus samples, including those too viscous or too scanty to be studied by the tube technique. When both in vivo and in vitro tests are negative, normal donor sperm may be tested in vitro against the partner's cervical mucus, and preovulatory mucus obtained from a fertile donor should be tested with the man's sperm. This crossmatch test will determine whether the sperm or cervical mucus is responsible for abnormal results.

CONCLUSIONS

The postcoital test determines the adequacy of sperm and receptivity of cervical mucus. It is the only test that evaluates the interaction between the sperm and the female genital tract fluids in vivo. The Sims-Huhner test should be an integral part of infertility investigation but must not be used as a substitute for semen analysis. Since cervical mucus accurately reflects the ovarian cycle, the postcoital test is also a useful indicator of endocrine preparation of the female reproductive system. Postcoital tests and in vitro sperm-cervical mucus penetration studies are also useful in evaluating the efficacy of a variety of barrier and hormonal contraceptives.

REFERENCES

Davajan V, Kunitake G (1969): Fractional in vivo and in vitro examination of post-coital cervical mucus in the human. Fertil Steril 20:197.

Elstein M, et al. (1973): Cervical mucus, present state of knowledge. In: Cervical mucus in human reproduction. WHO, Scriptor, Copenhagen, p 11.

Moghissi KS (1966): Cyclic changes of cervical mucus in normal and progestin treated women. Fertil Steril 17:663.

Moghissi KS (1973a): Sperm migration through the human cervix. In: Elstein M, Moghissi KS, Borth R, (eds): "Cervical Mucus in Human Reproduction." Scriptor, Copenhagen, p 128.

Moghissi KS (1973b): Composition and function of cervical secretion. In: Gripp R (ed): "Handbook of Physiology, Endocrinology II." Part 2, American Physiology Society, Ch. 31, p 25, Washington DC.

Moghissi KS (1977): Significance and prognostic value of postcoital test. In: Insler V, Bettendorf G (eds): "The Uterine Cervix in Reproduction." George Thieme, Stuttgart, p 146.

Moghissi KS (1984): The function of the cervix in human reproduction. Current Problems in Obstet Gynecol 7:2-58.

Moghissi KS, et al. (1964): Mechanism of sperm migration. Fertil Steril 15:15.

Moghissi KS, et al. (1972): A composite picture of the menstrual cycle. Am J Obstet Gynecol 114:405.

Tredway DR, et al. (1975): Significance of timing for postcoital evaluation of cervical mucus. Am J Obstet Gynecol 121:387.

Yurewicz EC, Moghissi KS (1981): Purification of human midcycle cervical mucin and characterization of its oligosaccharides with respect to size, composition and micro-heterogeneity. J Biol Chem 256:11895.

DISCUSSION

Henry Gabelnick: Just to follow-up on your last remark about the implications of dead sperm in the mucus, is it not possible that in sampling the mucus you are contaminating it with your aspirator and that would explain small numbers of immotile sperm in the mucus?

Kamran Moghissi: Not really. First of all, if you sample several hours after intercourse most of the vaginal spermatozoa would have been eliminated. Secondly, when you remove your device or whatever formulation, you try to clean up the surface of the cervix and get your sample of cervical mucus from the endocervix primarily. Then if you see a dead sperm in the cervix I think you should be a little concerned. I don't think that this would be contamination.

Erica Gollub: Can you comment on the extent and duration of the change in alkalinity of the vagina that occurs after exposure to pre-ejaculate as compared to full ejaculate?

Kamran Moghissi: We know that there is a change. That change is not to the point that it would help considerably the sperm survival in the vagina. It may change from a pH of 3 or 4 to perhaps 5, 5.5, or 6 even. The main contributor to pH change is the buffering capacity of the semen. You really do need a normal amount of seminal volume to provide that kind of a buffering.

James Koch: You mentioned that several caveats exist for people who are working with barrier contraceptive devices. I'm wondering if you might just briefly summarize those caveats and add any more that you think would be particularly relevant to people trying to use postcoital tests preceding actual efficacy tests with such methods.

Kamran Moghissi: It depends on what kind of a formulation you are using and whether it affects cervical mucus or sperm. I would try to start with an animal model, but not a rat or a mouse. Use an animal which does, in fact, have cervical mucus. Then I would go to the Phase I clinical studies, preferably *in vitro* tests, if the formulation tends itself to that kind of testing. Then I would go to Phase II during which you would do a postcoital test be on a limited number of patients. You want to make sure that the male and the female are both normal. In other words, the woman does in fact have an ovulatory cycle, does have cervical mucus that you have pretested, the male is normal and then take it from there.

James Koch: Thank you very much. I don't think I made my question entirely clear because I was talking about physical barrier devices rather than chemical means. I don't think those are practical in animal experiments.

Kamran Moghissi: I think a slide was shown yesterday which indicated that physical means alone probably are not sufficient to provide a complete barrier to sperm penetration into the cervical mucus. You're going to need, in addition to that, some form of spermicide. In other words, no matter how carefully you place your barrier or how nicely it is fitted to the vagina you're going to always have, during the sexual activities, some space that has not been covered through which the sperm can get into the cervical mucus. I think that is why you need a spermicide to actually destroy the sperm.

Shakti Upadhyay: Recent studies have shown that leukocytic populations in the vagina play a very important role in the fertilizing ability of the spermatozoa. There is a very significant leukocytosis after coitus and recent studies have shown that the presence of leukocytes has been correlated with infertility. Do you think that the number of leukoytes in the vagina should be one of the criteria for clinical evaluation in your selection of the cases?

Kamran Moghissi: Well, the vagina always contains some leukocytes. I would pay a lot more attention to leukocytosis within the cervical mucus. The normal mid-cycle cervical mucus has very few white cells in it. What you will find, however, in women who have some form of infection is a lot more white cells. The presence of the increased amount of white cells in the cervical mucus then may have some importance, but I don't think in the vagina that I would pay that much attention to it.

Claude Aguillaume: I was most interested in your finding of cyclic variations in the amount of spermatozoa in the ejaculate. Do you have any more studies that have looked at the function of nutrition or hormonal changes like FSH-LH or testosterone?

Kamran Moghissi: The study was published in the Journal of Andrology.

Zena Stein: When you test the sensitivity of the vaginal epithelium to a spermicide dose, it would make sense to do it in the presence and absence of seminal plasma, because the results would be completely different in the two situations.

Peter Mason: I was very interested in those changes in the composition of the cervical mucus. What is the factor that controls that? Is it the mucus-secreting cells themselves, or does something happen to the mucus after it has been secreted?

Kamran Moghissi: The mucus-secreting cells are regulated primarily by estrogen and progesterone. You can take a postmenopausal woman who is not menstruating, who has a dry vagina and cervix, and give her of estrogen and progesterone and cause exactly the same type of changes. This is very well-studied. It's purely the effect of estrogen and progesterone.

13

CONSIDERATIONS FOR THE CLINICAL EVALUATION OF CHEMICAL AND PHYSICAL BARRIER CONTRACEPTIVE EFFICACY

Roberto Rivera, Gaston Farr, Laneta Dorflinger, Carol Connell, Rosalie Dominik, Deborah Menius, Marilyn Metcalf-Whittaker, and Carolyn Reusché

Family Health International
P.O. Box 13950
Research Triangle Park, NC 27713

INTRODUCTION

The use of barrier methods for fertility regulation has been a common practice for many years. The condom was one of the earliest means of contraception. However, not until the last decade have other physical barrier methods and spermicides received much attention. Several factors have contributed to the current high degree of interest in this type of contraception. Many couples prefer to use barrier methods that do not have a systemic effect and are very safe to use. There also has been an increasing knowledge of the side effects and complications of some of the non-barrier methods, making barrier methods more appealing. In addition, family planning services have progressively recognized the need to offer a variety of contraceptive options, which has resulted in an increased availability of physical and chemical barrier contraceptives. However, the single most important factor that has caused the current interest in barrier contraceptive methods is the rising prevalence of sexually transmitted diseases (STDs), which has reached epidemic proportions in some societies. Many physical barrier contraceptives and spermicides have the dual effect of protecting against pregnancy and the transmission of STDs.

At present there is a great interest in learning more about the safe and effective use of barrier methods, particularly as it relates to protection against STDs. In addition, a variety of new products are currently in different stages of development. As a

Barrier Contraceptives: Current Status and Future Prospects, pages 185–200

consequence, the number of clinical trials designed to evaluate the safety and efficacy of barrier methods has increased substantially in recent years.

Family Health International (FHI) has as one of its main goals to improve women's reproductive health. One of FHI's programs focuses on the development, evaluation and introduction of new contraceptives that will enhance the freedom and abilities of individuals to choose voluntarily the number and spacing of their children. FHI has a long history of clinical research on physical and chemical barrier contraceptives, and has been responsible for the evaluation of these methods on a worldwide scale. In addition, it has participated in the development of new products in this category.

More recently, FHI has joined efforts with the Contraceptive Research and Development Program (CONRAD) and several pharmaceutical companies to work on the development of new barrier contraceptives. The experience of designing clinical trials for barrier methods has led to the recognition that there are a number of potential methodological problems associated with clinical trials designed for the evaluation of barrier contraceptive methods. Some of the criteria used in the design of barrier method clinical trials have been quite arbitrary, ill-defined and inconsistent across studies. Particularly, the varying criteria used for the evaluation of contraceptive efficacy has made it difficult to judge the relative efficacy of these products.

In 1991, CONRAD organized a meeting of public sector agencies involved in the research and development of physical and chemical barrier contraceptives to discuss problems encountered during the clinical evaluation of these methods. The agencies participating in this meeting (CONRAD, the World Health Organization, the Contraceptive Development Branch of the National Institutes of Child Health and Human Development, and FHI) agreed on the need to review the main methodological problems in the clinical evaluation of efficacy and safety of physical and chemical barrier contraceptives. FHI was designated to take the lead role and coordinate this inter-agency effort. A priority list of issues was developed with the participation of all the above-mentioned organizations. This paper represents FHI's initial considerations of some of these issues. Eventually, a much larger number of issues will be addressed, reviewed by the other participating agencies and made public as a proposed recommendation for the clinical evaluation of physical and chemical barrier contraceptives. The primary focus of this paper is to discuss methodologic issues in studies that evaluate clinical efficacy.

STUDY DESIGN

Inclusion/Exclusion Criteria

One should design a clinical study in which the composition of subjects is representative of the target population of users for whom approval will be requested. However, there are a number of factors, including age, marital status, parity, education and previous contraceptive experience, that may affect the safety and efficacy outcomes in contraceptive device and spermicidal studies. Center-by-center variability in these sociodemographic indicators raises potential concerns about the appropriateness of pooling data in multi-center trials.

Monogamy

Contraceptive efficacy studies are often conducted with mutually monogamous couples. This requirement is imposed, at least in part, to minimize the risk of women acquiring sexually transmitted diseases during the study. However, in barrier method studies designed to last 12 months or longer, it may not be feasible, or even desirable, to require monogamy. Barrier contraceptives, particularly those designed to prevent STDs, will be used frequently by women/couples who are not mutually monogamous. Non-monogamous individuals/couples may be more highly motivated to prevent pregnancy than monogamous married couples who almost certainly have different fertility desires. Furthermore, if multiple male partners are allowed in a clinical efficacy study, it may not be possible to document proven fertility of each male partner. This could result in lower pregnancy rates in this group of users. Despite these facts, it is not unreasonable to design studies to require monogamy, particularly in early studies with new products where efficacy in preventing STDs is unknown. In studies where monogamy is not required, safeguards should be added in the study design to follow and assess subjects for STDs.

Parity Status

It is ideal to recruit subjects with various parity histories into a clinical study and to perform stratified analyses or use multivariable models to determine whether or not parity history affects contraceptive efficacy or safety. Parity is often used as a proxy for level of fecundity and previous contraceptive failure [Trussell, 1991]. Whether or not equal numbers of parous and non-parous women should be required depends upon the sample sizes needed to obtain reasonably precise estimates of pregnancy rates for non-comparative studies or reasonably powerful tests of differences in efficacy rates between experimental groups for comparative studies within each parity stratum.

Fertility

Infertility is defined as one year of unprotected coitus without conception. This condition affects approximately 10 percent of couples of child-bearing age [Speroff et al., 1989]. By definition, individuals do not know they are infertile until the time at which they attempt conception. Having both partners with proven fertility should, theoretically at least, decrease the likelihood that couples are currently infertile. However, requiring proven fertility may confound the concern that clinical studies of new products should be conducted on populations that reflect the target population. A major target population of barrier methods is almost always young women with unproven fertility.

Other Demographics

Other baseline factors that should be considered are:

Age. Although contraceptive efficacy studies almost always recruit subjects between the ages of 18 and 35, some consideration should be given to extending enrollment to younger women. While it is difficult to conduct research with teenagers, this group of individuals is at high risk of pregnancy and STDs. Teenagers may have a different safety and efficacy profile than other users. Therefore, particularly when evaluating products with the potential for reducing STDs, information should be gathered on the safety, efficacy and acceptability of products in younger users when possible.

Education. In some studies, contraceptive failure rates have been shown to vary across educational levels. Education level may affect the individual's ability to comprehend written instructions and to use the product effectively. This is of particular concern for over-the-counter products. Therefore information should be gathered that would allow evaluation of user comprehension of written instructions for product use.

Race/Ethnicity. There is no theoretical reason that one would expect race or ethnicity *per se* to affect barrier method efficacy; however, some previous contraceptive studies have shown pregnancy rates to vary by ethnicity. This effect may relate most to clinic location or staff or other factors highly correlated with race or ethnicity. Therefore, information should be gathered on ethnicity and attempts made in controlled clinical studies to have a similar ethnic distribution in control and experimental groups.

Prior Use of Other Contraceptives

Because of the different mechanisms of action of various contraceptives, there may be differences in time to fertility following discontinuation of these various methods. Therefore, subjects may not be at equal risk of pregnancy if they discontinue different methods immediately prior to enrollment in an efficacy study. For non-hormonal methods such as copper IUDs, diaphragms or cervical caps, there is no delay in return to fertility with discontinuation. In contrast, some hormonal methods such as Depo-Provera, are known to produce delays in fertility following discontinuation.

For efficacy studies of barrier methods, it is recommended that users of injectables have at least two normal cycles before enrollment; however, Norplant®, combined oral contraceptives and progestin-only pill users need to have only one normal cycle before enrollment. In all cases, it would be ideal if enrollment could take place during menses (early follicular phase) to reduce the probability of an undetected pre-enrollment pregnancy.

It is commonly believed that users of certain barrier contraceptive methods become more experienced over time using their method and that better technique in use increases. When new barrier methods are compared with existing barrier methods, bias could be introduced. This bias occurs because subjects assigned to use the new product undergo a learning curve, whereas subjects assigned a product with which they have previous experience do not. Therefore users of the new product will potentially have a greater risk of failure than experienced users of the comparative method, biasing the failure rate downward for the established product.

The degree to which this bias is a problem will vary with device design and ease or difficulty of use. If a new device is sufficiently novel that a sponsor has concern about greatly elevated risk of pregnancy during early use, consideration should be given to allowing a training period in clinical studies. However, this should only be done in Phase II studies of products for which Phase III studies will be required. Phase III studies or Phase II studies of devices which qualify under the United States Food and Drug Administration guidelines should be evaluated during routine use, not ideal conditions, since this will best predict typical use effectiveness when marketed. In a Phase III study, calculating monthly failure rates should help assess whether there was a learning curve with the device under consideration. However, since the risk of pregnancy varies with coital frequency, level of fertility and compliance, a downward trend in monthly pregnancy rates may not necessarily be due to increasing experience. It may just mean that the women at the highest risk of pregnancy are leaving the trial earlier than other women.

Frequency of Intercourse

The FDA guidelines (Premarket Testing Guidelines for Female Barrier Contraceptive Devices Also Intended to Prevent Sexually Transmitted Diseases) require a history of sexual activity consistent with at least two to three coital episodes weekly. A coital frequency of two to three times weekly would almost certainly provide for intercourse each month on or close to the time of ovulation, thus ensuring a reasonable test of method effectiveness if the product is used consistently. Based on known and reported information from previous clinical studies, this coital frequency is not an unreasonable requirement at enrollment in terms of subject recruitment. However, it may not be reasonable to require that couples maintain that coital frequency throughout the duration of the study.

Subjects who fall below a predetermined minimum level of coital frequency during the study should be counseled to improve compliance. Investigators should attempt to verify that the subject wishes to continue participation in the trial. If it is determined that the subject does not expect to increase sexual activity to meet the minimum requirement for coital episodes per week, consideration should be given to discontinuing the subject from the study. The experience of these subjects is relevant to the safety analysis, even if the experience is quite limited. Rules for excluding some or all of the data for subjects with low coital frequency from the efficacy analysis should be included in the study protocol.

SAMPLE SIZE

Sample size should be determined in consultation with a biostatistician. The following usually needs to be provided:

1 the null hypothesis to be tested or parameter to be estimated
2 the expected level of outcome for the control product in a comparative trial or for the study product in a noncomparative trial
3 the smallest clinically significant difference between the new product and the control product for a comparative trial or an acceptable confidence interval around an estimated point for a noncomparative trial
4 the power one wants to have (usually 0.8, 0.9, or 0.95) to detect the smallest clinically significant difference between outcomes in a comparative trial
5 the chance one is willing to take of concluding that there is a difference in efficacy between two products in a comparative trial when there is no real difference (alpha level, usually 0.05)
6 the proportion of subjects predicted to drop out of the study for reasons other than the primary outcome measure; the predicted drop-out rate is usually estimated from prior clinical trial experience with the same or similar products.

In comparative studies of contraceptives, the null hypothesis of no difference in survival distributions is usually tested via the logrank test. Although the logrank test is usually the primary test of interest in contraceptive studies, sample size formulas for comparisons of simple binomial proportions (*i.e.*, percentages) can be used in planning studies. Since most of the subjects never experience the primary outcome (pregnancy), sample size requirements for tests of differences in proportions provide slightly conservative estimates of the numbers of subjects needed [Freedman, 1982].

The objective of a comparative trial may be to demonstrate that a new product is at least as effective as an existing product. An equivalence approach to sample size may be more appropriate in this situation. If standard hypothesis testing is planned for an equivalence study, high power ($\geq$ 0.90) is recommended. The confidence interval approach for equivalence studies, however, may be preferable to the use of standard hypothesis testing [Makuch et al., 1990]. If the planned analysis will involve a confidence interval approach, different procedures for sample size determination may be needed.

FOLLOW-UP SCHEDULES

The frequency of follow-up visits may affect the safety and efficacy results in a clinical trial. Therefore, follow-up visits must be strictly observed by all participating centers, and special care and procedures must be established to assure the compliance of the volunteers to the schedule. A follow-up schedule of one, three, six, nine, and 12 months seems to be adequate for most studies. However, more frequent follow-ups may help minimize loss to follow-up and discontinuation rates.

Regardless of the exact follow-up schedule, the number of clinic visits prior to admission must also be uniform in all centers. The procedures and contents of the pre-admission, admission and follow-up visits must be carefully considered as they may have an important effect on conclusion related to safety and efficacy. The addition of follow-up visits or procedures to be performed at follow-up visits not included in the protocol should be strongly discouraged because of the potential confounding effect on the study outcomes.

If the objective of the study is to estimate or compare six-month pregnancy rates, the last follow-up visit during product use should be at least six months after the date of first use of the product. An earlier visit would require that patients be censored from life tables before the end of month six. Similarly, if 12-month rates are of interest, the schedule should allow for 12 complete months of use. A post end-of study visit should be scheduled to confirm that subjects did not become pregnant while using the product, that an early pregnancy did not go undetected at the final study visit.

COUNSELING COMPONENTS

In any clinical trial, subjects should be counseled on the proper use of the study product and encouraged to ask questions. The counseling should include when to use the product, instructions on correct insertion/application and removal procedures and a review of possible side effects, if any are known or reasonably foreseeable. Subjects should also be told about the potential for failure of the study method. Subjects should be counseled about the importance of returning for all clinic visits. Good client-provider relations must be established to allow a free and comfortable communication. The continuity and availability of care must be understood by the volunteers, so they will feel confident to return or call the clinic any time a problem arises.

The degree to which the above-mentioned elements of counseling are addressed has a strong effect on both safety and efficacy. Subjects who are properly counseled have a more correct and continuous use of the study product, and are likely to have better continuation and efficacy rates.

Therefore, counseling should be consistent across all centers and throughout the duration of the study. The investigators must avoid introducing their personal biases for or against a particular product to the subjects in a clinical trial.

TIME SPAN TO ASSESS EFFICACY

Trussell *et al* address the issue of six versus 12-month rates in a paper in progress on the Reality female condom [personal communication, Trussell, 1993]. The authors suggest that there is no justification for requiring that comparisons of efficacy among products be based on 12-month rates and that life-table rates can be compared for any given duration as long as the duration is common across methods. Furthermore, since barrier studies typically have high discontinuation rates, large numbers of women must be enrolled to end up with enough subjects during month 12 to yield fairly precise estimates of the probability of pregnancy for the later months of follow-up.

A major advantage of barrier methods is that subjects can discontinue use any time they desire. Consequently, the kinds of women who might be interested in participating in a six-month study may be more like the target population than subjects interested in participating in a 12-month study.

Unless there is reasonable concern that a device degrades over time, the true efficacy of a product should not depend on the month of use. Pregnancy rates for barriers typically decline over time for a given sample of users. Although a learning effect may account for some of the decline, varying risks of pregnancy among women in a sample of users may produce most of the decline. Subjects with poorer compliance or higher coital frequency are more likely than other women to become pregnant early in a trial. They may also be more likely to decide a barrier method is not suitable for them and end up discontinuing for personal reasons. Therefore, differences in pregnancy rates between the first and last six months of a 12-month trial are more likely to be due to the changing composition of the sample than to the changing effectiveness of the product.

For these reasons, six months may actually be the preferred evaluation period for barrier contraceptive studies for efficacy. However, whether or not a six-month study will provide sufficient data for evaluating safety needs to be considered for each new method.

START DATE FOR EFFICACY ANALYSIS

Several dates have been proposed for the start date for pregnancy life-tables:

1 admission date - The date the subject is instructed to begin relying on the device for prevention of pregnancy.
2 date of first coitus - The date of first coitus after admission.
3 date of first use - The date the product is first used.

Admission date has generally been used as the start date for efficacy life-tables in barrier studies. This date is consistently used for the evaluation of efficacy of other contraceptives as well. USFDA Premarket Testing Guidelines for Female Barrier Devices Also Intended to Prevent Sexually Transmitted Diseases, however, recommend that the date of first coitus be used as the start date [USFDA, 1990]. If subjects are following a protocol which requires weekly coitus, use of the admission date or first date of coitus will lead to nearly identical life-table rates. Using date of first use however, would force the exclusion of pregnancies which occur due to noncompliance immediately following admission. The primary efficacy analysis should not exclude such events. Since we believe the intent is to measure the pregnancy rate during the time the subject is relying on the device as her method of contraception, we recommend that admission date be used as the start date for pregnancy life-table rates. The planned start date for contribution to life-tables should be included in the study protocol.

EVENT OR CENSORING DATES FOR EFFICACY ANALYSIS

The time at which a subject discontinues participation in a contraceptive efficacy trial has important implications for the analysis of efficacy. Subjects discontinue participation, and hence product use, for a variety of reasons including personal concerns, the desire to become pregnant, and so on. Rules for determining event or censoring dates should be included in the study period.

Ideally, a subject should be considered to have discontinued participation on the date the subject ceases relying on the method for contraception, regardless of the date of the last follow-up. Subjects who are discontinued due to pregnancy should contribute to life-tables until their estimated date of conception. Subjects who complete the required study period without pregnancy or an earlier termination should contribute to life-tables until the date of the final follow-up visit. Subjects who discontinue for other personal or medical reasons should contribute until their date of last use.

The standard convention for handling subjects lost to follow-up is to allow them to contribute to the life-table until their last visit date [Trussell, 1991]. However, Trussell argues that this is not appropriate, since subjects may have had an undetected pregnancy at their last visit. He recommends censoring subjects lost to follow-up three months prior to when they were last observed to assure that they were not pregnant at their last visit. We believe this rule is too conservative. We recommend that the end date for subjects lost to follow-up should be two to four weeks prior to their date of last visit or one week before the date of their last pregnancy test, whichever is later.

If an event (*e.g.*, accidental pregnancy) is detected after the subject has stopped using the method for some other reason (*e.g.*, medical reason), the date of discontinuation from the life-table should be the earliest of the two event dates, (*e.g.*, the estimated date of conception or the date of last use). The pregnancy would not be considered an event in the life-table pregnancy rate unless the estimated date of conception occurred before product use stopped.

Subjects who never use the product should contribute to the primary analysis of pregnancy rates until their discontinuation or until the date they report first using another method, whichever is earlier. This handling of data allows subjects to contribute to the pregnancy rate until there is some evidence that they are no longer considering the study product to be their method of contraception.

It should be noted that the guidelines above apply to final study analyses and not to interim analyses. At the time of an interim analysis, subjects still in the trial should be handled in a manner similar to the way subjects lost to follow-up are handled in a final analysis. The algorithms for determining how long each subject should contribute to an interim analysis of pregnancy and which events should be allowed to contribute to the rates can become very complicated. These details should be included in the study protocol if interim analyses are planned.

Whether subjects should be admitted to a study before they have gained a certain amount of experience with the method or only after they demonstrate that they know how to use the method correctly is discussed in an earlier section of this paper. Trussell and associates recommend that when there is a training phase in a clinical trial, investigators should report two sets of failure rates: one when entry into the analysis occurs at the start of the training phase and one when entry occurs at the start of the effectiveness phase [Trussell, et al., 1990]. The reasoning here is that, although efficacy cannot be adequately assessed until the subject has learned how to use the method, some failures will probably occur during the training phase when subjects are learning how to use the method. Furthermore, results from studies with a training phase would not be comparable to conventional studies without training unless all pregnancies and experience using the method are included in the analysis.

METHOD VERSUS USER FAILURE

The success of any compliance-based contraceptive method requires correct and consistent use every time the user has intercourse [Trussell, et al., 1990]. User failure has been defined as any accidental pregnancy that results from misuse of the method. Misuse includes not following product use instructions or failure to use the method during intercourse. Subjects may think they have used the method correctly and not report any misuse, even if misuse occurs. As a result, a pregnancy could be diagnosed by clinicians to be the result of method failure simply because there are no indications of user failure. There is often little or no information as to why the method failed and subjects may be unwilling or unable to return the method after it has failed. Also, for method failures, the investigator must determine if the reported date when the method was misused or failed coincides with the date of pregnancy, which itself is often difficult to determine accurately. User/method failure should be clearly defined prior to study initiation.

MEASURING CORRECT USE OF A PRODUCT

In order to calculate perfect use rates, each month or cycle of use must be classified as either a perfect or imperfect use interval. In order to minimize bias, this classification should be made without knowing whether or not a subject became pregnant during the interval.

A subject's use of a study product can be monitored through the use of a log of events which is completed by the subject, at home, during the study period. All subjects should record in the log all information pertinent to the study, such as dates of intercourse and whether the study product was used. If the product was not used, the subject should record what alternative method was used, if any. The subject should also be instructed to note any difficulties in using the product, any complaints or concerns about the product, and whether or not instructions for use were followed. The log should be returned to the clinic at each follow-up visit and reviewed by the investigator with the subject. In situations where there is a long period between follow-up visits, subjects should be encouraged, through phone contact, to keep accurate records on their coital logs. The log serves as the source document showing the subject's compliance with the protocol for the number of acts of intercourse and the number of times the product was used by the subject. The subject's log of events can also serve as a focus of discussion for the investigator to determine if the product was used correctly. For example, in discussing the log, the subject may reveal that she does not understand or is confused about how to use the product.

STUDY-RELATED PREGNANCIES

The usual rule for contraceptive efficacy studies is that any pregnancy is considered study-related if it occurs while the subject is still relying on the study method for contraceptive protection, and has not experienced a prior event that would disqualify her from the trial. A pregnancy that is not study-related, and therefore is not the reason for termination from the study, cannot be considered in the evaluation of efficacy. Yet, all pregnancies reported among study subjects during the course of their follow-up must be recorded and reported. Often investigators are provided with a standard rule that a diagnosed pregnancy with an estimated date of conception within seven days of last use of the method due to discontinuation from the study is to be considered a study-related pregnancy. But this rule of thumb may be too restrictive or too arbitrary in some cases.

If the subject has made a conscious decision to stop using the study method for contraceptive purposes, or has switched to another method of birth control, then a pregnancy occurring after this event should not be considered as study-related. In order to distinguish adequately between study-related pregnancies and pregnancies that are reported as events that occur during the course of the study, two categories for reporting pregnancy are recommended: a) all pregnancies diagnosed; and, b) pregnancies that occurred while the subject considered the product to be her method of contraception.

Pregnancies that are reported under the second category would form the basis for the analysis of method efficacy. Pregnancies reported under the first category would satisfy the requirements of reporting all events that occur during the course of a clinical trial.

DATA ANALYSIS

The primary analysis population should exclude subjects who never returned for a follow-up visit. However, women who never returned to follow-up should be described and total lost to follow-up and discontinuation rates should be adjusted to include their experiences. All data on subjects with protocol violations or deviations should be included in the primary analysis, otherwise bias may be introduced which favors one device group over another. This recommendation follows the practical definition of the Intent to Treat Population proposed by Gillings and Koch. In a comparative study, results should be presented categorized by the device used, whether or not it was the randomly assigned device. FHI refers to this population as the "Treated Population." The "Efficacy Population" excludes subjects with protocol deviations or violations, including subjects who never used the product. Following the recommendations of Gillings and Koch, a sensitivity analysis of the comparability of the conclusions from analysis for an Efficacy Population to the conclusions for the Treated Population is only necessary if more than five percent of the Treated Population is excluded from the Efficacy Population [Gillings and Koch, 1991].

PREGNANCY AND OTHER DISCONTINUATION EVENTS

Gross monthly and cumulative life-table rates using one month intervals should be calculated for each specific discontinuation reason, total method-related discontinuations (e.g., accidental pregnancy plus side effects), and total discontinuation. Kaplan-Meier event rates through the end of each month of study should also be produced for each category of discontinuation. The Kaplan-Meier rates provide a more exact estimate of the cumulative risk of pregnancy than event rates calculated using monthly intervals. Confidence intervals should be calculated for the cumulative event rates through the end of the study period [Kaplan and Meier, 1958].

The cumulative probability of pregnancy during perfect use (perfect use rates) should also be calculated [Trussell, 1991]. All months of perfect use should contribute to these event rates. For example, if a subject used the product perfectly during months 2 and 6 of a study and imperfectly during all other months, her month 2 and 6 data should contribute to the rates. Categorical rate models can be used to obtain these perfect use rates [Trussell and Guinnane, 1993]. The perfect use criteria should be clearly stated in the analysis plan of the study protocol.

CLINICAL TRIALS LIMITATIONS

The ability to generalize from the results of a clinical trial is dependent, in large part, on the composition of the population studied. Careful selection of study subjects is paramount if the trial results are to achieve the objectives of the trial. This is not easy in clinical trials research where the "real world" may not be easily replicated (e.g., OTC market).

Clinical trials of contraceptive methods are often conducted under artificial conditions. More specifically, some trials may be conducted in settings that are more conducive to making the trial successful, but may not be representative of the actual environment in which actual users will receive the services and in which the products will generally be used. For example, in order to insure that subjects are at sufficient risk of pregnancy, the study could limit participation only to subjects whose level of sexual activity is greater than many of the potential users of the method. Including subjects with infrequent sexual activity may be more reflective of the target population, but may not provide an adequate test of contraceptive efficacy.

Although clinical trials have their limitations, we hope that a continued effort to standardize methodologic approaches and study designs will lead to greater comparability of results across future studies of new barrier contraceptive methods. These issues are presented as a first step in what we hope will be an ongoing dialogue concerning the study of barrier contraceptive efficacy.

REFERENCES

Gillings D, Koch G (1991): The application of the principle of intention-to-treat to the analysis of clinical trials. Drug Inform J 25:11-24.

Freedman LS (1982): Tables of the number of patients required in clinical trials using the logrank test. Statistics in Med 1:121-129.

Kaplan EL, and Meier P (1958): Nonparametric estimation from incomplete observa tions. J Am Stat Assoc 53:457-481.

Makuck RW, Pledger G, Hall DB, Johnson MF, Herson J, Jiann-Ping H (1990): Statistical issues in drug research and development. Chapter 4. "Active Control Equivalence Studies". New York: Marcel Dekker, Inc.

Speroff L, Glass RH, Kase NG (1989): Clinical gynecologic endocrinology and infertility. Baltimore: Williams and Wilkins, pp 469-473.

Trussell J, and Guinnane T (1993): Techniques of event history analysis to be published in David Reher and Roger Schofield (eds): "Old and New Methods in Historical Demography". Clarendon Press:Oxford, 181-205.

Trussell, J (1991): Methodological pitfalls in the analysis of contraceptive failure. statistics in Med 10: 201-210.

Trussell J, Hatcher RA, Cates W Jr, Stewart FH, Kost K (1990): A guide to interpreting contraceptive efficacy studies. Obstet Gynecol 76(3):558-567.

USFDA, Center for Devices and Radiologic Health (1990): Premarket testing guidelines for female barrier contraceptive devices also intended to prevent sexually transmitted diseases.

DISCUSSION

Irv Sivin: I have just a couple of comments. One, you indicated a dichotomy between a large and representative target population and the possibility that in some way by enlisting a representative population one may introduce confounding factors. I think, however, that the emphasis is properly placed on having as close as possible to a representative population. That it is the only way in which one can indicate to the potential user what is likely to happen. To have a representative sample I think requires that there be several, certainly at least five, clinics or units in the study and this study should be large in absolute numbers. I think the important thing is the complexity of the group -- that it be as close to the world as possible. I also think, with all due respect, the important thing is what is the overall failure rate and not necessarily the division between correct and incorrect use.

The comment you made that perhaps we had paid too much attention to efficacy in the last 15 or 20 years in comparison with other factors, I think may be correct, but efficacy is a very important factor. Even in today's world, and for the next five decades at least, I think there will be large segments of the world's population that after a failure either will not use or will not choose to use, postcoital methods, menstrual regulation, or an elective abortion. The large segment requires that we have really effective methods.

Roberto Rivera: I agree entirely with what you said. I would only comment that one should not try to get efficacy information out of a study on issues or on subsets if the study cannot properly provide that kind of information.

Debbie Rogow: My question also pertains to the issue of sexual frequency, which raises a number of questions for me. One is that there is a tendency for many people to over-report sexual frequency which might account for why people are coming into your study easily and then falling by the wayside. Rather than their patterns changing, their pattern is emerging. That's one thing. Another is that, at least in one of the studies, which had an average frequency of 13 times a month or almost every other day not only is it perhaps not that widely representative of the population, but also using the results may be not comparable with other methods.

Roberto Rivera: I would only say that everything that you said makes sense and has to be considered. There is one biological aspect for the two to three times per week that I mentioned, i.e., trying to make sure that during ovulation there is a risk of exposure.

Debbie Rogow: I think that your last point is really the important one, which is, that we have a paucity of information on people's sexual frequency and behavior over the cycle. We might find people with monthly frequency which falls well below your standard but who have much higher frequency at the time of ovulation given the more recent data on increased libido at the time of ovulation, etc.

Mary Ann Leeper: This of course is a very pivotal and critical issue and has been for us for the last four years in terms of defining what is the best clinical study. If you look at our clinical study, motivation was the major issue. So your first point about having the right population I think is absolutely critical motivation and if you didn't use it you're going to get pregnant.

The second point that I wanted to make was this issue that you just talked briefly about earlier, about efficacy and defining efficacy. What is efficacy? And, in the real world, how important is it that we define it to the decimal point? If we know that the Reality female condom in the US market among different users averaged 12.4 failure rate, why should we spend millions of dollars more to make sure that it's 12.4 and not 12.7 or 11.3? The issue is the ballpark of where it is. Get it out and spend the millions of dollars on educating the people, the women and couples to use products and how important it is to use the product.

Roberto Rivera: Let me respond to your comment with another comment. I think we should have a look at efficacy particularly in barrier methods in a wider context. I like what Trussell said that we should know what are the failure rates during typical use, that we should know what are the failure rates during imperfect use, and which is the percentage of women who use a method perfectly. This kind of information is very important when you speak about the use of a new method.

Jacqueline Forrest: One of the things that we are really struggling with is this question of the typical user and typical use pregnancy rates. My question is, in your mind, do you think that it's even possible to design clinical trials that would have typical users of a method in them and if you don't think that is what we've already done, what do you use as your guide if you see a clinical trial result? Do you do a calculation in your head? I do one in mine to say that with this kind of method in a clinical trial, what do I think it's going to be like in a general population of users. I'm asking the questions because I am wondering how much more we should invest in trying to get typical-use rates out of clinical trials versus pushing more toward perfect use rates and then some other indication that we could give to potential users of the proportion of people who might be able to use them perfectly. Do you think we really can attain typical use rates from clinical trials and, if we don't do that, what do you think is the relationship between what we see in clinical trials and then what happens in real life?

Roberto Rivera: That is a very difficult question and I hope it will be addressed more extensively in the working groups. I certainly believe that we have to know more about the efficacy of barrier methods in typical users. Probably one thing we have to do is to define better what we believe is going to be the typical user of the method. If the typical user is not going to be the whole population but a given segment of the population according to the characteristics of the method, the outcome that we expect or the main use that is expected of the method, then I think it makes sense to focus the clinical trial or establish the admission and exclusion criteria so as to have a better definition of the intended population.

Zena Stein: I'm just wondering if you have experience with using the clinic as the unit rather than the individual? For instance, you mentioned counseling. People have mentioned here the providers as being key in getting compliance. Everyone knows that clinics as units have characteristics. They have people who believe and don't believe. They also have heterogeneity in ethnicity and so on. Have you had experience and have you considered the clinic as the unit and as opposed to the individual where you subject them to the different providers and the contamination of other people not in the trial?

Roberto Rivera: The message that I tried to present was the need that all these definitions and all these procedures should be established and should be consistent across clinics and across centers. Otherwise it becomes very difficult to compare the results from one site or from one clinic to another. To your final comment, yes, there is a big difference between the information that you get from a clinical trial and from what you get once the product is in actual or large scale use.

Jeffrey Spieler: One of my pet peeves is that we expect much more out of clinical trials than you really can get. They are imprecise tools, they are snapshots in time, and they frequently don't, as you've already said, reflect what is going to happen in the field. With all methods, particularly barrier methods, when the product gets on the market the clients are going to self-select. They are buying this product. They are not entering into a study. They're going to use that product as long as they have proper information on how to use it. They are going to use it as correctly or as incorrectly as they want. Therefore, when we design our clinical studies, I think we really have to design them much closer to what people actually do in terms of compliance and frequency of intercourse. These issues really have to be dealt with in order to deal with how people are going to behave but still recognizing that clinical trials are an imprecise tool. The other issue is that when we're studying products with dual mechanisms, STD prevention and contraception, until we get permission to actually study the method to prevent an STD transmission we end up recruiting people, in my personal view, that are totally unlike the people who will actually buy the product. The Reality female condom clinical trial which is a good study, recruited people who are probably not going to be the typical buyers of the product. The study involved only monogamous couples not at risk of STDs. Those people might be buying the product but I don't think they will be the majority of the buyers and we need a better idea of the majority of the buyers -- how they will perform when they use the device. Therefore, especially for products that are available and we want to evaluate them, we need a different tool than a clinical trial. We need some kind of a field-based study that's not nearly as rigorous as a clinical trial but that permits some minimal data collection to get an idea of how the product performs.

Roberto Rivera: I think we have to accept that there is also the need to know what is the efficacy of the method under perfect use conditions and you have to design a clinical trial that will allow you to do that. It will require you to have information to classify each month for each subject whether it was a month of perfect of imperfect use. That is when you run into study design problems.

PART IV: BARRIERS AND STDs

14

HIV, OTHER STDs, AND BARRIERS

Katherine M. Stone

Division of STD/HIV Prevention
National Center for Prevention Services
Centers for Disease Control and Prevention
Atlanta, GA 30333

INTRODUCTION

Effective collaborations between those in the fields of family planning and prevention of sexually transmitted diseases (STDs) will require a mutual understanding of the broad scope and heterogeneity of STD, impact of other STD on HIV transmission, similarities and differences between STD and unplanned pregnancy and the special burdens of STD on women.

OVERVIEW OF SEXUALLY TRANSMITTED DISEASES

STDs are a heterogeneous group of infections with various bacteria, protozoa and viruses [Holmes et al., 1990]. The term STDs as used herein includes the sexual transmission of HIV. Common sexually transmitted infections caused by bacteria include gonorrhea (caused by *Neisseria gonorrhoeae*), chlamydia (caused by *Chlamydia trachomatis*), syphilis (caused by *Treponema pallidum*), and chancroid (caused by *Haemophilus ducreyi*). Trichomoniasis is a common STD caused by a protozoan (*Trichomonas vaginalis*). Viruses that are commonly sexually transmitted include human immunodeficiency virus (HIV), herpes simplex virus (HSV), human papillomavirus (HPV), hepatitis B virus (HBV) and cytomegalovirus (CMV). Genital ulcer diseases include syphilis, chancroid and genital herpes. Many other species of bacteria and protozoa cause enteric infections spread by fecal-oral contact which may occur during sexual activity; these will not be considered here.

Barrier Contraceptives: Current Status and Future Prospects, pages 203–212
Published 1994 Wiley-Liss, Inc.

Bacterial and protozoal STDs are usually curable with antibiotics, whereas viral STDs are not. Because sexually transmitted viruses reside inside the host cell, killing these pathogens may necessitate killing the host cell as well. Primary prevention is the mainstay for the control of viral STDs. Many sexually transmitted viruses, most notably HSV and HPV, cause chronic subclinical infections which are far more common than the clinically apparent infections recognized as genital herpes and genital warts, and far more common than bacterial STDs.

Many sexually transmitted infections seem to potentiate HIV transmission because they disrupt genital epithelium or result in inflammation. The relationship between genital ulcer diseases -- i.e., syphilis, genital herpes, and chancroid -- and HIV transmission has been widely publicized. However, other nonulcerative STDs -- i.e., gonorrhea, chlamydia, trichomoniasis, and others -- may actually have an even larger impact on HIV transmission because they are so much more common than genital ulcer disease. Both ulcerative and nonulcerative STDs appear to increase risk of HIV transmission by approximately three to five fold [Wasserheit, 1992]. In addition, bacterial vaginosis (caused by overgrowth of various bacteria) and candidiasis (caused by a fungus, *Candida albicans* and other *Candida* species) are two other very common infections that involve the vulva and vagina and may facilitate HIV transmission. Because they result from alterations in endogenous vaginal flora, rather than contact with an infected partner, they are not considered to be STDs.

SIMILARITIES BETWEEN STDs AND UNPLANNED PREGNANCY

Efforts to prevent STDs and unplanned pregnancy are similar in several aspects [Cates and Stone, 1992]. First, sexual contact, usually sexual intercourse, is necessary for both. Thus, behavioral lessons learned about how to modify unsafe sexual practices can be applied to reducing risks of both STDs and unplanned pregnancy. Second, both STDs and unplanned pregnancy discriminate biologically against women [Aral, 1991, Aral and Guinan, 1984, Cates and Stone, 1992]. Due to anatomic differences, STDs are more easily transmitted to women, are more difficult to diagnose and are more frequently asymptomatic among women. Moreover, long-term complications in women are far more common and much more serious. Gender discrimination is more obvious for unplanned pregnancy, the entire biological burden of which falls on women. Third, those concerned with both STDs and unplanned pregnancy must contend with the power imbalance between genders that clearly favors men. In some cultures, both in the developed and the developing world, women frequently have little say about when, with whom, and under what conditions sexual intercourse occurs. Obviously this situation influences whether any prevention measures are used. Moreover, in many cultures a women's value to society is determined by her ability to bear children. Fourth, the groups at highest risk for STDs and unplanned pregnancy are similar, i.e., those under age 25 and those with lower incomes.

DIFFERENCES BETWEEN STDs AND UNPLANNED PREGNANCY

Despite the similarities, important differences exist between STDs and unplanned pregnancy [Cates and Stone, 1992]. Differences in health care providers include a microbiologic versus physiologic basic science background, directive versus nondirective approach to the patient, male versus female gender focus in clinic settings, and crisis (symptomatic) versus preventive health care emphasis. These aspects are discussed in greater detail elsewhere [Cates and Stone, 1992] and will not be elaborated here. Misperceptions that STDs occur primarily among commercial sex workers [women who have multiple sexual partners] and not the general population has resulted in a greater degree of negative social stigmatization for STDs compared to pregnancy. In the developing world where resources are scarce, policy makers, program planners and donor agencies consistently give low priority to control of STDs and other reproductive tract infections [Wasserheit and Holmes, 1992].

Several biologic differences are also important to consider [Cates and Stone, 1992] (Table 1). First, compared with the proportion of individuals able to conceive, a smaller percentage of men and women in most communities are capable of transmitting STDs at any one time, especially bacterial STDs. Second, the key determinant for STDs is the number of partners and partners' infection status, whereas for pregnancy, coital frequency and timing within the menstrual cycle are crucial factors. Thus, prevention foci are very different for the two outcomes. Third, STD transmission is bidirectional -- i.e., from male to female and from female to male -- whereas pregnancy is not; therefore, protection of the male partner should not be overlooked when planning STD prevention strategies. Fourth, the vehicles resulting in STDs include genital discharge and genital contact in addition to semen, whereas only semen is an important vehicle for spermatozoa resulting in pregnancy. Fifth, risk per coital episode for pregnancy varies from 0-20% [Trussell and Kost, 1987], depending on the stage of the menstrual cycle. This risk is generally lower than the risk of acquiring bacterial STDs and higher than the risk of acquiring viral STDs (Table 2). Sixth, potential portals of entry for STDs include the vaginal walls, vulva, urethra, anus and rectum in addition to the cervix, which is the sole "portal of entry" for spermatozoa (Table 3).

TABLE 1. Differences Between STD and Unplanned Pregnancy

Factor	STD	Unplanned Pregnancy
Percentage capable of transmitting	20% (bacterial) 70% (viral)	90%
Key determinants	partner selection	coital frequency, timing
Direction	bidirectional	male to female
Vehicle for causative agent	semen, genital discharge, genital contact	semen
Transmission risk per coital episode	0.1 - 90%	0 - 20%
Portals of entry	lower genital tract	cervix

RISK OF STD TRANSMISSION

Risk of transmission for many STDs is much higher from male to female than vice versa (Table 2). For most STDs, risk per coital episode has not been quantified or has been estimated based on limited data. Estimated risk of gonorrhea transmission per coital episode is 50 percent to 90 percent for women, clearly higher than the 20 percent risk for men [Platt, 1983; Hooper et al., 1978; Holmes et al., 1970; Judson, 1990.] Annual risk of HSV transmission to infected persons was 0.19 percent for women and 0.05 percent for men [Mertz et al., 1992]; assuming that couples engaged in an average of 100 coital episodes in one year, estimated risk per coital episode is 0.19 percent for women and 0.05 percent for men. For HIV, male to female transmission risk per coital episode has been estimated to range 0.1 percent to 20 percent, and female to male transmission has been estimated to range from 0.01 percent to 10 percent [Peterman et al., 1988; Padian et al., 1987; Brunham and Plummer, 1990; Anderson and May, 1988.]

TABLE 2. STD Transmission Risk Per Coital Episode

STD	Male to Female	Female to Male
Gonorrhea	50-90%	20%
Genital HSV	0.2%	0.05%
HIV	0.1% - 20%	0.01% - 10%

PORTALS OF ENTRY FOR STD PATHOGENS

Because STD pathogens attach only to susceptible epithelial cells, specific portals of entry differ among the various infections. Pathogens which cause STDs that primarily involve the male urethra (gonorrhea, chlamydia, trichomoniasis) or that are bloodborne (HIV, HBV) are generally carried in semen or other urethral secretions. Other infections (syphilis, genital herpes, chancroid, HPV) are spread by direct genital contact. Some pathogens (HSV, HPV, *T. Pallidum*) may occasionally be present in semen if the urethra is infected but are primarily spread through direct contact with infected penile epithelium. Some STDs (HIV? HBV?) perhaps are spread through semen *and* by direct genital contact. Although gonorrhea and chlamydia resemble conception in that the causative agents are carried in semen and enter the cervix, but not the vagina or vulva, they are more the "exception rather than the rule" in the universe of STDs. Most of the other common STDs, including the genital ulcer diseases and HIV infection, utilize other parts of the lower genital tract as portals of entry (Table 3).

TABLE 3. Portals of Entry for STD Pathogens

disease/agent	transmitted through semen?	cervix	vagina	vulva	anus/ rectum
conception/ sperm	+	+			
gonorrhea	+	+			+
chlamydia	+	+			+
trichomonas	+	+	+		
syphilis		+	+	+	+
chancroid		+	+	+	+
HIV	+	+	+	+?	+
HBV	+	+	+	+?	+
HSV		+	+	+	+
HPV		+	+	+	+

Specific barrier methods obviously differ in their ability to protect broad areas of the genital tract (Table 4). In addition to preventing cervical and vaginal contact with the penis and urethral fluids, condoms prevent vulvar contact with the penile shaft and glans, which are the major sites of clinically apparent and subclinical infection with syphilis, chancroid, HPV, and HSV in males. Other currently available barriers provide little or no protection for the vagina and vulva. Contraceptive methods which protect only the cervix will not be effective for preventing many STDs.

TABLE 4. Genital Areas Protected by Barriers

barrier	cervix	vagina	vulva	anus/ rectum
diaphragm/cap	+			
spermicides	+	+		
condom	+	+	+	+

LACK OF GENERALIZABILITY OF BARRIER EFFICACY

Because of the heterogeneity of STDs and subsequent differences in routes and risks of transmission, contraceptive barriers differ in the ability to protect against STDs. Contraceptive efficacy should not be generalized to STD prevention, and prevention efficacy for a specific STD can be generalized to another STD only if their transmission profiles are similar. The most effective contraceptive methods are those which do not rely heavily on user compliance, such as long-acting hormonal contraception and surgical sterilization. Unfortunately, these methods offer no protection against STDs. Moreover, users may not even perceive themselves to be at risk for STDs. Barrier methods of contraception provide variable degrees of protection against STDs, depending on coverage of the genital tract and user compliance. User compliance is a critical element in determining effectiveness. Compliance with barrier methods for contraception is notoriously low; however, little data are available to assess whether compliance among women using barriers solely or in part for STD prevention differs from compliance among women using barriers solely for contraception.

POTENTIAL MECHANISMS FOR STD PREVENTION

Potential mechanisms for barriers to prevent STDs differ somewhat from those to prevent pregnancy. Physical barriers must prevent contact with genital discharge or lesions in addition to semen. Chemical barriers must either kill pathogens and/or inhibit their attachment to susceptible epithelial cells, whereas an ideal spermicide should kill sperm on contact, inhibit sperm motility, and inhibit acrosin [Wooley, 1991]. Persistence of infectivity of pathogens after deposition into the female genital tract has not been determined, so the duration of action required to prevent infection is unknown. Ideally, a chemical barrier to prevent STDs should be effective against a broad range of pathogens (including bacteria, viruses, and protozoa) and should not impair local genitourinary defense mechanisms (including the genital epithelium and protective substances secreted by epithelial cells, as well as normal vaginal flora).

TABLE 5. Potential Mechanisms for STD Prevention and Contraception

Method	STD Prevention	Contraception
Physical barriers	Prevent contact with semen, discharge, lesions	Prevent contact with semen
Chemical barriers	Kill pathogens Inhibit attachment	Kill sperm Inhibit motility Inhibit acrosin

BURDEN OF STDs ON WOMEN

Women bear a greater burden from STDs for several reasons [Aral, 1991; Aral and Guinan, 1984, Cates and Stone, 1992]. First, common bacterial STDs are usually asymptomatic in women but not men and, therefore, are detected only by specific screening. For example, 30 to 80 percent of women but only 1 to 5 percent of men with gonorrhea are asymptomatic [Judson, 1990]; 85 percent of women [Judson, 1990] and perhaps 50 percent of men [Fish et al., 1989] with chlamydia are asymptomatic. Second, because the anatomy and normal flora of the lower genital tract are more complicated, STDs are generally more difficult and expensive to diagnose in women than in men. For example, a urethral swab and a Gram stain are sufficient to make a definitive diagnosis of gonorrhea in men, but a speculum exam and a specific culture for *N. gonorrhoeae* are required for women. Third, risk of transmission of many STDs to women is higher than risk of transmission to men. Mucosal epithelium is apparently more vulnerable than keratinized epithelium, and women have larger surface areas of susceptible mucosa (cervix, vagina, labia minora) than men (urethra). For example, annual risk of HSV acquisition among discordant couples was 19 percent for women and only 5 percent among men [Mertz et al., 1992]. Contact with semen or other urethral fluids is a likely explanation for higher male to female transmission risks for some STDs (e.g., gonorrhea) that are transmitted primarily by semen. Fourth, medical consequences of STDs are more frequent and more serious for women than for men. Pelvic inflammatory disease, which often results in infertility, is much more common than epididymitis and prostatitis. Many STDs frequently result in devastating female reproductive outcomes, such as spontaneous abortion, premature labor, fetal infection or neonatal infection. Genital (cervical, vulvar, anal, penile) cancers and precursors related to HPV, and perhaps other sexually transmitted infections, are much more common in women than in men in both developed and developing countries. Fifth, although advances have been made in many developed and developing countries over the last decade, a "double standard" for sexuality still exists in many cultures and places a greater psychosocial burden on women compared to men. Social values and attitudes related to sexuality, especially multiple partners, have not kept pace with changes in women's work roles and sexual behaviors [Aral, 1984]. For example, as recently as the late 1980's condoms were regarded as evidence of prostitution by vice squad workers in California [Cohen, 1988], one of the most liberal states in the United States [Cohen, 1988]. Development of new methods to protect

TABLE 6. Special Burden of STD on Women

1	Common bacterial STD usually asymptomatic in women but not in men
2	STD more difficult/expensive to diagnose
3	higher risk of STD transmission to women
4	Medical consequences of STD are worse for women
5	Psychosocial stigma; "double standard"

women from STDs, especially HIV, is clearly a high priority issue. Moreover, the assessment of safety and efficacy of new products should stimulate further research into the pathogenesis of various STDS, the protective role of local genitourinary mucosal defenses and the vaginal ecosystem -- research that is long overdue.

SUMMARY

STDs are caused by a variety of bacteria, protozoa and viruses with different portals of entry and specific modes of transmission. Viral STDs are not curable with antibiotics, and primary prevention is the mainstay for control. Prevention measures must consider specific modes of transmission and portals of entry for STD pathogens. Women clearly bear a larger burden than men from most STDs. Prevention and control of STDs will synergistically promote HIV prevention, family planning, child survival and safe motherhood [Wasserheit and Holmes, 1992].

REFERENCES

Anderson RM, May RM (1988): Epidemiological parameters of HIV transmission. Nature 333:514-519.

Aral SO (1991): Behavioral considerations of sexually transmitted diseases impact on women, mothers, and children [session 2036]. Presented at the 119th Annual Meeting of the American Public Health Association, November 12, 1991, Atlanta, Georgia.

Aral SO, Guinan ME (1984): Women and sexually transmitted diseases. In Holmes KK et al. (eds): "Sexually Transmitted Diseases" (1st ed.). New York: McGraw-Hill, pp 85-89.

Brunham RC, Plummer FA (1990): A general model of sexually transmitted disease epidemiology and its implications for control. Med Clin N Amer 74:1339-1352.

Cates W, Stone KM (1992): Family planning, sexually transmitted diseases and contraceptive choice: A literature update--Part I. Fam Plann Perspec 24:75-84.

Cohen J (1989): Condom promotion among prostitutes. In "Condoms in the Prevention of Sexually Transmitted Diseases: Proceedings of a Conference". Research Triangle Park, North Carolina: American Social Health Association, pp. 45-47.

Fish AN, Fairweather DV, Oriel JD, Ridgeway GL (1989): *Chlamydia trachomatis* infection in a gynecology clinic population: Identification of high-risk groups and the value of contact tracing. Eur J Obstet Gynecol Reprod Biol 31:67-74.

Holmes KK, Johnson DW, Trostle JH (1970): An estimate of the risk of men acquiring gonorrhea by sexual contact with infected females. Am J Epidemiol 91:170-174.

Holmes KK, Mardh PA, Sparling PF, Wiesner PJ, Cates W, Lemon SM, Stamm WE (eds) (1990): "Sexually Transmitted Diseases" (2nd ed.). New York: McGraw-Hill.

Hooper RR, Reynolds GH, Hones OG, et al. (1978): Cohort study of venereal disease. I: the risk of gonorrhea transmission from infected women to men. Am J Epidemiol 108:136-144.

Judson FN (1990): Gonorrhea. Med Clin N Amer 74:1353-1367.

Mertz GJ, Benedetti J, Ashley R, Selke SA, Corey L (1992): Risk factors for the sexual transmission of genital herpes. Ann Intern Med 116:197-202.

Padian N, Marquis L, Francis DP, Anderson RM, Rutherford GW, O'Malley PM, Winkelstein W (1987): Male-to-female transmission of human immunodeficiency virus. JAMA 258:788-790.

Peterman TA, Stoneburner RL, Allen JR, Jaffe HW, Curran JW (1988): Risk of human immunodeficiency virus transmission from heterosexual adults with transfusion-associated infections. JAMA 259:55-58.

Platt R, Rice PA, McCormack WM (1983): Risk of acquiring gonorrhea and prevalence of abnormal adnexal findings among women recently exposed to gonorrhea. JAMA 250:3205-3209.

Trussell J, Kost K (1987): Contraceptive failure in the United States: a critical review of the literature. Stud Fam Plann 18:237-283.

Wasserheit JN (1992): Epidemiological synergy: Interrelationships between human immunodeficiency virus infection and other sexually transmitted diseases. In Chen LC, Sepulveda J, Segal SJ (eds): "AIDS and Women's Reproductive Health". New York: Plenum Press, Ch. 5.

Wasserheit JN, Holmes KK (1992): Reproductive tract infections: Challenges for international health policy, programs, and research. In Germain A, Holmes KK, Piot P, Wasserheit JN (eds): "Reproductive Tract Infections: Global Impact and Priorities for Women's Reproductive Health". New York: Plenum Press, pp 7-33.

Wooley RJ (1991): Contraception--A look forward, Part I: New spermicides and natural family planning. J Am Board Fam Pract 4:33-46.

DISCUSSION

Gabriel Bialy: Dr. Stone, I was intrigued by your statement, albeit I agree with you, that you cannot take efficacy for one thing as a bona-fide measure for efficacy for some other disease. Could you expand just a tiny bit on this?

Katherine Stone: What I said was you can generalize prevention of one STD to another only if their transmission risk profiles are similar, i.e., the risk of transmission in general and also the portal of entry. For example, I think you can probably generalize prevention of hepatitis-B virus to HIV because they are spread in similar ways. Gonorrhea and chlamydia can maybe be generalized to each other because they have similarities but I'm not real sure of that because they have enough differences related to chlamydia being inside the cell. Those are a few examples. The point that I want to make is that you really can't generalize protection against gonorrhea and chlamydia. You cannot generalize that to HIV infection.

Dorothy Patton: One of your slides suggests that chlamydia was transmitted via semen and sperm. What is the direct evidence to show that?

Katherine Stone: My slide stated that the organisms were in semen. I didn't mean to imply that they are transmitted via spermatozoa, *per se*, although there are some individuals who feel that this is perhaps an important mechanism for pathogens ascending higher into the genital tract. There is some evidence for attachment of chlamydia to sperm. I think we really don't understand the magnitude of its importance as far as transmission goes, especially ascending infection.

Dorothy Patton: I am aware of those *in vitro* studies on the attachment of chlamydia to sperm but the evidence is very weak in those studies. *In vivo* we've been unable to show that there is a direct attachment of the organism to the sperm.

Katherine Stone: Yes, I agree the evidence is fairly weak and I think as far as pathogenesis of PID goes, people think sperm transport plays a very minor role if any.

15

USE OF THE SIV/RHESUS MACAQUE SYSTEM TO TEST VIRUCIDES DESIGNED TO PREVENT THE SEXUAL TRANSMISSION OF HIV

Christopher J. Miller

California Regional Primate Research Center
University of California Davis, Davis CA 95616

INTRODUCTION

As of March 1992, Centers for Disease Control and Prevention [CDC, 1992] reported that 213,641 cases of AIDS have been diagnosed in the United States, and approximately half of these patients have died as a consequence of the disease. In addition, between 1 million and 2 million Americans are infected with human immunodeficiency virus (HIV) but do not yet have symptoms of the infection. The World Health Organization estimates that 10 million to 12 million people are currently infected with HIV worldwide. The most conservative estimates predict that 30 million to 40 million people will be infected with HIV by the year 2000 and most of these people will be infected through heterosexual contact. This is particularly true in the urban centers of sub-Saharan Africa [Johnson, 1988; Piot, et al., 1988].

Despite its importance, little is known about the biology of this mode of transmission. In developed countries, most of the AIDS patients became infected with HIV through homosexual contact or the use of contaminated needles in intravenous drug abuse. However the vast majority of healthy HIV-infected individuals probably became infected by heterosexual contact and are unaware of their infection status. Because the mean incubation period from seroconversion to AIDS is about eight to 10 years [Bacchetti et al., 1989], these individuals have the potential, through sexual contact, to disseminate HIV widely before they become aware that they are infectious. Thus, AIDS has the potential to become a far more serious global epidemic than is currently the case.

Barrier Contraceptives: Current Status and Future Prospects, pages 213–223

BIOLOGY OF HETEROSEXUAL TRANSMISSION OF HIV

Rational strategies to prevent the sexual transmission of HIV must be based on an understanding of the biology of HIV transmission. Several underlying principles regarding the heterosexual transmission of HIV have emerged and much of our understanding is based on observations in the simian immunodeficiency virus (SIV)/ rhesus macaque model [Miller, et al., 1993]. The reproductive anatomy and physiology of the rhesus macaque are very similar to humans [Hendrickx, et al., 1987] making these animals an excellent species in which to conduct studies involving the reproductive tract. These similarities include regular menstrual cycles, 28-day cycle length and visible menstrual flow. The anatomic organization of the reproductive tract is virtually identical in the species. The main difference in the anatomy is that the cervical canal is more tortuous than the canal in women. In addition, the type and distribution of immune cells in the female genital tract of rhesus macaques and women is similar [Miller, et al., 1992b].

The identity and location of the target cells for HIV and SIV during sexual transmission are yet to be determined. In cervical biopsy material obtained from HIV-infected women, T cells and macrophages were determined to be the cell types infected with HIV [Pomerantz, et al., 1988]. These findings are consistent with the our results in SIV-infected monkeys. In chronically infected female rhesus macaques, SIV-infected cells are present in the uterus, cervix and vagina [Miller, et al., 1992c]. The majority of the SIV-infected cells are located in the submucosa of the ectocervix and vagina and have a morphology consistent with T lymphocytes and monocyte/macrophages. SIV-infected cells are also found within the stratified squamous epithelium of the vagina. Some of the infected cells in this location have a dendritic morphology consistent with Langerhans cells [Miller, et al., 1993; Miller, et al., 1992c]. Langerhans cells of the skin have been reported to be infected with HIV in AIDS patients [Tschachler, et al., 1987; Zambruno, et al., 1991]. The finding that cells in the vaginal epithelium are infected with SIV is the first indication that mucosal Langerhans cells might be infected with a lentivirus [Miller, et al., 1992c]. Presumably all the cells that were found to be infected in the genital tract of both human and rhesus express the CD4 molecule. There was no evidence that non-CD4+ cell types (epithelial cells) were infected in the reproductive tracts of either the monkeys or humans. SIV is efficiently transmitted to hysterectomized female macaques by inoculation of cell-free virus into blind vaginal pouches [Miller, et al., 1992d]. Since only a few CD4+ T cells are present in the submucosa of the vagina [Miller, et al., 1992b], the most likely target cells in the vaginal mucosa are macrophages or Langerhans cells [Miller, et al., 1992c; Miller, et al., 1992d].

The distribution of CD4+ Langerhans cells throughout the squamous epithelium of the vagina and ectocervix suggests that virus-target cells interaction can occur over a relatively large mucosal surface. Thus, any prophylactic strategy (vaccine, chemical or barrier) designed to prevent transmission of HIV must provide protection over the entire mucosal surface. A vaccine capable of preventing infection following contact with an infected individual offers the best hope of decreasing the incidence of HIV infection, but there are currently no candidate vaccines available that are likely to produce the type of genital and systemic immunity required. While research directed toward developing such vaccines continues, individuals at risk need strategies to reduce their chance of becoming HIV positive. Chemical prophylaxis through intravaginal application of virucidal compounds offers a direct and cost-effective method to protect women (and indirectly, men) from HIV transmission during sexual intercourse.

In order to test the efficacy of potential chemical compounds at preventing the sexual transmission of HIV an appropriate animal model challenge system must be used. The SIV/rhesus macaque model of AIDS [Miller, et al., 1989; Miller, et al., 1993] has emerged as a useful system in which to develop and test intravaginal virucidal compounds designed to prevent sexual transmission of HIV.

VAGINAL TRANSMISSION OF SIV

A genital mucosal challenge system has been developed using the SIV/rhesus macaque model of the heterosexual transmission of HIV [Miller, et al., 1990; Miller, et al., 1989; Miller, et al., 1993]. The transmission studies that will be discussed here utilized a stock of cell-free SIV_{mac} that was grown in human peripheral blood mononuclear cells. A detailed description of the virus stock has been previously published [Miller, et al., 1989]. The stock was titered in vivo by intravenous (IV) inoculation in a large number of animals. Intravenous inoculation with 1 ml of a 10^{-4} dilution of this stock infected five of five animals, while 0.5 ml of a 10^{-4} dilution of this stock infected two of two animals. Intravenous inoculation with 1 ml of a 10^{-5} dilution of this stock failed to infect two animals. Based on these results, 1 ml of 10^{-4} dilution is between 1 and 10 intravenous animal infectious doses (IVAID) of this stock of cell-free SIV_{mac} [Miller, et al., 1989].

The stock was next titered by genital inoculation of animals. The genital transmission studies in female monkeys were conducted by infusing the virus preparations through a 2.5 mm outer diameter (8 French) soft plastic, pediatric nasogastric feeding tube (American Pharmaseal, Valencia, CA) into the vaginal canal. The animals were immobilized with Ketamine HCl during the procedure. It is very unlikely that this inoculation technique causes significant trauma to the vaginal epithelium and any damage that occurs would be far less than caused by normal mating. The initial transmission experiments involved intravaginally inoculating the virus on multiple occasions over one menstrual cycle [Miller, et al., 1989]. Thus two animals were intravaginally inoculated with 1 ml of undiluted SIV_{mac} (10^4 IVAID) eight times over four weeks and two animals were inoculated with 1 ml of a 10^{-1} dilution of the virus stock (10^3 IVAID) eight times over four weeks. Both of the animals receiving the higher dose of virus became infected, one was viremic on day seven post-inoculation (PI) (after two inoculations) and the other was viremic on day 14 PI (after four inoculations). Of the two females receiving the diluted inoculum, one was viremic on day 14 (after four inoculations) and the other did not become infected. In addition to the adult animals described above, four juvenile female rhesus were intravaginally inoculated with undiluted cell-free SIV_{mac} (10^4 IVAID) twice a week for two weeks. All four of these animals became viremic and seropositive

[Miller, et al., 1989].

Subsequently, we attempted to infect mature female macaques with a single intravaginal inoculation of the cell-free virus stock [Miller, et al., 1990]. Thirteen female animals were intravaginally inoculated once with 1 ml of undiluted cell-free SIV_{mac} (10^4 IVAID). Nine of the 13 animals became infected. In addition, six mature female rhesus monkeys were intravaginally inoculated with 0.5 ml of the virus stock (5 x 10^3 IVAID) and three animals became infected. Thus it seems that 0.5 ml of the undiluted cell-free virus stock is a 50 percent genital animal infectious dose [Marthas, et al., 1992].

In order to locate the site of the SIV target cells in the female reproductive tract, four hysterectomized female macaques were intravaginally inoculated with cell-free SIV_{mac} [Miller, et al., 1992d]. The cervix and uterus of each animals was surgically removed and the vagina was formed into a blind pouch. The animals were allowed to completely heal prior to virus inoculation and culposcopic examination revealed that the vaginal epithelium completely lined the vaginal pouch. The animals were inoculated with 0.5 ml of the virus stock (5 x 10^3 IVAID) and two of the four animals were infected. Thus, the lack of the cervix and uterus did not affect the dose of cell-free virus required for the genital transmission of SIV.

PREVENTION OF SIV GENITAL TRANSMISSION BY SPERMICIDES CONTAINING NONOXYNOL-9

The results of experiments to test the ability of nonoxynol-9 containing spermicides to prevent the genital transmission have been previously reported [Miller, et al., 1992a; Miller, et al., 1990], but these studies will be briefly summarized here.

Effect of Nonoxynol-9 on SIV in Vitro

The in-vitro anti-HIV effect of nonoxynol-9 is similar for SIV, at nonoxynol-9 concentrations above 0.06 percent (v/v), no infectious SIV was detected in either SIV tissue culture supernatant/nonoxynol-9 mixtures or the SIV-infected rhesus peripheral blood mononuclear cell/nonoxynol-9 mixtures. These results are similar to those of Hicks et al., 1985 and Malkovsky et al., 1988, who showed that HIV was inactivated by 0.05 percent (v/v) nonoxynol-9.

Effect of Spermicides Containing Nonoxynol-9 on Genital Transmission of SIV

To determine if nonoxynol-9 containing spermicides were capable of preventing the genital transmission of SIV, 1 ml of contraceptive foam (Delfen® contraceptive foam, Ortho Pharmaceutical., Raritan NJ) containing nonoxynol-9 (12.5 percent v/v) was placed into the vaginal canal of six female rhesus macaques with a smooth glass pipette (5 mm outer diameter). Immediately following placement of the contraceptive foam, 1 ml of cell-free SIV was infused into the vaginal canal. The nonoxynol-9 foam infusion/SIV inoculation procedure was repeated four times over two weeks. We have previously shown that this dose of SIV produces infection in 100 percent of inoculated females (two adults and four juveniles) [Miller, et al., 1989]. Contraceptive foam containing nonoxynol-9 infused into the vaginal canal protected three of six animals from infection [Miller, et al., 1990].

In a second experiment, 1 ml of contraceptive gel (Gynol II Extra Strength, Ortho Pharmaceutical, Raritan NJ) containing nonoxynol-9 (3 percent by weight) was placed into the vaginal canal of an additional six female rhesus macaques prior to the infusion of 1 ml cell-free SIV suspension (10^4 intravenous animal infectious dose). The nonoxynol-9 gel infusion/SIV inoculation procedure was repeated four times over two weeks. The contraceptive gel containing nonoxynol-9 (3 percent by weight) infused into the vaginal canal protected two of six animals from infection following inoculation of 1 ml of cell-free SIV (10^4 intravenous animal infectious dose). Thus, nonoxynol-9 containing spermicides protect some animals against the genital transmission of SIV. There was no striking difference in the level of protection provided by the foam or gel preparations tested.

In the studies described above, commercially available spermicidal foam and gel preparations containing nonoxynol-9 protected some rhesus macaques against the vaginal transmission of SIV. The dose of SIV used in these studies was high, in previous experiments this dose had produced viremia in six of six animals; each dose of SIV contained approximately 10^4 more SIV than is required for intravenous transmission. It is possible that nonoxynol-9 containing preparations may be more effective against a lower dose of SIV. Semen from HIV-positive men usually does not contain this much virus and our finding of protection suggests that nonoxynol-9 may be effective at inactivating HIV in the ejaculate of infected men. We used two forms of nonoxynol-9, a foam and a gel, in order to determine if a particular formulation might provide more protection from genital transmission of SIV. The foam preparation (containing a higher concentration of nonoxynol-9) did protect one more animal than the gel preparation, however, due to the small size of the groups the significance of this difference is not clear. It is possible that the foam preparation was more effective because it more completely and rapidly coated the mucosa of the cervix and vagina.

OTHER STRATEGIES TO PREVENT SEXUAL TRANSMISSION OF HIV

It may be possible to reduce the sexual transmission of HIV by taking advantage of the sensitivity of the virus to acidic pH. Cell-free and cell-associated HIV is rapidly inactivated in-vitro at acidic pH [Ongradi, et al., 1990]. Cell-free SIV is also inactivated in-vitro at low pH [Miller, unpublished results]. In women, the vaginal pH is highly variable between individuals and can vary through the menstrual cycle. A recent report [Moller, et al., 1991] found that in sexually-active fertile women without gynecological abnormalities, the vaginal pH ranged from 3.4 to 6.9. The vaginal pH was highest (6.0 to 7.0) during menstruation, and decreased during intermenstrual period with no uniform pattern. In some women, the pH values throughout the menstrual cycle remained fairly constant and in others the vaginal pH dropped by as much as 2.1 pH units in the intermenstrual period. We recently undertook a small study to determine the vaginal pH of normal mature rhesus macaques [Miller, unpublished data]. Using a small pH electrode, the vaginal pH of 14 female rhesus macaques was determined once a week over a five week period. The menstrual cycles of the animals were monitored during this period. The pH values ranged from 4.8 to 8.0. In general, the pH of the rhesus macaque vagina was highest (5.6 to 8.0) around the time of menses. In most animals (10 of 14) the pH of the vagina varied less than 1 pH unit throughout the menstrual cycle. The mean vaginal pH in these animals was 5.9. In some animals (four of 14), the pH varied more than 1 pH unit at different stages of the menstrual cycle and in one animal the vaginal pH dropped by 2.1 pH units in the intermenstrual period. The mean vaginal pH in these four animals was 6.5. Thus the range of vaginal pH in women and female rhesus macaques is similar. It is possible that variations in vaginal pH play a role in the susceptibility of individuals to HIV transmission.

The pH of seminal plasma ranges from 7 to 8 and after ejaculation the pH of the vagina rises to 7.0 or higher [Fox, et al., 1973]. Cell-free HIV and SIV can survive for extended periods in these conditions. If preparations can be developed that can prevent the vaginal pH from rising after ejaculation, the sexual transmission of HIV may be reduced. In order to assess the importance of seminal plasma in the sexual transmission of the viruses, we have begun a series of vaginal transmission studies inoculating mixtures of seminal plasma and cell-free SIV into female macaques. Based on the above observations it is possible that the presence of seminal plasma may dramatically lower the dose of SIV required for the genital transmission of the virus. Using this approach we have identified a 100 percent genital animal infectious dose using a single inoculation of cell-free SIV. A large scale study is under way to determine the magnitude of effect of seminal plasma on the genital transmission of SIV.

A number of other chemical agents (including hydrogen peroxide, ethyl alcohol, Lysol and bleach) have been shown to inactivate HIV in vitro [Martin, et al., 1985]. In vitro studies have shown that spermicides containing benzalkonium chloride rapidly inactivate HIV in semen and vaginal secretions [Wainberg, et al., 1990]. Vaginal disinfection during labor with chlorhexidine has been shown to reduce the transmission of group B streptococci from mothers to infants [Burman, et al., 1992a; Burman, et al., 1992b]. Clinical trials have shown that spermicides containing nonoxynol-9 are capable of preventing the transmission of a number of STDs, including gonorrhea, chlamydia and trichomonas [Louv, et al., 1988; Niruthisard, et al., 1992; Rosenberg, et al., 1992]. Thus, chemical approaches can be used to prevent the vaginal transmission of a variety of microbial pathogens.

SUMMARY AND CONCLUSIONS

A vaccine capable of generating long lasting genital immunity offers the best hope of preventing the sexual transmission of HIV. However, such a vaccine is not currently available and the development of an effective vaccine may take many years.

The use of chemical agents to prevent the sexual transmission of HIV is an available and cost-effective method to reduce the sexual transmission of HIV. Chemical prophylaxis can be used by women without the cooperation of the male partner and virucidal preparations can be produced with or without spermicidal activity. The SIV/rhesus macaque model of genital transmission offers an excellent system in which to test candidate preparations designed to prevent the sexual transmission of HIV.

REFERENCES

Bacchetti P, Moss AR (1989): Incubation period of AIDS in San Francisco. Nature 338: 251-253.

Burman LG, Christensen P, Christensen K, Fryklund B, Helgesson AM, Svennsingsen NW, Tullus K (1992a): Prevention of excess neonatal morbidity associated with group B streptococci by vaginal chlohexidine disinfection during labour. Lancet 340:65-69.

Burman LG, Tullus K (1992b): Vaginal chlorhexidine disinfection during labour [letter]. Lancet 340:791-792.

CDC (1992): CDC Statistics. AIDS:343-345.

Fox C, Meldrum S, Watson BW (1973): Continuous measurement by radio-telemetry of vaginal pH during human coitus. J Reprod Fertil 33(1):69-75.

Hendrickx A, Cukierski M (1987): Reproductive and developmental toxicology in nonhuman primates. Prog Clin Biol Res 235:73-88.

Hicks DR, Martin LS, Getchell JP, Heath JL, Francis DP, McDougal JS, Curran JW, Voeller B (1985): Inactivation of HTLV-III/LAV infected cultures of normal human lymphocytes by nonoxynol-9 in vitro. Lancet 2:1422-1423.

Johnson AM (1988): Heterosexual transmission of human immunodeficiency virus. Br Med J 296:1017-1020.

Louv WC, Austin H, Alexander WJ, Stagno S, Cheeks J (1988): A clinical trial of nonoxynol-9 for preventing gonococcal and chlamydial infections. J Infect Dis 158: 518-523.

Malkovsky M, Newell A, Dalgeish AG (1988): Inactivation of HIV by nonoxynol-9 [letter]. Lancet 1(8586):645.

Marthas M, Sutjipto S, Miller C, Higgins J, Torten J, Unger R, Kiyono H, McGhee J, Marx P, Pedersen N (1992): Efficacy of live-attenuated and whole-inactivated simian immunodeficiency virus vaccines against intravenous and intravaginal challenge. In Brown F, Chanock RM, Ginsberg HS, Lerner RA (eds): "Vaccines 1992." Cold Spring Harbor Laboratory Press, 117-122.

Martin LS, McDougal JS, Loskoski SL (1985): Disinfection and Inactivation of the human T lymphotropic virus type III/lymphadenopathy-associated virus. J Infect Dis 152:400-403.

Miller CJ, McGhee JR, Gardner MB (1993): Mucosal Immunity, HIV transmission and AIDS. Lab Invest 68:129-145.

Miller CJ, Alexander NJ, Gettie A, Hendrickx AG, Marx PA (1992a): The effect of contraceptives containing nonoxynol-9 on the genital transmission of simian immunodeficiency virus in rhesus macaques. Fert Steril 57:1126-1128.

Miller CJ, McChesney M, Moore PF (1992b): Langerhans cells, macrophages and lymphocyte subsets in the cervix and vagina of rhesus macaques. Lab Invest 67:628-634.

Miller CJ, Vogel P, Alexander NJ, Sutjipto S, Hendrickx AG, Marx PA (1992c): Localization of SIV in the genital tract of chronically infected female rhesus macaques. Am J Pathol 141:655-660.

Miller CJ, Alexander NJ, Vogel P, Anderson J, Marx PA (1992d): Mechanism of genital transmission of SIV: A hypothesis based on transmission studies and the location of SIV in the genital tract of chronically infected female rhesus macaques. J Med Primatol 21:64-68.

Miller CJ, Alexander NJ, Sutjipto S, Hendrickx AG, Jennings M, Marx PA (1990): Effect of virus dose and nonoxynol-9 on the genital transmission of SIV in rhesus macaques. J Med Primatol 19:401-409.

Miller CJ, Alexander NJ, Sutjipto S, Lackner AA, Hendrickx AG, Gettie A, Lowenstine LJ, Jennings M, Marx PA (1989): Genital mucosal transmission of simian immunodeficiency virus: Animal model for heterosexual transmission of human immunodeficiency virus. J Virol 63:4277-4284.

Moller BR, Kaspersen P (1991): The acidity of the vagina. In Horowitz B, Marc PA (eds): "Vaginitis and Vaginosis." New York: Wiley-Liss Inc, pp 63-67.

Niruthisard S, Roddy R, Chutivongse S (1992): Use of nonoxynol-9 and reduction in rate of gonococcal and chlamydial cervical infections. Lancet 339:1371-1375.

Ongradi J, Ceccherini-Nelli L, Pistella M, Specter S, Bendinelli M (1990): Acid sensitivity of cell-free and cell-associated HIV-1: clinical implications. AIDS Res Human Retroviruses 6:1433-1436.

Piot P, Plummer FA, Mhalu FS, Lamboray J-L, Chin J, Mann JM (1988): AIDS: An international perspective. Science 239: 573-586.

Pomerantz RJ, de la Monte SM, Donegan CE, Rota TR, Vogt MW, Craven DE, Hirsch MS (1988): Human immunodeficiency virus (HIV) infection of the uterine cervix. Ann Int Med 108:321-327.

Rosenberg MJ, Davidson AJ, Chen JH, Judson FN, Douglas JM (1992): Barrier contraceptives and sexually transmitted diseases in women: a comparison of female dependent methods and condoms. Am J Public Health 82(5):669-674.

Tschachler E, Groh V, Popvic M, Mann DL, Konrad K, Safai B, Eron L, diMarzo Veronese F, Wolff K, Stingl G (1987): Epidermal Langerhans cells: A target for HTLV III/LAV infection. J Invest Dermatol 88:233-237.

Wainberg MA, Spira B, Bleau G, Thomas R (1990): Inactivation of human immunodeficiency virus type 1 in tissue culture fluid and genital secretions by the spermicide benzalkonium chloride. J Clin Micro 28:156-158.

Zambruno G, Mori L, Marconi A, Mongiardo N, De Rienzo B, Bertazzoni U, Giannetti A (1991): Detection of HIV-1 in epidermal Langerhans cells of HIV-infected patients using polymerase chain reaction. J Invest Dermatol 96:979-982.

DISCUSSION

Shakti Upadhyay: What was your control when you used nonoxynol-9 after inoculation with SIV? Any other spermatocidal agent or the base?

Christopher Miller: No we didn't use the carrier which would have been a great control. We just used a commercial foam that is 12.5 percent by volume in N-9 and a commercial gel that is 3 percent by volume. Our goal with those studies was just to try to demonstrate efficacy of currently available commercial products and not to, necessarily, prove that N-9 itself is the magic ingredient. So it was a very limited study aimed at just proving that a commercially available product could provide some efficacy.

Shakti Upadhyay: What is the time point when you did the inoculation after putting this nonoxynol-9?

Christopher Miller: Very short. We put it in, dropped the virus and then put that in, so it's less than five minutes. Usually between two and five minutes.

Jeffrey Spieler: I think it's worth mentioning again, as Chris Miller mentioned, that in those studies the inoculum was what he called a "whopping dose." It certainly was a super physiological dose and with such a large dose you could still prevent transmission. I think those results are actually very encouraging.

Christopher Miller: As a lot of people have said, any drop in transmission rates results in a lot fewer infected people. I think that is important.

Elizabeth Ngugi: Based on your results of the foreskin of the Macaque, would you be able to explain why it has been found that men with foreskin are much more likely to be infected with HIV?

Christopher Miller: Well, I can't give you a definitive answer but I certainly have my opinions. I'm very much of the opinion that Langerhans cells are very important in transmission and Langerhans cells are found in abundance in the foreskin, far more than in other sites. We've also localized virus in Langerhans cells in the foreskin in chronically infected animals. I would argue that the presence of those cell types in abundance in that site is one possible explanation for it.

Amy Pollack: I don't know if you can answer these questions. I just think that they are things we need to think about. This study that you did on the hysterectomized monkeys is really important. It has had great impact on how we think about transmission and my questions are: in the monkeys that you have, you've repeatedly anesthetized them I presume. What kind of impact does that have on the immune system in general? The other question is the actual local impact of having done major surgery on the monkeys and the impact of the vaginal cuff; you said you examined them three months after they were hysterectomized with a colposcope. I'm presuming that you looked with a colposcope and did the best job you could to determine that the cuff was completely closed, that there wasn't an opening so that there could be direct entry into the peritoneal cavity, and that the inflammatory response was completely gone. I've looked at women who are three months post-hysterectomy and there is frequently still some kind of inflammatory response or exposed mucosa that has not got stratified epithelium covering the mucosa. Then, of course, my next question would be if you are only looking at a couple of monkeys, and two out of four of them convert, what does that really mean for us in terms of numbers and the kind of dose that you're exposing them to. We all think it happens in the vagina but not in the cervix, and that's the key. It was a monkey study. I think we all need to be careful about that.

Christopher Miller: I agree. I don't think that you can say that target cells aren't in the cervix or in the peritoneal cavity. As far as the immune response to ketamine or immune suppression due to ketamine anesthesia, we've never looked locally to see what happens in the mucosal immune system with ketamine anesthesia. But systemically, you get a transient stress response as you always do when you squeeze-up an animal but you don't see any significant decline in immune parameters, such as CD-4 or CD-8. You just see very transient lymphocytosis and neutrophilia. After that everything is back to normal.

Kamran Moghissi: Could you comment on the *in vitro* effect of nonoxynol-9 over SIV?

Christopher Miller: It was very similar to the *in vitro* infectant in HIV, and thanks for asking that, 0.05 percent completely inactivated SIV as it does with HIV.

Laneta Dorflinger: It's interesting -- your data on enhanced vaginal transmission with dilution with seminal plasma. Have you looked at all at penile transmission by using seminal plasma dilutions?

Christopher Miller: No, we haven't and we don't immediately plan to do that. That's a good idea also. Some people reported that seminal plasma actually kills virus, so it's a fine line we're walking.

Debbie Rogow: I was interested in the work that you did looking at the impact of changes in the menstrual cycle on infection and I'm wondering, particularly given other findings -- yours included -- about the impact of the general ecological milieu, as well as specifically pH, on transmission infection rates, and whether you've thought of doing the same study but not at such high doses to see if there are any changes. Also, related to Amy's comment, are you sure that scar tissue is as capable of resisting transmission as normal vaginal mucosa?

Christopher Miller: I don't think we're sure of anything. As long as the epithelium is covering it, there is no direct exposure of granulation tissue, macrophages, or fibroblasts to the virus so it has to get through the epithelium somehow. This is an intact epithelium and the only CD4+ cells that are present in these epithelia are Langerhans cells. As far as looking at the stage of the menstrual cycle more with lower doses, one of the problems we have in the animal model that I think most people kind of intuitively understand is that we can't use doses that would provide 0.1 percent transmission as occurs with HIV. If we could get 1,000 monkeys and get one of them infected after this low-dose exposure then we could answer these questions of pH cycle without this overwhelming dose of virus. We really need to have a fairly consistent transmission to get any answers. I'd love to be able to do it, but there is a shortage of monkeys and money.

16

BARRIER METHODS FOR THE PREVENTION OF GONOCOCCAL AND CHLAMYDIAL INFECTION

Penelope J. Hitchcock

Sexually Transmitted Disease Branch
Division of Microbiology & Infectious Disease
National Institutes of Health
9000 Rockville Pike
Bethesda, MD 20892

INTRODUCTION

Since the advent of human reproductive behavior, it has been women (and their infants) who disproportionately suffer the consequences of sexually transmitted diseases (STDs). To what do we owe the recent political, social and medical awakening to the public health importance of STDs? Arguably, this change is attributable to the emergence of a new, fatal, *gender neutral* STD - human immunodeficiency virus infection. As part of this awareness, we have acknowledged the need for female-controlled barrier methods to protect women (and men) against these devastating infections. The need for preventive methods which can be implemented at the discretion of the woman reflects in high prevalences of non-consensual sex, sex without condom use, and risk behaviors that occur without partner knowledge.

One such approach would be to develop safe, effective intravaginal products such as foams, creams, or gels that would prevent infection. Such topical microbicides would be non-inflammatory chemical formations with virucidal and/or bactericidal activity designed for intra-vaginal use prior to, or immediately after, sexual intercourse. Ideally these products would not be inherently spermicidal, but could be so formulated for those interested in preventing pregnancy as well as infection.

Non-contraceptive barriers against infection would be extremely useful for those women who use one of the many safe, effective methods for contraception which either have no protective effect against infection or, arguably, exacerbate risk of infection. Indeed, a person's contraceptive choices may change over a lifetime. However, no matter

Barrier Contraceptives: Current Status and Future Prospects, pages 225–241
Published 1994 Wiley-Liss, Inc.

what an individual's current preference, so long as they are sexually active, they will always desire/require protection from sexually transmitted infections.

These products should be colorless, odorless, tasteless, non-inflammatory, fast yet long acting, stable, cheap, and readily available, and should not require the involvement of a health-care provider. Although it is recognized that correct and consistent use of these products will require the development and implementation of a behavioral research agenda, that subject will be developed in subsequent presentations.

This discussion will focus on 1) gonorrhea and chlamydial infection, the sexually transmitted diseases characterized by the presence of bacterial discharge; 2) the magnitude of the problem including the health and economic consequences; 3) the biology of the host/parasite interactions; 4) insights, based on laboratory data, into how we might alter the infectivity of the pathogen as well as the infectiousness of the partner; and 5) a brief overview of what we do or do not know about disease prevention efficacy with currently available barrier contraceptive methods.

SPECTRUM OF DISEASE:

Lower Reproductive Tract Infection

Gonorrhea. Acute lower tract infection with *Neisseria gonorrhoeae* in men manifests as urethritis. The chance of becoming infected with a single exposure is approximately 20 percent [Holmes, 1970]. Ninety-five percent of infected men will develop symptoms within 1-14 days, the mean being 3.4 days [Harrison, 1979; Handsfield, 1974]. Based on experimental gonococcal infection, the infectious dose is less than 1000 organisms [personal communication, Carolyn Deal, Walter Reed Army Institute of Research]; although infected men may shed as many as 10,000 viable gonococci per ejaculate [personal communication, Myron Cohen, University of North Carolina, Chapel Hill].

In women, gonorrhea manifests as cervicitis (and urethritis) after an incubation period of approximately ten days [Platt, 1983; Wallin, 1974]. Estimates of asymptomatic infections in women range from 60-80 percent [Pederson, 1971; McCormack, 1977; Upchurch, 1991]. If exposed, 50-90 percent of female partners of infected men will contract gonorrhea, however the number of exposures required is unknown [Holmes, 1970; Hooper, 1978]. In both men and women, gonorrhea can be a disease of the rectum and the oral pharynx [Wiesner, 1973; Klein, 1977; Kinghorn, 1979; Lebedeff,1980].

Chlamydial infection. Acute lower tract infection with *Chlamydia trachomatis* in men also manifests as urethritis. However compared to gonorrhea, this is a milder form of urethritis. Symptoms include dysuria and mucopurulent discharge. Estimates of asymptomatic infections in men range from 35-50 percent [McMillan, 1981; Stamm, 1984; Stamm, 1986]. The incubation period is seven to ten days. Transmissibility is poorly defined, with literature values ranging between 25-80 percent [Lycke, 1980]. Among heterosexual males estimates are that co-infection with gonorrhea occurs in 15-35 percent of cases [Bowie, 1978].

In women, the infection manifests as cervicitis, however only about 35 percent of women have symptoms [Harrison, 1985]. The disease is characterized by a mucopurulent discharge and hypertrophic ectopy. A correlation has been made between the presence of symptoms and the number of recoverable infectious elementary bodies [Hobson, 1980]. Increased susceptibility to chlamydial infection correlates with cervical ectopy (the presence of columnar epithelial cells), putting adolescents (60-80 percent of whom have cervical ectopy as a normal manifestation of puberty) and oral contraceptive users at increased risk for chlamydial infection. Up to 50 percent of women with cervicitis will also have urethritis [Paavonen, 1979; Brunham, 1981; Harrison, 1985].

Complications of Lower Tract Infection

Pelvic Inflammatory Disease. Pelvic inflammatory disease, or PID, is defined as upper tract infection which may include uterus, fallopian tubes, fimbriae as well as the surrounding peritoneal cavity. PID is the most common complication of gonococcal cervicitis - occurring in estimated 10-20 percent of cases [Eschenbach, 1975; Holmes, 1980]. The estimates are higher for chlamydial cervicitis; up to 50 percent of women will have endometritis [Sweet, 1983; Paavonen, 1985; Wasserheit, 1986]. The predominance of chlamydial PID reflects the frequency and the clinical manifestations of the infection. Women are either asymptomatic or have symptoms so indolent that they do not seek diagnosis or treatment [Harrison, 1985].

There are four significant consequences of PID which include 1) an increased risk for subsequent PID episodes; 2) chronic pelvic pain syndrome; 3) ectopic pregnancy; and 4) infertility. Tubal scarring mediates the latter two sequelae; if scarring is partial tubal pregnancy may occur, if complete tubal obstruction occurs infertility results [Reviewed, NIH Report on PID, 1991]. In the case of chlamydial infection, work in animal models has demonstrated a correlation with antibody responses to chlamydial heat shock protein, immune-mediated delayed hypersensitivity and scarring [Reviewed, Morrison, 1992]. Serologic studies in women who had a history of PID also demonstrated a correlation between antibody titers to this protein and tubal scarring. Women with subsequent ectopic pregnancy compared to those without had significantly higher titers to this antigen [Wager, 1990]. Whether this response mediates the pathogenesis of tubal scarring or is a surrogate marker for tubal scarring is unknown.

Adverse Outcomes of Pregnancy. Gonococcal and chlamydial uterine infections undoubtedly play a role in adverse outcomes of pregnancy, including spontaneous abortion, prematurity, chorioamnionitis, low birth weight and ophthalmia neonatorum [Reviewed, Cates, 1988; Edwards, 1978; Sweet, 1987; Brunham, 1990; Schulz, 1992]. The results of the Vaginal Infections and Prematurity Study implicated chlamydial infection as a risk factor for premature rupture of membranes [Nugent, 1990; Cotch, 1990]. In addition, chlamydial respiratory infection of the neonate, if untreated, may lead to chronic lung disease as well as otitis media [Reviewed, Alexander, 1983; Datta, 1988].

If one uses endometrial biopsies to assess whether the cervical barrier has been broken, there is evidence that uterine infection is quite common [Wasserheit, 1986]. Furthermore, animal model studies of chlamydial infection indicate that the organisms infect the uterus within seven days of vaginal inoculation [Rank, 1992].

Disseminated Disease. Disseminated gonococcal infection or DGI occurs in 0.5 to 3 percent of gonorrhea cases. This complication follows an asymptomatic infection for which the patient has not sought treatment. The disease manifests as an arthritis-dermatitis syndrome. Occasionally the disease may present as septicemia, although it is usually milder and more amenable to treatment compared to meningococcal septicemia. People with terminal complement deficiencies (C8 is the most common) may have repeated episodes of DGI. Gonococcal infection is the most common cause of septic arthritis in sexually active adults. Women are more commonly affected [Masi, 1981; O'Brien, 1983; Petersen, 1979].

STDs and HIV Infection. With respect to the epidemiological synergy between STDs and HIV, over 75 studies on the role of STDs in HIV transmission have been conducted. In 15, both ulcerative and non-ulcerative STDs were found to increase the risk of HIV transmission 3-5 fold, independent of behavior. Although the risk of transmission of HIV in genital ulcer diseases appears higher than in the discharge diseases, the high prevalence of discharge diseases results in a much higher population attributable risk [Reviewed, Wasserheit, 1992]. The mechanisms of action may involve 1) disruption of tissue integrity, including increased friability and bleeding; 2) recruitment of inflammatory cells including macrophages and lymphocytes, 3) loss of normal flora; 4) loss of mucous barriers; and 5) pH change.

Given that four bacterial, one viral and one protozoan STD alter the risk of HIV transmission, prevention of these diseases through use of effective microbicides would have enormous benefit, in terms of preventing the spread of both HIV and the STDs.

MAGNITUDE OF THE PROBLEM

Epidemiology and Economics

Epidemiology. It is very difficult to piece together the global burden of sexually transmitted diseases. The data from developing countries are insufficient to accurately depict the dimensions of the problem. Some cross sectional estimates have been made in a recent report from the Pan American Health Organization's 1991 AIDS/HIV/STD Annual Surveillance Report, as well as in the World Development Report, 1993. The World Health Association/ Global Program on AIDS estimates 14 million people are infected with HIV [Merson, 1993]. Arguably, since HIV is relatively non-infectious compared to other sexually transmitted pathogens, the number of other sexually transmitted infections is much larger. Sentinel studies from prenatal clinics in Africa have shown serologic evidence of syphilis in 1-15 percent, culture evidence of gonorrhea in 2-11 percent, bacteriologic evidence of chlamydial infection in 6-19 percent and microscopic evidence of trichomoniasis in 7-32 percent of attendees. The prevalences in STD clinic populations are much higher [Reviewed, Berkley, 1992].

In the United States and other developed countries, the data are more complete and may serve as a surrogate for the magnitude of the global problem, if not the trends. In 1992, there were approximately 10 million cases of STDs. Sixty-four percent of these infections were diagnosed in young people, less that 24 years old. In fact, three million occurred in teenagers. The total reflects one million cases of gonorrhea and four million cases of chlamydial infection. Adolescents cases represented 25 percent of cases of gonorrhea; high rates of chlamydial infection in adolescent populations were found in numerous studies and sentinel surveys [derived from CDC's Sexually Transmitted Disease Surveillance, 1991].

In 1992, the majority of the one million diagnosed cases of PID were attributed to chlamydial and gonococcal infection. Many other cases of undiagnosed PID are reflected in women with tubal factor infertility. As in developing countries, poverty and lack of access to health care not only contribute to the development of PID but to the untoward consequences of PID sequelae. Among minority teenagers who die due to pregnancy-related causes, the most common cause is ectopic pregnancy related to an earlier episode of PID. Twenty-five percent of women with a first episode of PID will have a subsequent episode, with chances of chronic sequelae increasing with each episode [Reviewed, NIH publication on PID]. In developing countries, PID-related infertility may threaten the marriage and the family stability since child bearing is often a non-negotiable, mandatory condition of the marriage [Fort, 1989].

A number of studies have tried to assess the impact of gonococcal and chlamydial infection on fetal wastage, low birth weight or prematurity, and congenital or perinatal infections. In general, these infections are *rarely* implicated in fetal wastage. In contrast, evidence of these infections has been found in 10-30 percent of low birth weight babies. Furthermore, congenital or perinatal transmission occurs in 40-70 percent of infected mothers. Pregnant adolescents are more frequently affected by these infections than pregnant adults [Reviewed, Brunham, 1990]

Economics. In addition to the costs associated with diagnosis and treatment of gonococcal and chlamydial cervicitis and urethritis, the most costly consequences of these infections reflect upper tract infection. However, the costs attributable to undiagnosed endometritis that result in premature birth and low birth weight are unknown; in 1980, the United States spent $460 million on neonatal intensive care. If the costs for prematurity were eliminated, a savings of $400 million would be predicted [Naeye, 1980]. Furthermore, chronic lung damage due to perinatal chlamydial respiratory infection represents additional health costs of unknown magnitude.

In the United Stated, we are able to estimate costs for pelvic inflammatory disease and chronic sequelae in <u>diagnosed</u> women. Estimates of the comprehensive costs of diagnosed PID exceeded five billion dollars in 1992. It is noteworthy that total costs associated with STDs represent six billion dollars for 1992. Furthermore, if the incidence of PID remains constant, at the current rate of inflation, projected costs of PID for the year 2000 will be almost $10 billion [Washington, 1991]. Sadly these figures do not account for infertility due to undiagnosed post-PID tubal scarring.

THE BIOLOGY OF HOST/PARASITE INTERACTIONS

Developing successful strategies to prevent infection with topical microbicides is critically dependent upon an understanding of the infectious process - the molecular basis and chronology of the host/parasite interaction. It is from this vantage point that the "pressure points" for effectively interrupting infection will be apparent.

The Surface of the Problem

Chlamydia trachomatis is a gram negative bacterium with two morphologically and metabolically distinct forms: the elementary body, the stable infectious particle; and the reticulate body, the fragile, metabolically active, replicative, intracellular form. Based on tissue culture and animal model studies, the infectious elementary body will bind to the surface of the host cell and induce uptake. This may occur within the first five minutes of exposure. Once sequestered inside the phagocytic vacuole an unidentified surface component of the elementary body prevents fusion with the lysosome; this would otherwise be a lethal event for the organism. A process of differentiation follows, during which the elementary body is transformed into the metabolically-active, replicative reticulate body. Large chlamydial inclusions form within the cell; these rupture within 48 hours, releasing infectious elementary bodies into the cervical, vaginal, and urethral secretions.

Although a true gram-negative organism, *C. trachomatis* has many unusual features including an obligate intracellular niche, the absence of a peptidoglycan in the cell wall, and a non-toxic lipopolysaccharide. The rigidity of the bacterial cell wall appears to be the result of disulfide binding between proteins, including the major outer membrane protein, MOMP [Reviewed, Schachter, 1991]. Recently a carbohydrate-mediated attachment mechanism has been described [Zhang, 1992].

In addition to the morphological heterogeneity, there are ocular and genital biovars of chlamydia. Although there is some overlap, one of the ocular biovars - Biovar A - has never been isolated from a genital site [Grayston, 1975; Kuo, 1983]. Whether the tissue tropism reflects specific receptor/ligand interactions or metabolic differences in tissue types is unknown.

Neisseria gonorrhoeae is a classical gram-negative bacterium; it does not undergo a developmental cycle. However, the surface of the gonococcus is quite complex. During the infection, any organism, at any moment, is likely undergoing numerous, independent, simultaneous changes. Furthermore the gonococcus is facultatively intracellular, spending at least part of its existence sequestered within host cells [Reviewed, Sparling, 1990].

Attachment to target cells is mediated by several surface components including pili, hair-like appendages which protrude from the cell surface [Swanson, 1971; Jephcott, 1971]. These form rope-like bundles which function in attachment to host cells and in the formation of microcolonies [Brinton, 1978; Todd, 1984]. The major outer membrane protein is one of the few stable surface components. It functions as an ion transport channel and probably intercalates into host cell membranes as part of the early steps in the infectious process [Young, 1983; Lynch, 1984; Haines, 1988]. Protein II is involved in attachment as well, perhaps in conjunction with lipopolysaccharide, LPS [Dekker, 1990; personal communication, Milan Blake, Rockefeller University]. This complex of fat and sugar is critical for maintaining the integrity of the bacterial cell wall as well as camouflaging the organism [Guymon, 1978; Perry, 1977]. Although infection does not confer long term immunity, protective antibodies specific to protein I and LPS are elicited. These function in complement-dependent bactericidal assays and phagocytic assays [Rice, 1980; Apicella, 1986].

The ability to avoid the consequences of the host's immune response to infection is attributed to the organism's propensity for surface alteration. At least a half dozen surface components are able to undergo phenotypic variation including phase variation, antigenic variation, or both, or they exhibit some heterogeneity within or between strains [Reviewed, Meyer, 1989]. In simple terms, phase variation is analogous to either wearing a sweater or not, antigenic variation is analogous to wearing sweaters of different colors. In addition, gonococci exhibit microheterogeneity - the organisms infecting an individual may be quite different from the ones next to them. If one examines gonococci in a urethral exudate using gold-tagged antibody particles that are specific for a given surface antigen, only some of the bacteria will have this molecule on the surface [Hitchcock, 1986; Apicella, 1987].

Much of what we know about the infectious process of gonorrhea has been gleaned from the human fallopian tube model and early histopathology studies done by Harkness and colleagues [Harkness, 1948]. In the fallopian tube, gonococci probably attach to the non- ciliated columnar epithelial cells while simultaneously delivering a toxic insult to ciliated cells. Isolated lipopolysaccharide or peptidoglycan fragments have been shown to effectively stop ciliary action and mediate selective extrusion of these cells from the mucosal surface. Gonococci induce uptake by non-professional phagocytes, perhaps through Protein I and/or Protein II or other "invasins" which appear to be induced when gonococci are incubated with target cells [Haines, 1988; Shaw, 1983; Chen, 1991]. Once inside the cell, gonococci replicate within vacuoles. The bacteria-laden vacuoles are transported to the basal aspect of the cell and the bacteria are exocytosed into the subepithelial space [Taylor-Robinson, 1974; McGee, 1978].

Experimental Approaches to Altering Chlamydial and Gonococcal Infectivity

Based on experimental evidence, a number of interventions <u>decrease</u> the infectivity of infectious elementary bodies for tissue culture cells or in animal models. These include mild heat treatment, reduction of disulfide bonds, heparin or dextran sulfate treatment, incubation with specific antibodies, and neuraminidase treatment. On the other hand, a number of interventions <u>increase</u> infectivity including estrogen therapy, the addition of DEAE- dextran to host cells prior to infection, irradiation, IUDR or cycloheximide treatment of cells, progesterone treatment of mice (prior to infection), and co-infection with *N. gonorrhoeae* [Reviewed, Moulder, 1991].

In the case of gonorrhea, decreasing infectivity can be accomplished by incubating the organisms with specific antibodies [Ward, 1978; Rice, 1980; Schneider, 1982], divalent cation chelators, such as EDTA [unpublished data, PJH] and pyocins (bacterial viruses) [Morse, 1982]. In addition, patients taking hormonal contraceptives appear to be at decreased risk for infection [McCormack, 1982; Louv, 1989]. Bacteria that have been recently isolated from humans are more infectious those grown in the laboratory [personal communication, Carolyn Deal, Walter Reed Army Institute].

MICROBIOLOGICAL EFFICACY OF CURRENTLY AVAILABLE BARRIER METHODS

The efficacy of spermicides, male condoms, and female condoms to prevent gonorrhea and chlamydial infection have been reviewed by Dr. Stewart (refer to Chapter 7, this volume). A few points should be emphasized. Although studies have been done, differences in study design, small sample sizes, high frequencies of sexual activity, problems with compliance, including measurement of compliance, and self-reported data limit our ability to draw firm conclusions or extrapolate the findings to different populations.

There are clearly trends in these results - spermicides and the male condom, if used consistently correctly, appear to reduce the risk of gonorrhea and chlamydial infection [Reviewed, Cates, 1992]. The diaphragm, the sponge, and the cervical cap deserve further study before such a statement could be made. Concerning the female condom, we have no data on prevention of gonococcal and chlamydial infection.

Unfortunately, the testing of barrier methods in populations at risk for STD and HIV presents some serious ethical issues, especially if the device is incompatible with the male condom. However, for several reasons, the prevention of pregnancy is not an acceptable surrogate for prevention of infectious diseases. First, the window of susceptibility is much smaller with pregnancy than with infection (three to five days each month versus thirty days). Second, STD pathogens are quite variable with respect to infectivity. For example hepatitis B and human papillomavirus are quite infectious, while herpes simplex virus is not, nor is HIV. Third, STD pathogens vary enormously in size. One protozoan is larger than spermatozoa (7um head and 15um tail), but most are many times smaller. The protozoan that causes trichomoniasis is the size of a white blood cell (10-20um), whereas hepatitis B, the smallest sexually transmitted agent, is 42 nm. HIV is 100 -120 nm, depending upon the presence of the envelope. Fourth, since the eukaryotic cell membrane (i.e. the sperm cell membrane) is quite different from the bacterial cell membrane or the viral capsid in non-enveloped viruses (i.e., HPV). Therefore, it is not possible to predict microbicidal efficacy based on spermicidal efficacy. Fifth, if a given compound has spermicidal activity it is unlikely to be specific for sperm. Cell membranes of vaginal and cervical cells could also be damaged, and toxicity and inflammation could mitigate the protective effects of the microbicide by decreasing infectious dose.

SUMMARY

The development of effective barrier products is a global priority of the highest order. Governmental agencies, non-governmental agencies, industry and research institutions must work together to support, coordinate, and synergize the individual and collective efforts in this endeavor. Simultaneously we must implement three critical, complementary strategies including basic research, product evaluation research, and clinical evaluation research. Basic research will be maximized if it is done in the context of multidisciplinary efforts. Product evaluation should be expedited and limited to evaluation of microbiological efficacy and preclinical safety testing. Clinical evaluation should include early evaluation of products for safety as well as larger efficacy trials. Evaluation of N-9 containing spermicides would enable the development of study designs, safety evaluation and efficacy. Such an evaluation would permit the establishment of relevant, effective regulatory guidelines for the approval of microbicides.

It is clear that we have both rationale and strategy for the development of topical microbicides that will neutralize viruses, bacteria, and/or protozoans. This task will not be easy. As obligate pathogens of humans, the etiologic agents of these products are realistic, feasible and attainable. Because of the innate differences between bacteria and host cells, there is a large window in which to find effective compounds which are not toxic to host cells. In addition, because we do not have to worry about cell-associated transmission, safe and effective bactericides may be easier to develop. Although the task is difficult, the rewards are enormous - gonorrhea and chlamydial infection and PID rates are ALL likely to be reduced. Furthermore, prevention and control of bacterial STDs is a critical component of HIV prevention. It is clear that the risk of becoming infected, or infecting others with HIV, is substantially increased if one has an STD. Development and implementation of an STD/HIV prevention strategy based on female controlled barrier methods is a logical step in preventing the spread of HIV. The question we will be answering is not *whether* this will prevent HIV infection, but *what proportion* of HIV infections will it prevent?

REFERENCES

Alexander ER, Harrison HR (1983): Role of *Chlamydia trachomatis* in perinatal infection. Rev Inf Dis 5:713-719.

Apicella MA, et al. (1987): Phenotypic variation in epitope expression of the *Neisseria gonorrhoeae* lipooligosaccharide. Infect Immun 55:1755.

Apicella MA, et al. (1986): Bactericidal antibody response of normal human serum to the lipooligosaccharide of *Neisseria gonorrhoeae*. J Infect Dis 153:520.

Berkeley S (1992): The public health significance of sexually transmitted diseases for HIV infection in Africa. In Chen LC, Sepulveda J, Segal SJ (eds): "AIDS and Women's Reproductive Health." New York: Plenum Press., pp 73-84.

Bowie WR, et al. (1978): Etiologies of post-gonococcal urethritis in homosexual and heterosexual men:Roles of *Chlamydia trachomatis* and ureaplasma urealyticum. Sex Transm Dis 5:151.

Brinton CC, et al. (1978) Uses of pili in gonorrhoea control: Role of bacterial pili in disease, purification and properties of gonococcal pili and progress in the development of a gonococcal pilus vaccine for gonorrhea. In Brooks G et al (eds.): "Immunobiology of *Neisseria gonorrhoeae*." Washington, DC: American Society for Microbiology, pp 155-178.

Brunham RC, Holmes KK, Embree JE (1990): Sexually transmitted diseases in pregnancy. In Holmes KK, Mardh, PA, Sparling PF, Wiesner, PJ. "Sexually Transmitted Diseases" (2nd ed). New York: McGraw-Hill, pp 771-801.

Brunham R, et al. (1981): Epidemiological and clinical correlates of *C. trachomatis* and *N gonorrhoeae* infection among women attending at STD clinic. Clin Res 29:47A.

Cates W, Alexander ER (1988): Sexually transmitted diseases and the fetus. In: Knudsin RB, Falk L, Hopp SS, (eds): "Impact on the Fetus of Parental Sexually Transmitted Disease." New York: New York Academy of Sciences, pp 1-6.

Cates W, Stone KM (1992): Family Planning, sexually transmitted diseases and contraceptive choice: a literature update-- Part I. Fam Plann Perspec 24:75-84.

Chen JC, Bavoil P, Clark VL (1991) Enhancement of the invasive ability of *Neisseria gonorrhoeae* by contact with HecIB, an adenocarcinoma endometrial cell line. In: Achtman M, Kohl P, Marchal C, Morelli G, Seiler A, Theisen B, (eds): "'Neisseriae 1990." Berlin: deGruyter, pp 567-572.

Cotch MF, Vaginal Infections and Prematurity Study Group (1990): Carriage of *Trichomonas vaginalis* is associated with adverse pregnancy outcome. In "Proceedings of the 30th ICAAC." Atlanta: ICAAC, abstract 681.

Datta P, Laga M, Plummer A, et al. (1988): Infection and disease after perinatal exposure to *Chlamydia trachomatis* in Nairobi, Kenya. J Infect Dis 158:524-528.

Dekker NP, Lammel CJ, Mandrell RE, Brooks GF (1990): Opa (protein II) influences gonococcal organization in colonies, surface appearance, size and attachment to human fallopian tube tissues. Microbiol Path 9:19-31.

Edwards LE, Barrada MI, Hamann AA, Hakanson EY (1978): Gonorrhea in pregnancy. Am J Obstet Gynecol 132:637-641.

Eschenbach DA, et al. (1975): Polymicrobial etiology of acute pelvic inflammatory disease. N Eng J Med 293:166.

Expert Committee on Pelvic Inflammatory Disease (1991): Pelvic inflammatory disease: Research directions in the 1990s. Sex Trans Dis 18:46-64.

Fort AL (1989) Investigation of the social context of fertility and family planning: A qualitative study in Peru. Fam Plan Persp 15:88-94.

Grayston JT, Wang SP (1975): New knowledge of chlamydia and the diseases they cause. J Infect Dis 132:87.

Guyman LF, et al. (1978): Altered outer membrane components in serum-sensitive and serum-resistant strains of *Neisseria gonorrhoeae*. In Brooks G, et al. (eds) "Immuno biology of *Neisseria gonorrhoeae*." Washington, DC: American Society for Microbiology, p 139-141.

Haines KA, Yeh L, Blake MS, Cristello P, Korchak H, Weissman G (1988): Protein I a translocatable ion channel from *Neisseria gonorrhoeae* selectively inhibits exocytosis from human neutrophils without inhibiting O2 generation. J Biol Chem 263:945-951.

Handsfield HH, et al. (1974): Asymptomatic gonorrhea in men: Diagnosis, natural course, prevalence and significance. N Engl J Med 290:117.

Harkness AH (1948): The pathology of gonorrhoea. Br J Vener Dis 24:137.

Harrison HR, et al. (1985): Cervical *Chlamydia trachomatis* infection in university women: Relationship to history, contraception, ectopy and cervicitis. Am J Obstet Gynecol 153:244.

Harrison WO, et al. (1979): A trial of minocycline given after exposure to prevent gonorrhea. N Eng J Med 300:1074.

Hitchcock PJ, Boslego J, Joiner KA, Robinson EN Jr (1988): Analyses of the immuno accessibility of H8 antigen and the functionality of H8 specific monoclonal antibody 10. In Poolman JT et al. (eds) "Gonococci and Meningococci." Dordrecht, the Netherlands: Kluwer Academic Publishers, pp 543-552.

Hobson D, et al. (1980): Quantitative aspects of chlamydial infection of the cervix. Br J Dis 56:156.

Holmes KK, et al. (1970): An estimate of the risk of men acquiring gonorrhea by sexual contact with infected females. Am J Epidemiol 91:170.

Holmes KK, et al. (1980): Salpingitis - overview of etiology and epidemiology. Am J Obstet Gynecol 183:893.

Hooper RR, et al. (1978): Cohort study of venereal disease: 1. The risk of gonorrhea transmission from infected women to men. Am J Epi 108:136.

Jephcott AE, et al. (1971): Brief report - *Neisseria gonorrhoeae:* III. Demonstration of presumed appendages to cells from different colony types. Acta Path Micro Scand (B)79:487.

Kinghorn GR, Rashid S (1979): Prevalence of rectal and pharyngeal infection in women with gonorrhea in Sheffield. Br J Vener Dis 55:408.

Klein EJ, et al. (1977): Anorectal gonococcal infection. Ann Intern Med 86:340.

Kuo C-C, et al. (1983): Immunotypes of *Chlamydia trachomatis* isolates in Seattle, Washington. Infect Immun 41:865.

Lebedeff DA, Hochman EB (1980): Rectal gonorrhea in men: Diagnosis and treatment. Ann Intern Med 92:463.

Louv WC, et al. (1989): Oral contraceptive use and the risk of chlamydial and gonococcal infections. Am J Obstet Gynecol 160:396.

Lycke E, et al. (1980): The risk of transmission of genital *Chlamydia trachomatis* infection is less than that of genital *Neisseria gonorrhoeae* infection. Sex Transm Dis 7:6.

Lynch EC, et al. (1984): Studies of porins: Spontaneously transferred from whole cells and reconstituted from purified proteins of *Neisseria gonorrhoeae* and *Neisseria meningitidis*. Biophy J 45:104.

Masi AT, Eisenstein BI (1981): Disseminated gonococcal infection (DGI) and gonococcal arthritis (GCA): II. Clinical manifestations, diagnosis, complications, treatment and prevention. Semin Arthritis Rheum 10:173.

McCormack WM, et al. (1977): Clinical spectrum of gonococcal infection in women. Lancet 2:1182.

McCormack WM, et al. (1982): Effect of menstrual cycle and method of contraception on recovery of *Neisseria gonorrhoeae*. JAMA 247:1292.

McGee ZA, Melly MA, Gregg CR, Korn RC, Taylor-Robinson D, Johnson AP, McCutchan JA (1978): Virulence factors of gonococci: Studies using human fallopian tube organ cultures. In Brooks G et al. (eds) "Immunobiology of *Neisseria gonor rhoeae*." Washington, DC: American Society for Microbiology, p 259-262.

McMillan A, et al. (1981): Chlamydial Infection in Homosexual Men: Frequency of Isolation of *Chlamydia trachomatis* from the urethra, ano-rectum, and pharynx. Br J Vener Dis 57:47.

Merson MM (1993): Slowing the spread of HIV: Agenda for the 1990s. Science 260: 1266-1268.

Meyer TF, van Putten JPM (1989): Genetic mechanisms and biological implications of phase variation in pathogenic neisseriae. Clin Micro Rev:S139-145.

Morrison RP, et al. (1992): Immunology of *Chlamydia trachomatis* infections - immunoprotective and immunopathogenic responses. In Gallin J, Fauci AS, Quinn TC (eds) "Adv Host Def Mech: Sexually Transmitted Diseases." 8:57-84.

Morse SA, Apicella MA (1982): Isolation of lipopolysaccharide mutant of *Neisseria gonorrhoeae* using an R-type bacteriocin from *Pseudomonas aeruginosa*: An analysis of the antigenic and biologic differences. J Infect Dis 145:206-218.

Moulder JW (1991): Interaction of chlamydia and host cells in vitro. Microbiol Rev 55: 143.

Naeye R, Kassane JM (1990): Perinatal diseases: A neglected area of the medical sciences in perinatal diseases. In Naeye RL, et al. (eds) "International Academy of Pathology Monograph." Baltimore: Williams and Wilkins, p 1.

Nugent R, Martin D, Cotch MF, Rettig P (1990): Chlamydial infection and adverse pregnancy outcome: Interim results from a multicenter prospective study. In Bowie WR, Caldwell HD, Jones RP, et al. (eds.): "Chlamydial Infections: Proceedings of the Seventh International Symposium on Human Chlamydial Infections." New York: Cambridge University Press, pp 344-347.

O'Brien JA, et al. (1983): Disseminated gonococcal infection: A prospective analysis of 49 patients; and a review of pathophysiology and immune mechanisms. Medicine 2: 395.

Paavonen J, et al. (1985): Prevalence and manifestations of endometritis among women with cervicitis. Am J Obstet Gynecol 152:280.

Paavonen J (1979): *Chlamydia trachomatis* induced urethritis in female partners of men with nongonococcal urethritis. Sex Trans Dis 6:69.

Pedersen AHB, Bonin P (1971): Screening females for asymptomatic gonorrhea infection. Northwest Med 70:255.

Perry MB, et al. (1977): Lipopolysaccharides of *Neisseria gonorrhoeae*. In Roberts, RB (ed): "The Gonococcus." New York: Wiley, p 285.

Peterson BH, et al. (1979): *Neisseria meningitidis* and *Neisseria gonorrhoeae* bacteremia associated with C6, C7, or C8 deficiency. Ann Intern Med 90:917.

Platt R, et al. (1983): Risk of acquiring gonorrhea and prevalence of abnormal adnexal findings among women recently exposed to gonorrhea. JAMA 250:320.

Rank RG, Sanders MM (1992): Pathogenesis of endometritis and salpingitis in a guinea pig model for chlamydial genital infection. Amer J Path 140:927-936.

Rice PA, et al. (1980): *Neisseria gonorrhoeae* surface antigens: Their interaction with human sera. In Danielsson D, Normark S (eds.) "Genetics and Immunobiology of Patho genic Neisseria." Umeo, Sweden: University of Umeo, p 255.

Rice PA, et al. (1980): Natural serum bactericidal activity against *Neisseria gonorrhoeae* isolated from disseminated, locally invasive and uncomplicated disease. J Immunol 124: 2105-2109.

Schachter J (1990): Biology of *Chlamydia trachomatis*. In Holmes KK, Mardh, PA, Sparling PF, Weisner, PJ. "Sexually Transmitted Diseases" (2nd ed). New York: McGraw-Hill, pp 167-180.

Schneider H, et al. (1982): Immunological basis of serum resistance of *Neisseria gonorrhoeae*. J Gen Micro 128:13-22.

Schulz KF, Schulte J, Berman S (1992): Maternal health and child survival: Opportunities to protect both women and children from the adverse consequences of reproductive tract infections. In Germain A, Holmes KK, Piot P, Wasserheit JN (eds) "Reproductive Tract Infections: Global Impact and Priorities for Women's Reproductive Health." New York: Plenum Press, pp 145-182.

Shaw JH, Falkow S (1983): Model for invasion of human tissue culture cells by *Neisseria gonorrhoeae*. Infect Immun 56:1625.

Sparling PF (1990) Biology of *Neisseria gonorrhoeae*. In Holmes KK, Mardh, PA, Sparling PF, Weisner, PJ. "Sexually Transmitted Diseases" (2nd ed). New York:

McGraw-Hill, pp 131-147.

Stamm WE, Cole B (1986): Asymptomatic *Chlamydia trachomatis* urethritis in men. Sex Trans Dis 13:163.

Stamm WE, et al. (1984): *Chlamydia trachomatis* urethral infections in men: Prevalence, risk factors, and clinical manifestations. Ann Intern Med 100:47.

Swanson J, et al. (1971): Studies on gonococcus infection 1. Pili and zones of adhesion: Their relation to gonococcal growth patterns. J Exp Med 184:886.

Sweet RI, et al. (1983): Failure of beta-lactam antibiotics to eradicate *Chlamydia trachomatis* in the endometrium despite apparent clinical cure of acute salpingitis. JAMA 250:241.

Sweet RI, Landers DV, Walker C, Schachter J. (1987): *Chlamydia trachomatis* infection and pregnancy outcome. Am J Obstet Gynecol 156:824-833.

Taylor-Robinson D, et al. (1974): Effect of *Neisseria gonorrhoeae* on human and rabbit oviducts. Br J Vener Dis 50:279.

Todd WJ, Wray GP, Hitchcock PJ (1984): Arrangement of pili in colonies of *Neisseria gonorrhoeae*. J Bact 159:312-320.

Upchurch DM, Weisman CS, Shepherd M, Brookmeyer R, Fox R, et al. (1991): Interpartner reliability of reporting of recent sexual behaviors. Am J Epi 134: 1159-1165.

Wager EA, Schachter J, Bavoil P, Stephens RS (1990): Differential human serological response to two 60,000 MW *Chlamydia trachomatis* antigens. J Infect Dis 162:922-927.

Wallin J (1974): Gonorrhea in 1972: A 1-year study of patients attending the VD unit in Uppsala. Br J Vener Dis 51:41.

Ward ME, et al. (1978): The surface properties of *Neisseria gonorrhoeae*: Determinants of susceptibility to antibody complement killing. J Gen Micro 198:205-212.

Washington AE, Katz P (1991): Cost of and payment source for pelvic inflammatory disease: Trends and projections 1983 through 2000. JAMA 266:2565-2569.

Wasserheit JN (1992) Epidemiological synergy: Interrelationships between human immunodeficiency virus infection and other sexually transmitted diseases. In Chen LC, Sepulveda J, Segal SJ (eds.): "AIDS and Women's Reproductive Health." New York:

Plenum Press, Chapter 5.

Wasserheit JN, et al. (1986): Microbial causes of proven pelvic inflammatory disease and efficacy of clindamycin and tobramycin. Ann Intern Med 104:187.

Wiesner PJ, et al. (1973): Clinical spectrum of pharyngeal gonococcal infections. N Engl J Med 288:181.

Young JD-E, et al. (1983): Properties of the major outer membrane protein from *Neisseria gonorrhoeae* incorporated into model lipid membranes. Proc Natl Acad Sci USA 80:3831.

Zhang JP, Stephens RS (1992): Mechanism of attachment of *Chlamydia trachomatis* to eukaryotic host cells. Cell 69:861-869.

DISCUSSION

Felicia Stewart: I had the privilege of hearing Kathleen Toomey from the CDC talk about this subject last week and she said something which I found extraordinarily moving. She said that if we could control chlamydia we could potentially reduce the incidence of new cases of HIV new cases by as much as 40 percent. In my view, that was just flabbergasting since I don't know of any other way to prevent HIV by 40 percent. There could be a one-month sexual abstinence festival every year in the world and at the very beginning of it an azithromycin day. Could you comment on what you think the impact of these approaches might be?

Penelope Hitchcock: Felicia, your question is much better than my answer. Let me say that, first of all, I completely agree with the sentiment of Kathleen's assessment. David Sokal at the AIDS meeting in Amsterdam using both a stochastic and a nonstochastic model, postulated that within 20 years we could decrease the prevalence of HIV between 46 and 75 percent. I think it is extraordinary. As I said, I think it's clear that this is the most efficient, responsible, and immediate thing we can do to prevent the spread of HIV. How much of it would we prevent? This is a relatively difficult infection to transmit, which is true of the naturally occurring veterinary models as well. There are three things that seem to be driving the epidemic. One is anal intercourse. The CD4-like receptors on the rectal epithelium are extremely efficient in transmitting this bug, especially with the trauma that comes from anal intercourse. Intravenous exposure is also an extremely efficient way to transmit the infection. In the United States, the epidemic is growing fastest among intravenous drug users and their partners. The third factor driving this epidemic is other sexually transmitted disease. One problem with mass treatment has to do with the antibiotic resistance of neisseria gonorrhea. We already have azithromycin-resistant GC. We are planning to do a very large mass treatment study in Africa looking at control of trichomonas in three different villages in three countries. For a number of reasons we will not look primarily at STDs in that scenario, but we will look at parameters related to mass treatment. I'm hoping that we can have home diagnostic screening in which a urine sample is used to give someone an initial indication of infection and then it is followed by confirmatory testing and treating. In developing countries we may be simply looking at mass treatment. I think that this needs to be done right away. Immediately.

Gabriel Bialy: Penny, when we develop spermicides, we have one target to kill. When you go after a number of targets, obviously the strategy will change. I was very much pleased to hear that a strategy is being evolved by the working group. Could you very briefly state some of the crucial elements of this strategy?

Penelope Hitchcock: These organisms are different and yet there are similarities. Because of the difference of the bacterial cell surface and the host cell surface, the window between killing these bugs and hurting the host cell is much larger than it is for either trichomoniasis, which has a eukaryotic cell membrane, or for a virus that is cell-associated. We can probably neutralize free virus fairly easily but it will be an enormous challenge to develop something to kill virus inside cells and not kill non-target eukaryotic cells. I think that we're going to have a bactericide much sooner than we're going to have a virucide. I also think that the development of this will allow us to protect women against gonococcal and chlamydial infection, and, given the prevalence of these diseases, to impact on HIV transmission. Although it's disappointing to think that we may not be able to get the very virus that we are all most worried about, I don't think that means that we are immobilized technologically. Again, I think this partnership with the private and the public sector is a critical one and we've got to nurture it and move forward with research and product development.

PART V: CLINICAL STUDIES ON THE EFFECTIVENESS OF BARRIERS IN PREVENTING STDs

17

FOUR FAMILY HEALTH INTERNATIONAL STUDIES ON BARRIER METHODS AND SEXUALLY TRANSMITTED INFECTIONS

Ronald E. Roddy

Contraception and STD/HIV Unit
Contraceptive Use and Epidemiology Division
Family Health International
Research Triangle Park, NC 27709

This paper will present a summary of four studies conducted by Family Health International from 1986 through 1992 concerning condom use, nonoxynol-9 (N-9) use, and sexually transmitted infections (STI). For two of the studies, the primary interest was cervical infection with *N. gonorrhoeae* and *C. trachomatis*. The major focus for the other two studies was HIV infection. We were also able to determine the rates of yeast infections in all four studies and genital ulcers in three. Rate ratios (RR) are used to show the effect of a particular method on preventing infection. It is calculated by dividing the rate of infection for the barrier method group (number of infections/number of person-time) by the rate of infection for the reference group. A rate ratio less than 1 indicates a protective effect, a rate ratio of 1 indicates no effect, and a rate ratio larger than 1 indicates that the barrier group was more likely to become infected.

The first study was a randomized controlled trial conducted in 1986 in Bangkok, Thailand among 492 massage parlor workers [Rosenberg, et al., 1987]. Gonorrhea and chlamydial infection were considered separate outcomes. Women were randomly assigned to either an N-9 (1 gm) sponge group or a control group. The women were examined weekly for six weeks. They were instructed to use a sponge during every sexual intercourse. The results indicate that the N-9 sponge provided some protection against gonorrhea and chlamydial infection. For gonorrhea, the RR was 0.31 and the 95 percent confidence interval (CI) was (0.16, 0.8) or about a 70 percent reduction in the rate of infection. For chlamydial infection, the RR was 0.67, 95 percent CI: (0.42, 1.1),

Barrier Contraceptives: Current Status and Future Prospects, pages 245–248

about a 30 percent reduction in the rate. The study also demonstrates an increase in the rate of vaginal yeast infection; RR=2.76, 95 percent CI: (0.96-7.98). There was no reported increase in genital ulcers.

The second study was also a randomized controlled trial conducted in Bangkok, Thailand among 343 massage parlor workers [Niruthisard, et al., 1992]. The women were randomized to receive condoms and N-9 films (70 mg) or condoms and placebos. The women were examined weekly for nine weeks. The results indicate that the N-9 film provided some protection from cervicitis (gonorrhea and chlamydial infection combined). The overall reduction in the rate of infection was about 25 percent; RR=0.75, 95 percent CI: (0.5, 1). However, when comparing women using N-9 for more than 75 percent of their coital acts with women who used placebo with a similar frequency, N-9 provided better protection and reduced the rate of infection by approximately 40 percent; RR=0.6, 95 percent CI: (0.38, 0.95). Use of condoms was also a part of the study, and their use was distributed similarly in the N-9 and placebo groups. Adjusting for condom use did not change the estimates for N-9 protection. Because placebo condoms were not used, the comparison to assess condom protection was between poor (0-50 percent) condom use and better (>75 percent) condom use. Comparing poor condom use with better condom use demonstrated a reduction in the cervicitis rate of about 55 percent; RR=0.45, 95 percent CI: (0.31, 0.65). It must be remembered that this estimate is not made from randomized groups, and there is no placebo with which to compare. Therefore, the estimate is probably affected by two biases. Selection bias would cause overestimation by an unknown amount, and having no placebo to compare with would cause an underestimation by some unknown amount. There was no increase in the rate of yeast infections or genital ulcers during the study.

A cohort study was conducted in Lusaka, Zambia among HIV-discordant couples with one partner HIV infected and one partner uninfected [Feldblum, et al., 1992]. The outcome of interest was HIV infection. Condoms and N-9 products (gel, film, suppository) were made available to couples. About 114 couples contributed follow-up time to the study. Only 13 seroconversions occurred among the couples. Comparing couples that used N-9 for every intercourse with couples who used N-9 less than 100 percent of the time, the RR was 0.5, 95 percent CI: (0.1, 3.9) for women and RR=2.2, 95 percent CI: (0.4, 11.6) for men. Condom use was also evaluated using the same comparisons, and none of the women who used condoms during each intercourse became infected. However, men who reported 100 percent condom use appeared to have an increase in the rate of infection; RR=1.4, 95 percent CI (0.2, 11.4). We think the apparent increase in the rate of infection for men for both N-9 and condom use is the result of sexual activity apart from that of the couple. The men may have used condoms or N-9 at home with their partner they knew to be HIV infected, but may not have used them with other sexual partners of unknown serostatus. The rates of yeast infection and genital ulcers were not increased by N-9 use in the study.

Another cohort study was conducted among 273 HIV seronegative women with multiple partners in Yaounde, Cameroon [Zekeng, et al., 1993]. In this study, condoms and N-9 (100 mg) suppositories were available. Each woman chose for herself which products to try and no randomization took place. There were 19 seroconversions during the study. The results indicate that N-9 reduced the rate of infection considerably for women who used N-9 for at least 67 percent of their coital acts when compared with women who were less consistent users; RR=0.2, 95 percent CI: (0.1, 0.7). Condom protection was also estimated. When comparing women who used condoms for at least 63 percent of their coital acts with women who used condoms less frequently, the reduction in the rate of seroconversion was 20 percent; RR=0.8, 95 percent CI: (0.2, 2.9). Women who were already using condoms at admission apparently received more protection than women who were not regularly using condoms at the time of their enrollment; RR=0.5 and RR=1.4, respectively. Yeast infections and genital ulcers were not associated with N-9 use during the study.

TABLE 1. Summary table of the effects of nonoxynol-9 (N-9) and condoms on cervicitis, HIV, yeast infection, and genital ulcers.

Reference	Cervicitis	HIV	Yeast	Ulcers
Rosenberg 1987	GC, RR=0.31 CT, RR=0.67		RR=2.76	
Niruthisard 1992	<u>N-9</u> RR=0.75 >75%, RR=0.6 <u>Condoms</u> >75%, RR=0.45		No increase	No increase
Feldblum 1992		<u>N-9</u> ♀, RR=0.5 ♂, RR=2.2 <u>Condoms</u> ♀, RR=0.0 ♂, RR=1.4	No increase	No increase
Zekeng 1993		<u>N-9</u> RR=0.2 <u>Condoms</u> RR=0.8	No increase	No increase

RR = rate ratio, GC = gonorrhea, CT = chlamydial infection, >75 percent = method used for >75 percent of coital acts, ♀ = female, ♂ = male.

In summary, these four studies conducted by FHI have demonstrated some reduction in the rates of cervical gonorrheal and chlamydial infection and HIV infection among women who use condoms or intravaginal nonoxynol-9. The evidence is more solid for protection from bacterial infection than for viral infection. Further randomized controlled trials are needed to better measure the effect of N-9 on the reduction of HIV infection in women.

ACKNOWLEDGEMENTS

This work was supported in part by Family Health International and the US Agency for International Development. The contents do not necessarily reflect USAID policy.

REFERENCES

Feldblum PJ, Hira SK, Godwin S, Kamanga J, Mukelabai G (1992): Efficacy of spermicide and condom use by HIV-discordant couples in Zambia. VII International Conference on AIDS/III STD World Congress, Amsterdam, The Netherlands 19-24 July 1992, Abstract WeC 1085.

Niruthisard S, Roddy RE, Chutivongse S (1992): Use of nonoxynol-9 and reduction in rate of gonococcal and chlamydial cervical infections. Lancet 339:1371-1375.

Rosenberg MJ, Rojanapithayakorn W, Feldblum PJ, Higgins JE (1987): Effect of the contraceptive sponge on chlamydial infection, gonorrhea and candidiasis: a comparative clinical trial. JAMA 257:2308-2312.

Zekeng L, Feldblum PJ, Oliver RM, Kaptue L (1993): Barrier contraceptive use and HIV infection among high-risk women in Cameroon. AIDS 7:725-731.

18

PREVENTION OF VAGINAL TRICHOMONIASIS BY COMPLIANT USE OF THE FEMALE CONDOM*

David E. Soper, Donna Shoupe, Gary A. Shangold, Mona M. Shangold, Jacqueline Gutmann, and Lane Mercer

Department of Obstetrics and Gynecology
Medical College of Virginia
12th and Marshall Streets
Richmond, VA 23298 (DES)
University of Southern California (DS)
University of Chicago (GAS, LM)
Hahnemann University (MMS)
Yale University (JG)

INTRODUCTION

The female condom consists of a soft, loose-fitting polyurethane sheath and two diaphragm-like, flexible polyurethane rings. One of the rings lies inside the sheath and serves as an insertional mechanism and internal anchor. The other ring forms the outer edge of the sheath and remains outside the vagina once the condom is in place. This outer ring also protects the woman's labia and the base of the penis, providing more physical coverage for both the man and the woman than the conventional male condom.

A number of laboratory studies have found that male condoms provide an impervious barrier to most sexually transmitted pathogens. In experimental settings, male condoms have been shown to be effective barriers against herpes simplex virus, *Chlamydia trachomatis*, cytomegalovirus and the human immunodeficiency virus (HIV) [Cates and Stone, 1992]. The female condom, likewise, has been shown to be impermeable to both cytomegalovirus and HIV [Drew, et al. 1990]. Several studies suggest that the male

*Reprinted from Sexually Transmitted Diseases, May-June 1993, Published by JB Lippincott Company, Vol 20, No 3, pages 137-139.

Barrier Contraceptives: Current Status and Future Prospects, pages 249–253
Published 1994 Wiley-Liss, Inc.

condom protects women against some sexually transmitted diseases including herpes, cervical gonorrhea, cervical chlamydia, vaginal trichomoniasis and genital ulcers [Cates and Stone, 1992].

The purpose of this study was to determine if appropriate use of the female condom decreased the rate of recurrent vaginal trichomoniasis in previously diagnosed and treated women.

MATERIALS AND METHODS

Sexually active women at least 18 years of age with documented vaginal trichomoniasis were recruited from the gynecology outpatient clinics of the participating urban medical centers. Subjects attending these clinics were predominantly indigent minority women. Subjects underwent pelvic examinations prior to enrollment in the study and again 45 days later at the completion of the study. The diagnosis of vaginal trichomoniasis was made by microscopic examination of a saline slide of the vaginal secretions showing motile flagellated organisms slightly larger than white blood cells. The slide was prepared by dipping a cotton tipped applicator into the vaginal secretions and suspending the secretions in a drop or two of saline previously placed on the glass slide. Microscopy using 400X magnification was used to scan at least four high power fields. Patients were excluded from the study if they relied upon another barrier contraceptive, were pregnant or had a clinical diagnosis of pelvic inflammatory disease. All subjects were treated with a single two gram oral dose of metronidazole while in the physician's office. They were advised about the risk of reinfection and offered the option to have their partners evaluated, treated and counseled that barrier protection should be used against reinfection.

Like male condoms, to be effective, the female condom must be put on prior to genital contact, must remain intact and, most importantly, must be used consistently and correctly [CDC, 1988]. Since the objective of the study was to evaluate whether or not the device, if properly used, protected against reinfection, it was important to enroll subjects into the user group that would consistently use the device. For this reason, subjects were not randomly allocated to the user and control groups. Each subject was shown the WPC-333 female condom and asked if she would use it every time she had intercourse during the study period. If she responded that she would be compliant, the subject was enrolled into the user group. If the subject felt that she could not be compliant, she was enrolled into the control group. The user group was instructed to use the WPC-333 female condom with each act of intercourse over the next 45 days. All subjects kept a diary of the number of coital episodes in which they participated. In addition, they noted the number of times genital contact occurred prior to the insertion of the female condom.

The frequency of vaginal trichomoniasis in each group at the end of the study was compared using a single-tailed Fisher's exact test. The median test was used to compare the number of genital contacts prior to insertion of the condom between infected and noninfected non-compliant users.

RESULTS

One hundred twenty-six subjects were entered into the study. Of these, 22 subjects did not satisfactorily complete the study for the following reasons: 19 (12 controls, 7 users) were lost to follow-up; two users did not have a sexual encounter during the study period; and one user reported that her partner used male condoms at the same time as she used the WPC-333 female condom.

One hundred four subjects completed the study. For analyses, the subjects were divided into three groups: controls, compliant users, and non-compliant users. Noncompliant users had unprotected intercourse at least once during the study period (average number unprotected genital contacts = 5.1). Fifty controls and 54 users completed the study. Of the 54 users, 20 were compliant (used the WPC-333 female condom with each act of sexual intercourse throughout the study period).

There was no difference among the three groups for the following variables: age, number of partners during the study period, total number of episodes of sexual intercourse during the study period, and likelihood of their partners receiving treatment for trichomoniasis (Table 1). Patients compliant with their use of the WPC-333 female condom were less likely to become reinfected with trichomonas than patients with either noncompliant use of the female condom (p = 0.09) or the control group (p = 0.08) (Table 2). Interestingly, we noted that the number of genital contacts (including pre-ejaculatory coitus) prior to insertion of the female condom was significantly higher in the noncompliant group of women reinfected with trichomonas when compared to those noncompliant women not reinfected (12.25 ± 11.98 vs 4.23 ± 4.58 [p = 0.03]).

TABLE 1. Group Characteristics

Variable	Controls (n=500	USERS Non-compliant (n=34)	 Compliant (n=20)
Age (years)*	31.4 ± 8.9	28.8 ± 7.3	29.5 ± 6.6
No. of Partners*	1.1 ± 0.3	1.2 ± 0.5	1.1 ± 0.3
Total No. Coitus*	12.1 ± 10.3	15.9 ± 11.7	20.5 ± 14.6
Partners treated	17 (34%)	10 (29.4%)	6 (30%)

*mean ± SD

TABLE 2. Recurrent Vaginal Trichomoniasis by Group

	Controls (n=50)	Non-compliant (n=34)	Compliant (n=20)
Number reinfected	7 (14%)	5 (14.7%)	0

DISCUSSION

No barrier device under the woman's control currently exists that is approved as effective in the prevention of sexually transmitted diseases (STDs). Only the latex male condom is recognized as effective against the transmission of STDs. This study is only the second to evaluate the utility of condoms in the prevention of vaginal trichomoniasis. Rosenberg, et al., in a case-control study have shown that use of the male condom decreases the risk of vaginal trichomoniasis in women (adjusted odds ratio = 0.33). Our investigation confirms the utility of a condom, in this case, the female condom, in preventing reinfection with trichomonas. Not only is perfect compliance protective against infection, but the risk of infection seems to be related directly to the number of times that one is exposed to sexual contact without the device. This last observation suggests that, despite no difference in the reinfection rates between nonusers and noncompliant users, a "dose response" relationship exists in that infected partial compliers had significantly more genital exposures than uninfected partial compliers. Indeed, the number of genital exposures (mean = 12.25) of the infected noncompliers was similar to the total number of unprotected coital episodes in the controls (mean = 12.1), suggesting that their risk of reinfection was similar.

Our study has several limitations. We could have conceivably chosen women at lowest risk for reinfection by assigning only those subjects likely to be compliant users of the female condom to the user group. However, variables that should be associated with the risk of reinfection (number of sexual partners, treatment of sexual partners and number of sexual acts) were not significantly different among the three groups (Table 1). Although many times analyses are based upon an "intention to treat," we preferred to evaluate outcome based upon the compliant use of the device. Reinfection rates were similar in the control and user groups (14 percent versus 9.3 percent) if compliance was not considered. Although we did not reevaluate the subjects after treatment and before further sexual activity, we would expect only an occasional patient to have failed treatment with the single two gram dose of metronidazole, reported to have a median cure rate of 96 percent [Lossick, 1982]. In addition, examination of the saline wet prep has a sensitivity of approximately 60 percent and, therefore, we may have missed patients reinfected with a smaller number of microorganisms.

While not statistically significant in the traditional sense ($P < 0.05$), the female condom, when appropriately used, was effective in preventing women from becoming reinfected with trichomonas. Further study of the female condom in the prevention of not only vaginal trichomoniasis but also gonorrhea and Chlamydia, as well as other STDs, should be performed.

REFERENCES

Cates W, Stone KM (1992): Family planning, sexually transmitted diseases and contraceptive choice: A literature update - Part I. Fam Plan Perspect 24:75-84.

CDC (1988): Condoms for prevention of sexually transmitted diseases. MMWR 37:133.

Drew WL, Blair M, Miner RC, Conant M (1990): Evaluation of the virus permeability of a new condom for women. Sex Transm Dis 17:110-112.

Lossick JG (1982): Treatment of *Trichomonas vaginalis* infections. Rev Infect Dis 4(Suppl):S801-S818.

Rosenberg MJ, Davidson AJ, Chen JH, Judson FN, Douglas JM (1992): Barrier contraception and sexually transmitted diseases in women. Am J Pub Health, 82: 669-74.

19

CONDOM USE EVALUATION IN HOTEL AND MOTEL ROOMS

Ernesto Guerrero, Santo Rosario, and Lydia Bond

Latin American Union Against STD/AIDS
ULACETS
Apartado Postal 1139
Santo Domingo, Dominican Republic (EG)
Centro de Orientacion e Investigacion Integral
COIN (SR)
Pan American Health Organization
PAHO (LB)

INTRODUCTION

The Dominican Republic is one of several countries that began programs several years ago for the promotion, advertisement and distribution of condoms for the prevention of sexually transmitted diseases (STDs) and AIDS. The promotional strategy focused on a number of different population groups and used a number of different approaches including:

- wide-scale promotion and advertising, through mass media such as television, radio, press, billboards and distribution of printed material
- free distribution of condoms among groups practicing high risk behaviors, such as women prostitutes, homosexuals, prisoners, and the resident sugar plantation population
- KAP usage and methodology research on the proper use of condoms carried out among different population groups
- a 1987 health ordinance requiring hotels and motels to place condoms in rooms, irrespective of any request by the customer.

Barrier Contraceptives: Current Status and Future Prospects, pages 255–259

Due to cultural and economic reasons, there exist in Santo Domingo (the capital city of the Dominican Republic) more than 100 hotels and motels used primarily for having casual sex. A typical hotel averages 50 clients a day. Prices are charged by the hour or by the day, and there is great variation in facilities and commodities available (air conditioned, jacuzzi, etc.). Clients do not have to register; in the motels there is no direct visual contact, all transactions such as services, beverages, etc. take place through a small window in the wall.

Because the sanitary sewer system has been deficient for years, most people never throw things into the toilet (including used condoms). Consequently, used condoms can be retrieved from trash for study. Since 1988, a sample of these establishments has been surveyed in order to determine condom use prevalence and condom breakage among their clients.

OBJECTIVES

The objectives of the research activity were as follows:

- determine the prevalence of condom usage among the customers of hotels and motels in the city of Santo Domingo
- develop a profile of condom users
- ascertain trends in condom usage
- determine the effects of the presence of printed materials explaining the proper use of condoms on usage prevalence rates
- compare breakage rates for condoms with different manufacturing dates.

METHODOLOGY

The study population consisted of the customers of five hotels and five motels in the city of Santo Domingo. The identity of the clients was completely unknown to the researchers and the clients were unaware that they were participants in this study. The selection of sites was made with an eye toward securing business establishments that would represent a range of room rates, as well as businesses from different geographic locations.

To encourage optimal cooperation, the owners or managers of the establishments received a letter followed by visits. One week prior to the observation, condoms were distributed to the businesses selected. On the day preceding the activity, printed materials (instructional literature and/or stickers) on proper condom use were randomly distributed. The proprietors were instructed to place the condoms next to the soap and towel in each room, in accordance with the health ordinance.

Preparation of the Instrument. In order to record the information observed, a data collection instrument was developed, along with instructions regarding its use. Another document was prepared with instructions relating to the encoding process.

Selection and Training of Surveyors. A two-member team, consisting of one health inspector and one health professional (educator or social worker), was assigned to each of the establishments. All team members received training in the use of the data gathering instrument. Every effort was made to ensure that the health inspector would be in charge of carrying out the inventory of the condoms and that he and the health professional would reach an agreement on the data observed. The health professional was to transcribe the information onto the form.

Pre-test. In order to standardize the completion and validation of the instrument, a pre-test was carried out in which surveyors received training in the actual hotels and motels.

Information Gathering. The information gathering activity in Santo Domingo was to be carried out over a four-day period. (As a result of logistical problems, one business was surveyed on the fifth day.) Each selected establishment was visited and surveyed twice over a period beginning at 4 p.m. and ending at midnight. Once the rooms were vacated and the customers had left, the survey team would enter the room before the cleaning staff and carry out a detailed inspection, in order to ascertain what had happened to the condoms that had been placed in the room. If two rooms were vacated at the same time, the team would split up. If there were opened condom packets in one of the rooms, surveyors were required to reach a consensus. Prior to beginning data collection, surveyors visited empty rooms to ensure that the two condoms of the specified brands were being placed in the rooms. They were also to ensure that the printed materials explaining the proper use of condoms were actually being distributed and placed in the business establishments and to replace such printed materials when missing or damaged.

Information Verification. On days when observation activities were carried out, one supervisor and one coordinator would visit the surveyors in the establishments and verify the information being collected. During the data encoding process, the interviewers were consulted on any particular information requiring clarification.

Selected Variables. All of the questions on the survey instrument were cross-tabulated against a profile consisting of the variables listed in Table 1.

TABLE 1. Independent Variables

CONDOMS:	Used
	Not Used
	Broken
	Taken
	Left behind

A total of 19 other variables were researched; they are shown in Table 2.

TABLE 2. Other Variables Studied

Country
Business establishment (hotel/motel)
Type of condoms (manufacture date)
Time of observation
Day of observation
Social category (price per room)
Presence of instructional materials
Length of stay
Consumption of alcoholic beverages
Number of condoms or wrappers
Presence of other (non-study) condoms
How far open condom was unrolled
Open condoms that were broken
Open condoms that were used
Open condoms that were not used
Condoms taken
Condoms left behind intact
Presence of semen
Knot at the (open) end of condom

RESULTS

The total number of surveys carried out for this study was 949. The total number of rooms available among all establishments selected for the survey was 385 (202 in hotels and 183 in motels).

Usage Prevalence

The prevalence of condom usage was 12.0 percent for the customers of the hotels and motels of Santo Domingo.

Of the condoms placed in the rooms, 60.2 percent were left intact.

User Profile

Some type of alcoholic beverage was consumed by 60 percent of the users.

Usage prevalence varied among establishments, from 4.4 percent to 27.8 percent.

Customers in the cheaper hotels had a lower rate of usage prevalence than the customers at more expensive hotels or motels.

One out of every three users of condoms used more than one.

Printed Materials

One out of every four customers took the instructional pamphlets away with them.

In those establishments not provided with any information, the percentage of usage was lower.

Condoms

The average rate of breakage was 8.6 percent of the total number of open condoms.

A very low percentage of users tied a knot in the condom before throwing it away.

Condom breakage was more frequent among customers of the lower social categories.

Customers preferred to use the brand of condom found in the room. Only 8.6 percent used their own condoms.

Breakage rates were 10 times higher among condoms manufactured in 1985 than among those manufactured in 1989.

Trends

The prevalence of condom usage in Santo Domingo is decreasing:

18 percent in 1988 and 1989
12 percent in 1990

Conclusion

The method of condom use evaluation in hotels and motels could be implemented in other developing countries with similar characteristics. The methodology may serve to:

- establish a system of epidemiological surveillance
- improve condom quality control
- evaluate impact of mass media campaign
- evaluate educational materials
- determine STD prevalence (through the lab-examination of used condoms found)
- compare with self report.

20

EFFICACY OF NONOXYNOL-9 CONTRACEPTIVE SPONGE USE IN PREVENTING HETEROSEXUAL ACQUISITION OF HIV IN NAIROBI PROSTITUTES

Elizabeth Ngugi, Joan Kreiss, King Holmes, Jackoniah Ndinya, Peter Waiyaki, Pacita L. Roberts, Irene Rumijo, Rose Sajabi, Joyce Kimata, Thomas R. Fleming, Aggrey Anzala, Donna Holton, and Francis Plummer

University of Nairobi, Departments of Community Health and Medical Microbiology, Nairobi (EN, JN, IR, RS, AA)
University of Washington, Departments of Epidemiology and Biostatistics, Seattle, WA (JK, KH, PR, TF)
Kenya Medical Research Institute, Nairobi (PW, JK)
University of Manitoba, Microbiology Department, Winnipeg, Canada (DH, FP)

Sexual transmission is the most frequent route of transmission of HIV worldwide. Given the apparent limitations of educational programs in controlling this epidemic, other options should be explored. The male condom is one such option. However, condoms are not always used due to religious, cultural, economic or administrative issues, or due to a lack of gender equality. In addition, condoms do not offer 100 percent protection. Condom breakage rates with vaginal intercourse are frequently reported to be between 1 and 5 percent, although higher rates have been reported.

The objective of this study was to determine the efficacy of the N-9 contraceptive sponge in preventing sexual acquisition of the human immunodeficiency virus (HIV) in a heterosexual population. The study design was a prospective randomized placebo-controlled trial enrolling commercial sex workers attending Pumwani clinic in Nairobi. A total of 138 women were recruited, with 74 assigned to N-9 and 64 assigned to the

Barrier Contraceptives: Current Status and Future Prospects, pages 261–264

placebo. The sample size and duration of follow-up were expected to detect a 75 percent decrease in HIV infection in N-9 users compared with those using placebo, with a power of 80 percent (α= 0.05).

Woman were randomly assigned to the two groups. Physical examination were carried out on all subjects in both groups at entry. They were then requested to come back at monthly intervals, or more frequently if necessary, for follow-up. There was no significant difference between the groups in baseline, demographic data, duration of prostitution, mean number of sexual partners per week, number of partners who used condoms, sex during menses, and oral contraceptives use (Table 1).

TABLE 1. Comparison of the N-9 sponge and placebo groups at enrollment

	N-9 Sponge (n=74)	Placebo (n=64)
Mean age (years)	29.4	30.1
Mean duration of prostitution (years)	4.6	4.3
Mean no. of sex partners per week	42	56
Mean % of partners who use condoms	54	51
Number who had sex during menses (%)	10 (14%)	3 (5%)
Number using oral contraceptives (%)	21 (28%)	14 (22%)

The prevalence of genital ulcers at any site and on the vulva specifically was found to be lower in the placebo group than the N-9 groups. However, vulvitis, genital warts, vaginal discharge, cervical discharge, cervical ectopy, and uterine adnexal tenderness were similar in both groups. All were effectively treated for any of these STDs. There was no difference in sexual practices and other risk exposure in the two groups (Table 2). The women were counselled to use condoms with every sexual partner properly and refuse to have sex with clients who declined, regardless of their group membership. Condoms were provided free of charge.

RESULTS

There was a 19 percent loss to follow-up in the N-9 group and 13 percent in the placebo group. The most frequent reason for not returning for follow up was having left Nairobi. In this group of women who were in a randomized, placebo-controlled trial and are at high risk of exposure to HIV, N-9 was ineffective in preventing HIV transmission even though there was 60 percent condom use. There was 3.3 fold increased risk of both vulvitis and genital ulcers in N-9 users compared to placebo. It is noted that the incidence of candida vaginitis was increased in a study among prostitutes in Bangkok, Thailand [Ngugi, 1982].

TABLE 2. Sexual Practices and Other Risk Exposures in N-9 Sponge and Placebo Groups During Follow-up

	N-9 Sponge (n=60)	Placebo (n=55)
Mean duration of follow-up in months (range)	14 (1-45)	17 (1-46)
Mean number of visits	9	9
Mean number of sex partners per week	34	34
Mean percentage of partners using condoms (range of percentages)	58 (0-100)	63 (0-100)
Number having sex during menses (%)	5 (8)	7 (13)
Number who complained about N-9 or placebo (%)	28 (47)	4 (7)
Mean percentage who complied with N-9 sponge or placebo use (range)	81 (17-100)	90 (29-100)
Number receiving blood transfusion (%)	none	1 (2)
Number who received injections at other clinics (%)	13 (22)	19 (35)
Number using oral contraceptives (%)	19 (32)	19 (35)

CONCLUSION

Genital ulcers and vulvitis occurred with increased frequency in N-9 sponge users. We were unable to demonstrate that N-9 sponge use was effective in reducing the risk of HIV infection among highly exposed women. There is an urgent need to develop safe and effective HIV prevention methods for women in view of the explosive spread of HIV infection and its subsequent effect on the social and economic welfare of our communities.

REFERENCES

Ngugi EN, Plummer FA, Simonsen JN, et al. (1982): Prevention of HIV transmission through health education and promotion of condoms among high risk prostitutes. Lancet 2:887-890.

Rosenberg MJ, Rojanapithayakorn W, Feldblum PJ, Higgins JE (1987): Effect of the contraceptive sponge on chlamydial infection, gonorrhoea, and candidiasis: a comparative clinical trial. JAMA 257:2308-2312.

Hooton TM, Hillier S, Johnson C, Roberts PL, Stamm SE (1991): *Escherichia coli* bacteriuria and contraceptive method. JAMA 265:64-69.

21

PROTECTAID, A NEW VAGINAL CONTRACEPTIVE SPONGE WITH ANTI-STD PROPERTIES

Alexandre Psychoyos

Laboratoire de Physiologie de la Reproduction,
Hôpital Bicêtre, Bâtiment INSERM,
80 rue Général Leclerc
Kremlin-Bicêtre, France 94276

INTRODUCTION

Contraceptive methods that also prevent sexually transmitted diseases (STDs) are at the present urgently needed. This crucial need emerges not only from the AIDS pandemic but also from the tremendous recrudescence of STDs, estimated at 250 million new cases per year [Crosignani, et al., 1992]. Genital infections by STD pathogens are considered to promote the ability of microorganisms to adhere to mucosa cells and to increase further the risk of an HIV contamination.

To date, the only method that has been proven to be effective for the prevention of STDs, including AIDS, is the male condom. However, the condom is unacceptable to many men and women. It can be used only for one act of sexual intercourse, is placed just before coitus, and women have little control over its use.

Vaginal sponges impregnated with spermicidal compounds that also have anti-STD properties show several advantages. They are entirely under the woman's control, can be placed several hours before coitus, and allow repeated acts of sexual intercourse for a 24 hour period.

Two contraceptive vaginal sponges have been commercially available for several years: the TODAY sponge (VLI Co., CA) containing as spermicide 1 g of nonoxynol-9 (N-9); and the PHARMATEX sponge (Pharmelac, France) containing a cream with 60 mg of benzalkonium chloride (BZK). Both of these sponges have a high content of active ingredients that may cause vaginal or penal irritation (pruritus), particularly if used often.

The TODAY sponge may also cause discomfort to the partner due to the presence

Barrier Contraceptives: Current Status and Future Prospects, pages 265–270

of a string. The PHARMATEX sponge has no string and this makes its retrieval difficult. Also, its polyvinilic foam is not soft enough to allow a good diffusion of the active cream.

THE PROTECTAID SPONGE

PROTECTAID is a new vaginal sponge under investigation as contraceptive and protective agent against several STDs. It is impregnated with F-5 Gel, which combines the inherent spermicidal and anti-STD properties of sodium cholate (Na cholate), BZK and N-9.

This new product, compared with other methods of local protection, shows several advantages:

- The sponge itself, by its flexibility, adapts and fits each woman's particular size and shape. The male partner may not detect its presence during sexual intercourse.

- Insertion and removal of the sponge are simple. To insert the sponge the user places the index finger into one of the slots and pushes the sponge into place. To remove the sponge, she places the index finger into the vagina to find one of the slots and pulls it gently out.

- The PROTECTAID sponge is made from polyurethane foam with interconnecting "open" cells. This allows the entire sponge to be impregnated with F-5 Gel. Not only is the vaginal wall covered or lubricated with F-5 Gel immediately during insertion, but additional F-5 Gel is released as the sponge is pressed during sexual intercourse.

- F-5 Gel contains a low concentration (25 mg/5 g) of each one of the active ingredients (Na cholate, N-9, and BZK). The N-9 content is 40 times lower than that of the TODAY sponge, and the BZK content is 2.5 times less than that of the PHARMATEX sponge.

- F-5 Gel also contains polydimethylsiloxane, a dispersing agent, which by its coating and anti-adhesive action allows the gel to cover and protect the entire vaginal mucosa.

Clinical Evaluation

A first study was performed in collaboration with Professor G. Creatsas and Dr. E. Hassan at the First Department of Obstetrics and Gynecology of the University of Athens, Greece (Psychoyos, et al., 1993). The aim of this study was to evaluate the PROTECTAID sponge, provided by Axcan, Ltd. (Mont St. Hilaire, Canada) essentially for its acceptability and contraceptive effectiveness.

The PROTECTAID sponge impregnated with F-5 Gel, was given to 20 young women aged 19-25 years, mostly nulliparous and high school graduates, for a period of one year. Pretreatment examination showed normal ovulation and intact internal genitalia. All women had a single sexual partner with normal sperm. Cervical and vaginal cultures for isolation and identification of aerobic-anaerobic pathogens including chlamydia, mycoplasma and fungi species, were performed before treatment and at six and 12 month intervals.

A written consent was obtained from each woman separately before enrollment in the study. Pregnant women and those with positive cervical cultures were excluded from the study, as well as men with positive urethral cultures. All women were instructed on the way to insert the sponge, i.e., to insert it within the 12 hours preceding the first act of sexual intercourse regardless of the day of the cycle, and to remove it four to six hours after the last intercourse and in no case to leave the sponge in place for more than 24 hours following insertion.

During 12 months of use in conjunction with at least three acts of intercourse per week, the contraceptive efficacy of the PROTECTAID vaginal sponge was 100 percent. Cervical cultures at the six month interval showed the presence of *Mycoplasma hominis* and *Candida albicans* in one and two cases respectively. These patients were properly treated, due to minor clinical symptoms. At the end of the study, cervical cultures revealed the presence of E. coli and Candida albicans in one case respectively.

No side effects were reported. One woman complained of discomfort at the beginning of treatment which was relieved thereafter. All women used the PROTECTAID sponge during the 12 month period with no discontinuation.

Spermicidal Effect of F-5 Gel

The spermicidal activity of F-5 Gel, provided by Axcan Ltd., was evaluated by the Sander-Cramer test as modified under the recommendation of I.P.P.F (1964). Briefly, F-5 Gel dilutions in 0.9 percent saline were mixed with 0.2 ml of liquified semen, vortexed for five seconds and examined microscopically within 15 seconds. In the absence of any sperm motility the mixture was then re-examined after the addition of 2 ml of an alkaline (pH 8.1) 5 percent solution of glucose and incubation for 30 min at 37°C. All spermicidal dilutions were tested with at least three different semen samples showing normal sperm concentration and percent motility.

In all semen samples, sperm motility was instantaneously inhibited by the addition of 1:10-1:100 dilutions of F-5 Gel. Sperm immobility was definitive and not modified by the addition of an alkaline solution of glucose and incubation at 37°C.

Ultrastructural analysis under transmission electron microscopy (TEM) and scanning electron microscopy (SEM) show that spermicidal dilutions of F-5 Gel led to profound ultrastructural alterations of spermatozoa, such as removal of the plasma membrane, absence or fragmentation of the acrosomal complex, rupture of the flagellum, etc.

In Vitro Inactivation of HIV-1 by F-5 Gel

Numerous laboratory studies have shown that two of the active ingredients of the F-5 Gel (BZK and N-9) can inactivate most STD pathogens in vitro, including HIV-1 [Malkovsky, et al., 1988; Wainberg, et al., 1990]. In our own studies we have examined the in vitro inactivation of HIV-1 by sodium cholate or F-5 Gel. These studies were conducted in collaboration with Professors A. Gravanis and V. Georgoulias at the Department of Pharmacology, School of Medicine of the University of Crete, Heraklion, Greece [Psychoyos, et al., 1993].

In fact, both sodium cholate and F-5 Gel exert a dose dependent in vitro inhibitory effect (a) on the activity of HIV-1 associated reverse transcriptase in an acellular system and (b) on the potential of HIV-1 to efficiently infect human lymphocytes.

Sodium cholate (choleic acid) and F-5 Gel, at doses higher than 2.5 2mM and 10^{-4} v/v respectively, also inhibit the proliferation of normal human lymphocytes and CEM cells originating from a T-cell leukemia.

Consequently, F-5 Gel appears useful in the selective destruction of HIV-infected cells, as well as in the inhibition of HIV-1 itself. This possibility becomes more apparent in cases where the HIV-1 virus is present in cavities, such as the vagina.

In Vitro Effect of F-5 Gel on *Chlamydia trachomatis* and *Trichomonas vaginalis*

As known [Crosignani, et al., 1992], these two STD pathogens account every year for 70 percent of new cases of STDs. The in vitro effect of F-5 Gel on chlamydia trachomatis serotypes C, D, H was studied by Baskin, et al. [1992]. In this study, toxicity results were observed by microscopic observation using a fluorescein conjugated monoclonal antibody directed at the chlamydia trachomatis antigen. The chlamydia control exhibited infectivity, whereas F-5 Gel completely inactivated Chlamydia trachomatis serotypes C, D, and H at a concentration of 10^{-3}.

In addition, data communicated by Axcan Ltd. [1992] show that F-5 Gel exhibits in vitro a trichomonacidal activity at concentrations of the product as low as 0.5 to 1 percent.

The Risk of Local Irritation

Local application of compounds with cytotoxic-detergent properties entails the risk of local irritation of the vaginal mucosa, as well as systemic effects following their absorption and transfer into the circulation. This has particularly been observed with nonoxynol-9 used in certain barrier contraceptives at high concentrations [Niruthisard, et al., 1991]. However, no local irritation was noticed by visual observation or reported by the women who participated in our study using the PROTECTAID vaginal sponge throughout a one year period.

As stated above, the PROTECTAID sponge is impregnated with F-5 Gel that contains a low concentration (25 mg/5 g) of N-9. Wistar rats exposed to F-5 Gel for up to 72 hours showed no trace of irritation upon visual examination and histopathological evaluation. Those exposed to a similar gel, incorporating an amount of N-9 proportional to that contained in the TODAY contraceptive sponge (1 g per sponge) displayed signs of vaginal irritation [Panisset et al., 1992].

CONCLUSION

The present STD/AIDS pandemic requires the intensified use of vaginal methods of protection. The combined spermicidal and anti-STD (including HIV-1) properties of F-5 Gel used in the PROTECTAID sponge offers a new and highly promising protective method. With the urgent needs for such methods, this low-dose spermicide vaginal sponge deserves particular attention. It is important to confirm by prospective clinical trials its effectiveness in preventing the heterosexual transmission of STDs, including HIV.

Our pilot clinical study revealed high contraceptive efficacy and acceptability of the PROTECTAID sponge, as well as an absence of side effects. I would therefore strongly encourage the undertaking of prospective epidemiological investigations within the framework of family planning programs in order to evaluate the anti-STD properties of this new contraceptive method.

REFERENCES

Axcan Ltd. (1992): "Investigator's Brochure. Protectaid contraceptive sponge with F-5 Gel". Mont St-Hilaire, Canada.

Baskin BL, Wetherall NT, Brady CS (1992): Microbial activity of Protectaid F-5 Gel on Chlamydia Trachomatis. In Axcan Ltd: "Investigator's Brochure. Protectaid contraceptive sponge with F-5 Gel". Mont Saint-Hilaire, Canada.

Crosignani PG, Diczfalusy E, Newton J, Rubin B (1992): Sexually transmitted diseases. Human Reproduction 9:1330-1334.

I.P.P.F. (1964): "Medical Handbook" Kleinman RL (ed). London; IPPF.

Malkovsky M, Newell A, Dalgeish AG (1988): Inactivation of HIV by nonoxynol-9. Lancet 1:645.

Niruthisard S, Roddy R, Chutivongse S (1991): The effects of frequent nonoxynol-9 use on vaginal and cervical mucosa. Sex Transm Dis 18:1761-1779.

Panisset JC, Greselin E, Bouthillier L, Denniel C, Audet MA (1992): Vaginal tolerance of F-5 Gel in rats: a controlled study. VIII International Conference on AIDS/III STD World Congress, Amsterdam, Abstract Pub 7059.

Psychoyos A, Creatsas G, Hassan E, Georgoulias V, Gravanis A (1993): Spermicidal and antiviral properties of a new vaginal contraceptive sponge (PROTECTAID). Human Reproduction (in press).

Wainberg MA, Spira B, Bleau G, Thomas R (1990): Inactivation of Human Immunodeficiency virus type 1 in tissue culture fluid and in genital secretions by the spermicide Benzalkonium chloride. J Clin Microbiol 1:156-158.

22

DEVELOPMENT OF VAGINAL MICROBICIDES

Marc Karam

Global Programme on AIDS
World Health Organization
Geneva, Switzerland

The purpose of this chapter is to present the perspective of the World Health Organization's Global Programme on AIDS on barrier methods to prevent HIV transmission and other sexually transmitted diseases. I would like to start by giving a few data to outline the magnitude of the problem. The current number of HIV infections resulting from vaginal intercourse is estimated to account for approximately 75 percent of all HIV infections worldwide. The relative proportion of heterosexually and homosexually transmitted HIV infections varies from one part of the world to the other. In North America, HIV has mainly been transmitted homosexually, while in other parts of the world, such as in Africa, the male to female sex ratio of HIV-infected people indicates a predominance of heterosexual transmission.

Over 5 million women are infected with HIV. In some large cities of sub-saharan Africa, the proportion of women being HIV-infected is growing rapidly: up to one-third of women attending antenatal clinics are infected in some urban areas. In the absence of effective preventive measures, this number will continue to increase.

Now, what do we have to prevent the transmission of HIV and other STDs? Methods available today range from sexual abstinence to the use of condoms. Despite the effectiveness of condoms as a barrier method, significant failure rates for prevention of STDs have been reported, most of them user-related. We need to be innovative and develop a product that can be used by women. Various options are presented in other chapters and particular attention was paid to the use of intravaginal products. Experience acquired in other programmes in WHO and elsewhere in family planning tells us that if a contraceptive method is under the control of the female, compliance with its use will be greater and prevention of pregnancy will be improved. There is every reason to believe that the same rule would apply to the prevention of HIV and other STDs should a female-controlled method be available. There is, therefore, a great need to develop

Barrier Contraceptives: Current Status and Future Prospects, pages 271–272

such a method. Innovative ways of adapting existing contraceptive barriers should be sought. The ideal product does not exist and a lot of work is needed to develop a safe and effective female-controlled method for the prevention of HIV and other STDs, especially one that is acceptable and affordable to women in the developing world. This work cannot be achieved by a single agency and requires the commitment and the partnership of universities, research institutions, the public and private sectors, and international institutions. WHO/GPA is prepared to play an important role in coordinating this effort.

What should be done? Compounds available on the shelf should be screened for their anti-microbial activity, including anti-HIV activity, and for their safety. Innovative formulations for intravaginal use should be sought and new vehicles should be identified. Research should be undertaken for the identification of new molecules for intravaginal use as soon as possible. Suitable animal models should be identified and safety trials of existing products should be undertaken in a population at low risk of HIV infection. WHO/GPA is prepared to collaborate with all parties involved in this research, to review the methodological approach and to organize clinical trials of candidate compounds, allowing comparison of the results of different studies. To this end, WHO/GPA will organize later this year a workshop for the standardization of colposcopical examination and reporting of findings. We also feel that there is a need to prepare the ground for efficacy trials for the time when a safe product becomes available. The development of appropriate clinical trial sites and the organization of appropriate cohorts of population are crucial steps in this preparation and will require coordinated mobilization of considerable efforts and resources of all kinds.

23

PRANEEM POLYHERBAL CREAM AND SUPPOSITORIES

G.P. Talwar, Sanjay Garg, Ritu Singh, P.L. Sharma, Veneeta Dhar, Vibha Taluja, Suman Dhawan and S.N. Upadhyay

National Institute of Immunology
New Delhi-110067, India (GPT, SG, RS, VD, VT, SD, SNU)
Postgraduate Institute of Medical Education and Research
Chandigarh-160012, India (PLS)

Two polyherbal preparations have been formulated with the dual objectives of achieving contraception and promotion of genital health. The preparations have three active constituents; purified neem seed extract (Praneem); saponins from reetha (*Sapindus mukerossi*) and quinine hydrochloride. Neem (*Azadirachta indica*), a tree native to India but now grown in many parts of the world, is recognized as having many medicinal properties. In the cream, neem seed extracts are incorporated as a water-washable emulsion using pharmacopoeially acceptable ingredients, antioxidants and preservatives. An applicator is provided for self-administration of 3 to 5 ml of the cream deep in the vagina. The fragrance and vehicles employed may undergo further refinement to make them more attractive to potential users. The shelf life of the cream at room temperature is estimated to be 18 months based on the accelerated stability studies carried out at 45° C. The suppositories have been made by incorporation of the same constituents into three different grades of polyethylene glycol.

The choice of the constituents is based on the following rationale. Neem seed extracts have components with spermicidal properties [Riar, et al., 1993]. There are, in addition, compounds with a wide range of antimicrobial [Patel and Trivedi, 1982], antiviral [Rao, et al., 1969], antifungal [Pant, et al 1986] and antiparasitic [Khalid, et al., 1986] activities. Many of these compounds have been purified and their chemical structure determined.

Barrier Contraceptives: Current Status and Future Prospects, pages 273–276

Sapindus mukerossi (Reetha or soapnut) extracts have been used by communities in India for washing of hair and for soft woolen clothes. Reetha extracts contain saponins with mild action. Quinine hydrochloride has been included for its anti-HIV action. Quality control specifications have been worked out for these ingredients.

Besides spermicidal and antimicrobial components, neem seed extracts have two other interesting classes of compounds. The water soluble fraction has polysaccharides endowed with the property of inducing the production *in situ* of cytokines, in particular, of interferons, which can act on intracellular infections. These may also be effective against cancers [Kaisha, 1985], presumably by activation of NK cells as well as by direct action. The nonpolar fraction has a compound, which appears to stimulate cell mediated immunity locally at the site at which it is applied. Its application in the uterus impairs fertility without impairing ovarian function or hormone levels [Kaushic, 1992]. The effect is highly localized in the rat: The untreated uterine horn allows unhindered normal reproductive functions of implantation and embryonic development while the treated horn has no implantation sites [Upadhyay, et al., 1990]. Our group has recently purified this compound to near homogeneity, and its chemical structure is being determined.

We have opted at this stage to retain the natural mixture of these compounds in our preparations, not only to keep the costs low, but also to dispense together an array of useful compounds for the diverse purposes of counteracting sperm, a variety of microbes and viruses in the genital tract. The use of Praneem does not entail toxicity, and no undesirable side effects have been observed.

TOXICOLOGY STUDIES

Acute and subacute toxicology studies were conducted with the Praneem Polyherbal Cream (PPC) in monkeys at the Postgraduate Institute of Medical Education and Research. Four Rhesus monkeys (*Macacca mulata*) received daily application of the therapeutic dose of the PPC for 12 weeks, four other monkeys received a fivefold higher dose for the same duration. The four control monkeys received petroleum jelly in the same way for the duration of the study. There was no statistically significant difference in the weight changes of animals in the three groups and no change in any of the hematological and biochemical parameters compared to pretreatment values. Autopsy revealed no gross changes, and histopathology of tissues of all animals was normal.

SKIN IRRITABILITY

Skin irritability was tested using the Draize test on normal and abraded skin of rabbits, and by 21 day cumulative skin sensitivity testing on human volunteers of both sexes. In both tests, the PPC was devoid of sensitization potential and irritability.

The effect on the cervix was tested in monkeys. Twenty-one daily applications did not cause apparent inflammatory reactions.

SPERMICIDAL ACTION

Spermicidal action of each of the three constituents (P, R, Q) was evaluated individually and in combination by the Sander Cramer Test. Each one of these had spermicidal properties. Of interest was their synergistic effect, which enabled the reduction in concentrations of these ingredients in the final formulation to attain 100 percent sperm immobilization and death within 20 seconds.

CONTRACEPTIVE EFFICACY

Contraceptive efficacy of the polyherbal preparations was tested in rabbits and monkeys. None of the 26 rabbits became pregnant on mating with males of proven fertility within 30 minutes of application of the PPC. The fertility rate in controls was 87 percent. Bonnet monkeys of proven fertility were also largely protected. Only one pregnancy took place in 44 cycles of cohabitation with males of proven fertility after application of the cream. Control monkeys treated with peanut oil became pregnant in the colony within two to three cycles under similar conditions.

ANTI-MICROBIAL ACTION

Limited studies have so far been done in the laboratory on the antimicrobial action. The constituents of the polyherbal preparations have demonstrated a growth inhibiting effect at concentrations utilized in the formulation on pathogenic strains of *Candida albicans* and *staphylococcus aureus*. They also inhibit the growth of *Chlamydia*.

Praneem Polyherbal Cream and Suppository have recently been approved by the Drugs Controller of India and the Institutional Ethics Committees for clinical trials. In the Phase I clinical trials, the acceptability and lack of side effects of the preparations will be studied, which will include the effect of the PPC on microbial flora. Praneem polyherbal preparations are also under study in Nigeria, Kenya, China, Dominican Republic and Brazil by the South to South Cooperation in Reproductive Health.

REFERENCES

Kaisha TK (1985): An Interferon Inducer, Japanese Patent, Sho 60-45521.

Kaushic C (1992): Ph.D. Thesis, JNU, India.

Khalid SA, Farouk A, Geary TG, Jansen JB (1986): Potential anti-malarial candidates from African plants: an *in vitro* approach using *Plasmodium falciparum.* J Ethno pharmacol 15:201-209.

National Research Council (1992): Neem: A tree for solving global problems. National Academy Press, Washington, DC.

Pant N, Garg HS, Madhusudanan KP, Bhakuni DS (1986): Sulfurous compounds from *Azadirachta indica* leaves. Fitoterapia 57:302-304.

Patel RP, Trivedi BM (1962): The *in vitro* antibacterial activity of some medicinal oils. Ind J Med Res 50:218.

Rao AR, Kumar SSU, Paramasivam TB, Kamalakshi S, Parashuraman AR, Shantha M (1969): Study of antiviral activity of tender leaves of margosa tree (*Melia azadirachta*) on vaccinia and variola virus a preliminary report. Ind J Med Res 57:495-502.

Riar SS, Sawhney RC, Ilavazhagan G, Roy JB, Kain AK, Thomas P, Singh R, Singh B, Devkumar C, Singh M, Sawhney RC (1993): Neem as a contraceptive. Proceedings of World Neem Conference, Bangalore, India. In Press.

Siddiqui S, Siddiqui BS, Faizi S, Mahmood T (1988): Tetracycline Triterpenoids and their derivatives from Azadirachta indica. J Nat Products 51:1, 30-43.

Sinha KC, Riar SS, Tiwari RS, Dhawan AK, Bardhan J, Thomas P, Kain AK, Jain RK (1984): Neem oil as a vaginal contraceptive. Ind J Med Res 79:131-136.

Upadhyay SN, Kaushic C, Talwar GP (1990): Anti-fertility activity of neem (*Azadi rachta indica*) oil by single intrauterine administration: A novel method for contracep tion. Proc R Soc Lond B 242:175-179.

24

SOUTH TO SOUTH EXPERIENCE WITH POLYHERBAL CREAM IN THE TREATMENT OF VAGINAL INFECTION

O.A. Ladipo

South to South Cooperation in Reproductive Health
Rua Caetano Moura, 35 - FEDERAÇÃO
40210-350 Salvador, Bahia
Brazil

INTRODUCTION

Vaginal discharge is the commonest gynecological complaint of sexually active women. Although not all vaginal discharge is abnormal, the majority of cases are attributable to either bacterial, fungal, protozoan or viral causes. The first three groups account for more than 80 percent of cases of vaginitis.

In resource-poor settings around the world, reproductive tract infections are extremely common, and the consequences for the health and social well-being of women and their children are frequent and potentially devastating [Brunham and Embree, 1992, Wasserheit, 1989]. Management of these cases depends upon identification of the causative pathogens, either by direct examination or by culture of the discharge and treatment with appropriate therapeutic agents.

Recently, Prof. G.P. Talwar and colleagues at the National Institute of Immunology of India in New Delhi developed a polyherbal cream containing three active substances: the purified extracts of neem (Praneem), extracts of the spindus plant and quinine hydrochloride. Together, the three have antimicrobial and antiviral activities. These components are dispensed in a water-washable cream base utilizing pharmaceutically approved excipients. The cream is physically stable for up to 1 1/2 years at room temperature. The South to South Cooperation in Reproductive Health investigated in a Phase I study the acceptability and therapeutic effect in patients with vaginal discharge.

Barrier Contraceptives: Current Status and Future Prospects, pages 277–282

METHODOLOGY

The South to South centers that participated in this study were the Fertility Research Unit, University of Ibadan, Ibadan, Nigeria; Profamilia, Santo Domingo, Dominican Republic; Kenyatta National Hospital, Nairobi, Kenya; and CEPARH, Salvador, Brazil. Fifty-two patients attending the gynecology clinic with symptoms suggestive of vaginitis were recruited for a Phase I clinical trial of Praneem vaginal cream. The inclusion criteria are age range of 18-40 years, regular menstrual cycle for the previous three months, age of last child more than one year, and good physical health. Exclusion criteria were suspected pregnancy, history of abortion, allergic disorders, autoimmune/endocrine disease and any malignancy or current use of other medications.

Age and parity were recorded, and after clinical and bacteriological evaluation, voluntary consent to participate in the trial was obtained. The patients were carefully instructed as to how to apply the cream and were given tubes with instructions to apply daily for seven days. The applicators supplied with the cream were designed to release about three grams at each application. All subjects returned between the 8th and 10th day post treatment for clinical review. In 10 patients venipuncture was performed before and after treatment for evaluation of clinical chemistry and hematological status.

RESULTS

The mean age of volunteers was 33.9 ± 4.1 years, while the mean parity was 3.3 ± 1.1.

Microbiology

Table 1 illustrates the organisms isolated prior to treatment. *Candida albicans* occurred in almost one-third of the patients while *gardnerella* and *trichomonas* infections were diagnosed in 23.1 percent and 19.2 percent of the patients, respectively.

TABLE 1. Organisms Isolated Prior To Praneem Therapy

Organism	Frequency	Percentage Frequency
Candida Albicans	17	32.7
Gardnerella Vaginalis	12	23.1
Trichomonas Vaginalis	10	19.2
Staphylococcus aureus	14	26.9
E. Coli	2	3.8
Klebsiella aeruginosa	2	3.8
Pseudomonas aeruginosa	1	1.9
Micro-coccus species	1	1.9

NB - More than one organism isolated in seven cases

Laboratory Studies

All hematological and biochemical parameters evaluated before treatment with Praneem cream were within normal range (Tables 2 and 3). There was no significant change in any parameter after treatment.

TABLE 2. Hematological Parameters Before and After Treatment with Polyherbal Vaginal Cream (N = 10)

Test	Normal Range	Before Treatment	After Treatment
White blood cell count X 10^3	5 - 10/mm^3	7.85 ± 0.34	7.55 ± 0.24
Red blood cell count X 10^6	3.8 - 5.2/mm^3	4.42 ± 0.08	4.4 ± 0.04
Haemoglobin gm%	11 - 18	12.2 ± 0.30	11.9 ± 0.32
Hematocrit %	35 - 47	39.5 ± 0.83	39.5 ± 0.92
ESR (1st hour)	0 - 22/mm	11.2 ± 2.30	17.1 ± 2.80

NB - No statistically significant difference in all parameters

TABLE 3. Biochemical Parameters Average Before and After Treatment with Polyherbal Cream (N = 10)

Test	Normal Range	Before Treatment	After Treatment
Renal Function			
Urea (mg/dl)	15 - 40	27.7 ± 2.0	27.9 ± 1.6
Sodium (meq/l)	135 - 150	143.2 ± 1.7	138 ± 0.9
Potassium (meq/l)	3.5 - 5	4.08 ± 0.09	4.2 ± 0.1
Chloride (meq/l)	97.6 - 105	100.6 ± 1.2	99.8 ± 0.7
Creatinine (mg/dl)	0.4 - 1.4	0.9 ± 0.04	0.9 ± 0.09
Uric Acid (mg/dl)	1.5 - 6.0	3.6 ± 0.29	3.7 ± 0.18
Liver Function			
Bilirubin (mg/dl)	1.2	0.6 ± 0.08	0.5 ± 0.08
SGOT (u/L)	4 - 36	27.5 ± 1.38	23.7 ± 0.82
SGPT (u/L)	3 - 32	24.0 ± 1.46	21.6 ± 0.82
Total Protein (g/dl)	6 - 8	7.1 ± 0.22	7.28 ± 0.20
Albumin (g/dl)	4.5 - 5.5	4.56 ± 0.19	4.92 ± 0.12
Globulin (g/dl)	1.5 - 2.5	2.54 ± 0.2	2.36 ± 0.2
Others			
Calcium (meq/l)	4.5 - 5.7	5.12 ± 0.06	4.94 ± 0.07*
Glucose (mg/dl)	70 - 110	92 ± 3.4	92 ± 3.3

* $p < 0.05$

Clinical Response

At the follow-up visit, the clinical response to therapy was assessed (Table 4). Complete clinical response was noted in 35 subjects (67.2 percent), while 12 (23.1 percent) had persistence of symptoms and 2 (3.9 percent) had partial response. In three patients (5.8 percent), symptoms became worse. One patient had very profuse vaginal discharge one day after commencement of therapy, but this subsided gradually over the subsequent five days while still on therapy. One other patient had severe pruritus vulvae following daily application of the cream on the first three days. Two patients in whom symptoms became worse terminated due to severe allergic reaction.

TABLE 4. Clinical Response to Praneem Cream

Clinical Response	Frequency	Percentage Frequency
Complete	35	67.2
Partial	2	3.9
Persistence of symptoms	12	23.1
Became worse	3	5.8
TOTAL	52	100

A repeat culture after therapy indicated a 52.9 percent cure for *candida albicans* infection, while for *gardnerella* and *trichomonas* infections the therapeutic cure rates were 80 percent and 58 percent respectively. *Staphylococcus aureus* was identified in only 14 percent of subjects in the repeat culture, while 15 percent of subjects had *E. coli* detected in post-therapy culture (Table 5).

TABLE 5. Microbiologic Response to Praneem Cream

Organism	Results of repeat culture (% cure)
Candida Albicans	52.9%
Gardnerella Vaginalis	80 %
Trichomonas Vaginalis	58 %
Staphylococcus aureus	86 %
E. Coli	85 %

Aesthetic Assessment (n = 10)

Forty percent of the patients thought the color of the Praneem Cream was acceptable while 10 percent found it displeasing and 50 percent were indifferent. The consistency of the cream was considered good by 50 percent of patients while 50 percent expressed no opinion. The odor of the cream was acceptable to 30 percent of the patients, while 40 percent thought it was displeasing and 30 percent were indifferent.

DISCUSSION

In this preliminary Phase I study of the therapeutic effect and acceptability of the Praneem cream in patients with abnormal vaginal discharge, the results suggest an antimicrobial and antibacterial effect with two-thirds of patients becoming asymptomatic after one week of treatment. *E. coli* and *Candida albicans* were the principal organisms isolated in those patients classified as having partial or no clinical response to the Praneem cream.

Further studies are needed on the use of the cream in patients with pathological discharge. Of importance in future studies would be the need for *N. gonorrhea* and *Chlamydia trachomatis* cultures. Some of the patients who had partial or no response may have had infection with one of these organisms. Preliminary results from India suggest a therapeutic effect on cervical Chlamydia infection. A longer period of follow-up is also recommended in future studies to identify later treatment failures.

During this preliminary study, 5.8 percent of subjects experienced itching and/or irritation. Whether this reaction was to the vehicle or the active ingredients is not known. The use of a placebo cream in future studies will clarify the issue.

Although the contraceptive effectiveness of the Praneem cream is yet to be confirmed in clinical trials, studies on the bonnet monkeys and rabbits indicate effectiveness as a vaginal preparation [Talwar, 1992]. The ingredients immobilize and kill 100 percent of sperm in Sander Cramer testing. Praneem cream with its potential contraceptive, bactericidal and virucidal properties may in the future be one choice of methods provided for family planning.

ACKNOWLEDGEMENT

The author acknowledges the contributions of Drs. A. O. Adekunle, J.K. Mati, P. Spinola, V. Brache to the South to South Praneem Study. Prof. G.P. Talwar supplied the Praneem cream from the National Institute of Immunology, New Delhi, India.

REFERENCES

Brunham RC and Embree JE (1992): Sexually Transmitted Diseases: Current and future dimensions of the problem in the third world. In Germain A. et al, (eds), "Reproductive Tract Infections: Global Impact and Priorities for Women's Health", New York, Plenum Press p 35-38.

Talwar GP (1992): Praneem Polyherbal Cream. Presentation at the 11th South to South meeting, Salvador, Bahia, Brazil, May 20-25.

Wasserheit JN (1989): The significance and scope of reproductive tract infections among third world women. Int. J. Gynecol. Obstet. suppl 3:145-68.

25

PRANEEM POLYHERBAL CREAM PHASE I CLINICAL TRIALS IN INDIA

Sanjay Garg, Rajinder Kaur, S.N. Upadhyay, G.P. Talwar[1]

National Institute of Immunology
Shaheed Jeet Singh Marg
New Delhi-110067, India

With the increase in prevalence of AIDS and other sexually transmitted diseases, the need for contraceptives that have an anti-microbial effect is high priority. It is believed that reduction in prevalence of sexually transmitted diseases will reduce the occurrence of AIDS [Chantler, 1992].

Praneem polyherbal cream, a formulation with active ingredients of plant origin [Talwar et al., 1993] is undergoing clinical trials in India for contraceptive efficacy and for its effect on pathogenic vaginal microorganisms.

With the permission of the Drug Controller of India and Institutional Ethical Committees, clinical trials have started on the Praneem polyherbal cream in three centers in India, Postgraduate Institute of Medical and Research, Chandigarh; Safdarjung Hospital, New Delhi; and Maulana Azad Medical College, New Delhi.

In Phase I studies, the objectives are to determine the safety and acceptability of the cream and its noninterference with intercourse. The study is in progress in symptomatic women for cervicitis/vaginitis attending the gynecology clinic of the above hospitals. Most patients enrolled had symptoms of vaginal discharge, dyspareunia and pain in the lower abdominal region. The patients were of reproductive age group (22 to 35 years) with regular menstrual cycles. Besides clinical examination, swabs were taken for determination of the microorganism(s) responsible for the infection. The diagnostic tests employed are listed in Table 1 [Ripa and Mardh, 1977; Mittal et al., 1993]. In addition

[1]In Clinical Collaboration with: A. Mittal and S. Kapoor at the Institute of Pathology, New Delhi, 110029, India. S.K. Das, S. Suri and S. Gupta at the Safdarjung, New Delhi, 110029, India. S. Gopalan, J. Kaur and V. Jain at the Postgraduate Institute of Medical Education & Research, Chandigarh, 160012, India.

Barrier Contraceptives: Current Status and Future Prospects, pages 283–285

to evaluating the cream's safety and acceptability, the effect of the cream on vaginitis due to *Chlamydia trachomatis*, *Candida albicans*, Trichomonas, *G. vaginalis*, and *N. gonorrhea* was investigated. A separate protocol has been written to determine the effect of the cream on human papilloma virus associated with dysplasia and cervical cancers.

TABLE 1. Diagnostic Methods Used In The Study

C. trachomatis	Smear - GIEMSA Antigen detection
T. vaginalis	Wet smear Whittington media
N. gonorrhea	Smear - Gram stain Culture - Thayer Martin media
G. vaginalis	Smear - Gram stain Whiff test Culture - Blood agar media

In the ongoing study, results are available on 28 patients who had cervicitis due to *Chlamydia trachomatis* and on 24 patients with infections due to other organisms. Twenty-two out of the 28 patients with cervicitis ascribable to *C. trachomatis* infection recovered clinically and microbiologically after seven to 21 daily applications of the cream [Garg, et al., 1993]. Six subjects were lost to follow-up. The formulation was well tolerated in all subjects.

In patients with other infections, follow-up has been completed in one patient with *T. vaginalis* infection, 14 patients with *C. albicans* infections and in 11 patients with *G. vaginalis* infection. The patient with *T. Vaginalis* infection recovered both clinically and microbiologically. Nine out of 14 candida-positive patients showed complete clinical recovery while they were still harboring the microorganism as per culture. Eight of the 11 patients with *G. vaginalis* recovered clinically and microbiologically after 14 to 21 daily applications. One patient, who reported irritation with the cream, showed no reaction on applications of the cream on the skin. A study using placebo cream containing the vehicle alone without any active ingredients is in progress. The ethical committee considers it unethical to give a placebo cream to patients with pathogenic infection. For this reason, the number of controls is much less than required for a complete clinical study.

To study the acceptability of the Polyherbal cream and pessaries, both were given to couples of reproductive age in which either the wife or husband had undergone prior elective tubal or vasal ligation. Neither preparation was reported to cause interference with intercourse. Thirty percent of the subjects, however, disliked the odor of the cream. Most subjects preferred the suppositories owing to their ease of insertion.

Prior to undertaking of Phase II contraceptive efficacy trials, post-coital test with both the Praneem polyherbal cream and suppository are being conducted in women using IUDs. Early results are encouraging.

Although much more work remains to be done, preliminary results with polyherbal preparations are promising. These preparations may be an alternative to the present preparations based on nonoxynol-9 and will be a novel combination of a contraceptive with prophylactic/therapeutic agents for genital infections.

REFERENCES

Chantler E (1990): New and existing spermicides with virucidal properties. In Alexander NJ, Gabelnick HL, and Spieler JM (eds): "Heterosexual transmission of AIDS." New York:Wiley-Liss, Inc., pp 303-11.

Garg S, Mittal A, Kapur S, Gupta S, Upadhyay SN, Talwar GP (1993): Clinical studies with Praneem polyherbal cream in Chlamydial cervicitis. Proceedings of World Neem Conference, Bangalore, India, In press.

Mittal A, Kapoor S, Gupta S, (1993): Chlamydial cervicitis, role of culture enzyme, immunoassay and GIEMSA cytology in diagnosis, APMIS, In press.

Ripa KT, Mardh PA, (1977): New simplified culture technique for Chlamydia trachomatis, nongonococcal urethritis and related infections. Am Soc Microbiol 6:323-327.

Talwar GP, Garg S, Singh R, Kaur R, Dhar V, Taluja V, Dhawan S, Upadhyay SN (1993): Praneem polyherbal cream and suppositories. In Mauck C, Cordero M, Gabelnick HL, Spieler JM, and Rivera R (eds): "Barrier contraceptives: current status and future prospects." New York: Wiley-Liss, Inc., In Press.

GENERAL DISCUSSION

Howard Price: This question is for Dr. Guerrero. Were all of the condoms used in your study prelubricated and do you know what extra lubrication was used in some cases?

Ernesto Guerrero: All of the condoms were lubricated and we didn't find any visual evidence that they were using an additional lubricant. It could be that if they used extra lubricant, they took the lubricant away with them.

Debbie Rogow: My questions are for Sanjay Garg, about the trials of neem and in particular about the control subjects. First, for those who did return for follow-up, was there also some return to asymptomatic status? Second, what were they told when they were subsequently treated about their participation in the trial? Third, how many controls were lost to follow-up without treatment after having come in?

Sanjay Garg: These were very preliminary studies. We do not have any data on the patients who were lost to follow-up. As far as controls are concerned, we are improving the protocol for the next study.

Henry Gabelnick: I wanted to ask Ron Roddy a question. You said with a certain degree of understatement that you're a little surprised about the ratio of 0.8 in the condom use between high and low users. I'm wondering if you could speculate a little bit more than you did about why that result was found.

Ronald Roddy: We have lots of speculations and I warned you about the overestimate bias' that were inherent in those, but there were also some bias' that would make you underestimate some things. One of them is the comparison groups we were choosing. For instance, if you're comparing 100 percent condom users to 99 percent condoms users that's going to underestimate the effect. In my opinion, from my study, a little bit of spermicide use is not very protective, but a lot is. So the difference there is going to be greater whereas for the condom use, people who are moderately using condoms compared to people who were using them perfectly, the difference there is going to be less, so that would underestimate also. Not only did we not have any perfect users in most of the studies but we didn't have any perfect non-users. A lot of the women wanted to use either one or both, but a lot of the combinations were not allowed by the clients of these women. All the women used something sometimes and so we didn't have a zero use group to compare to.

Kamran Moghissi: This question is also directed to Dr. Roddy. Can you tell us how you collected your data regarding compliance? Was it by recall or was it perspective recording of the data of coital activities?

Barrier Contraceptives: Current Status and Future Prospects, pages 287–292

Ronald Roddy: Paul Feldblum of FHI and I have discussed the problem of compliance in these studies for some time. We have come up with what we call a pictorial coital log so even illiterate women can use it. What they do is whenever they have sex -- it is prospectively recorded -- we give them the log, explain how to use it and show some examples on the log for them to look at. As they are going along, if they have sex they put an X in one of the boxes and then we have a picture of a condom and a picture of whatever else we are using in the study and they put an X over those. As you know, people's sexual activities are difficult to validate but we have used these logs in three different studies. In the HIV and gonorrhea/chlamydia studies that I presented, the compliance seems to fit with the biological way that we would look at things. We try to validate the results by interviewing the couples separately. We would interview the men and ask them about their sexual activity and then do the same with the women. We have tried to look at it several ways and I personally feel that looking at it retrospectively, asking a woman in the last month how many times did she use this or use that, is not good. So, we are trying to follow this prospectively. I know we get misclassification and it is not perfect information but I think it is probably better than asking them what they did in the last month.

Alfred Shihata: This question is addressed to Elizabeth Ngugi. Did I understand you correctly when you said that 85 percent of the candidates for the study were HIV positive and you selected the 15 percent who were HIV negative?

Elizabeth Ngugi: I didn't give that information, but at the recruitment time the seropositivity was in that region. We selected those who were willing to participate from the 15 percent.

Alfred Shihata: What about the incubation period between the HIV infection and the development of seropositivity? For very active sex workers in areas where HIV is highly endemic, whether you give them the sponge or condoms or a placebo, it is a matter of time before every one of them will become seropositive. The study is actually biased towards the people who use the sponge. They have more ulcer disease and the ones using the placebo did not use a sponge. Can you address the incubation period, and whether this single study should change all of our knowledge despite overwhelming evidence that nonoxynol-9 has antiviral and anti HIV activity?

Elizabeth Ngugi: We can't. We don't like to generalize from the study. You are quite right also in saying that where there is high exposure of HIV infection, even with 100 percent use of condoms it is just a matter of time because of the condom bias and so forth. So even when we were using the N-9 we did not know at that time if they were really seronegative. However, if one can delay the infection, one has prolonged the life of that person. Also, if one delays the infection then transmission to other members of society is reduced.

Alfred Shihata: The basic issue here is that the women you selected probably had the infection to start with. Can you tell if they already had the HIV infection and did not seroconvert after you tested them?

Ronald Roddy: I would like to address that. One approach would be to throw out the people who seroconverted early in the study. The other is to remember that this was a randomized study, so seroconverters should have been randomized to both groups. Since they are looking at rate ratios, it should not affect the results.

Michael Burnhill: This is directed at Dr. Guerrero. I am totally swept away by your ability and your workers' ability to do this kind of study. I don't think it could be done in the United States. I am also amazed that you could get condoms into hotel rooms. I wonder what would happen if international hotel chains would stop putting sewing kits and shower caps in the rooms and would supply condoms? Did you see in 1988 or 1989 and 1990 any impact on gonorrhea and chlamydia when you supplied condoms? Is there a diminution in your infection rate now that you are providing condoms?

Ernesto Guerrero: The major large hotel chains were also asked to place condoms next to the bible. We had some meetings on this issue and even though it was required by law, they refused to comply. Maybe the law should apply to hotels or motels that have a rate for periods of less than 24 hours. We knew that those hotels are used for casual sex most of the time. We are still pushing to have those condoms there and to keep that ordinance regarding hotels and motels in place. Unfortunately, many of them are still not complying. Two or three weeks ago we did a quick survey and less than 75 percent were complying. About 90 percent of the owners of these smaller hotels and motels are foreigners, mostly Chinese. They all have the condoms now. Some of them don't want to put them in the room, but if you say through the intercom that you want condoms they will bring them to you. When we asked those people why they were not putting the condoms in the rooms, they had many excuses.

Michael Burnhill: The other part of my question was whether providing condoms had any effect on chlamydia or gonorrhea rates.

Ernesto Guerrero: In the STD clinics we have seen some decline in the rates of gonorrhea but not chlamydia. Chlamydia rates seem to be climbing. But I don't know how these rates relate to providing condoms in motels.

Jeffrey Spieler: At the last CONRAD international workshop, which was on heterosexual transmission of HIV, we had the hotel put a rubber ducky condom rather than a mint on everybody's pillow when the beds were turned down in the evening. We actually got a very interesting and positive response from many of the participants. Everybody was talking about it and said that it should be done routinely.

Ernesto Guerrero: We usually talk about condoms whenever we have a meeting like this. We ask the people how many of them are carrying condoms right now. Since this workshop is about condoms and barriers, how many of you are carrying condoms?

Jeffrey Spieler: The manufacturers only. I don't know how many people came to Santo Domingo expecting a sexual encounter.

Ernesto Guerrero: That's always the answer. That's always the excuse. No, I didn't come for that. It just happened.

Katherine Laguardia: My questions are addressed to Dr. Soper. I have two questions concerning the study that you did with the female condom. Did you look at other microbiologic parameters? And could you comment on the relative value or weakness of studying recurrence of infection with these sorts of methods as outcome measures?

David Soper: We did not study other bacteriological infections. What we were trying to do in this study was to find a disease or an infection that has a relatively high incidence so that we could see if the device would decrease the rate of infection. In the literature, the reinfection rate of women who are exposed to untreated males is between 6 and 20 percent. In our study it was 14 or 15 percent which allowed us to do a study with about 120 subjects in two arms or 180 in three arms and achieve statistical significance.

Penelope Hitchcock: I have a comment and then a question for the group at large. My comment has to do with the pitfall of using symptoms as a measure of response to treatment. These diseases, especially gonococcal and chlamydial infections, are often asymptomatic or have symptoms that wax and wane. Microbiological testing of the effect of these compounds would be extremely helpful. There is a paucity of diagnostic tests that are rapid, inexpensive, and easy to use that would aid in this. This may be an opportunity for us to collaborate and develop some protocols which would be useful.

I would like to take this opportunity to ask you as a group what the issues are with respect to testing devices of unknown efficacy in populations at high risk for STD and HIV. How are you coming to terms with that in your study designs and in your communication with regulatory agencies?

David Soper: A high percentage of gonococcal and chlamydial infections and all STDs, frankly, are asymptomatic. What we're learning now is that the so-called "typical" presentations of STDs are probably actually less typical and that the atypical presentations are more typical. Certainly, at least in our population, two-thirds of the chlamydial infections and one-third of the gonococcal infections seen in our ER receive no treatment whatsoever. So they either cause symptoms that are not recognized as associated with these infections or they are asymptomatic.

Jeffrey Spieler: The next question was related to testing chemicals or devices of unknown efficacy in the field. I would like to ask Lisa Rarick to comment on new things like the Proctectaid Sponge which has a trivalent approach with a base, and Praneem which has three major components. In order to evaluate multivalent products like these, what would we need to do to get approval to do Phase I and Phase II studies?

Lisa Rarick: It gets very complicated to do studies in the United States on combination products. You would have to start by doing comparison studies with multiple arms to show that each component of the product contributes to its overall effect. I am not sure how that would be done with the polyherbal combination. With Protectaid, there would have to be some evidence that each of the three major components is effective for something that was going to be claimed in the label. That is how we would normally look at it. It is not easy.

Jeffrey Spieler: Let's go back to Penny's question about testing these in the field.

G.P. Talwar: The best way to do field studies would be to have a placebo-controlled trial.

Jeffrey Spieler: If I understand Dr. Hitchcock's question properly, we don't know if the experimental product is going to work or not. To expose people to it when we don't know whether there are advantages is questionable unless we go through a proper development process and that takes a long time.

G.P. Talwar: Our product has gone through preclinical toxicology studies in good centers, and then we started the Phase I clinical studies to see whether it is well-tolerated and whether there are any side effects before efficacy studies are undertaken. The data that we presented on microbiology applies to the product's main indication which is contraception. We hope contraception can be achieved without the side effects that are associated with the currently available spermicides. In the Phase I study we looked for microbiological effects. In the hospital that we are using, clinicians say that about 30 to 40 percent of the women who come have vaginitis and cervitis. With informed consent, they recruited them for these Phase I studies to determine whether cream affects clinical symptoms and the microbiology of vaginal infections. The chlamydial study was done by a doctor in India who was trained at the CDC and who is probably one of the few who has a qualified center for culture of chlamydia. Our data on the effect of polyherbal cream on chlamydia looks pretty good but we would appreciate greatly having further cross-checks by other investigators.

Marc Karam: At WHO we have two steps. First, we look at the safety of the product. If a product contains multiple molecules it would be important to look at each molecule independently. And also to look at not only the active compounds but the carriers. As for the efficacy of these compounds I think it is a very difficult question and one best addressed by studying them with concomitant use of condoms.

Malcolm Potts: The ethical approach is to counsel everybody to use condoms and then randomly allocate to your compound and a placebo. I think that is what Dr. Roddy has been doing at FHI. I would like to ask Dr. Psychoyos a question. I think the idea of a low concentration of several chemical entities is an interesting one. What is the pH of Protectaid?

Alexandre Psychoyos: The pH is 4.4. It also contains acetic acid used to adjust the pH. Maybe any odor present is due to the acetic acid.

Jeffrey Spieler: I made the comment that when I showed the product to several people they commented on the chemical odor. This is something that Nancy Williamson spoke about in her presentation on acceptability. The product should not be malodorous.

Alexandre Psychoyos: I just wanted to have the opportunity to say that the combination of these three substances was done to synergize the spermicide effect, allowing much lower dilutions than with other spermicidal products. We also want to cover a large spectrum of pathogens which could have different sensitivities to each one of the three ingredients. That was the main reason of the combination.

Clark Taylor: The Protectaid sponge may initially smell a little, but in use it doesn't smell and, more importantly, it has no taste. That's very important to notice. My question is for Dr. Talwar. Have you done any studies to see if the cream is compatible with condoms?

G.P. Talwar: That has not been tried yet.

Michael Burnhill: Praneem may never make it through the approval process in the U.S. because as a phytotherapeutic and probably or possibly unpatentable product, nobody may ever choose to invest the amount of money that would be required to get it through the FDA. There are a whole host of traditional and botanical products that have antibacterial and antiviral properties, some of which have been in use for thousands of years, and yet they have to go through a full drug protocol if they want to make any claims. I wonder if the FDA shouldn't have a separate division for folk medicine and phytotherapeutics.

G.P. Talwar: A patent for Praneem has already been granted. Some of the best drugs of the century have come from plants.

Jeffrey Spieler: WHO had a task force on plants for fertility regulation. As soon as we applied the scientific method to isolating the active ingredient and determining activity, we could not find either in almost any of the plants that were studied. That doesn't mean they don't work when used traditionally. Dr. Talwar's approach is to actually isolate the active ingredients which would permit a proper analysis by the FDA of the individual and combined components. I think the public sector would be very interested in supporting novel approaches to both STD prevention and contraception if we can work out a development plan that is reasonable. Thank you all.

PART VI: CONTRACEPTIVE APPLICATION OF BARRIER METHODS WORKING GROUPS

26

PRECLINICAL STUDY DESIGN WORKING GROUP REPORT

David Katz (Moderator)
Douglas S. Colvard (Rapporteur)

Department of Obstetrics and Gynecology
University of California, Davis
Davis, CA 95616 (DK)
Contraceptive Research and Development Program
1611 North Kent Street, Suite 806
Arlington, VA 22209 (DSC)

This Working Group focused upon the sequence of issues that must be addressed before a compound that is topically contraceptive, and/or prophylactic, can be evaluated in a clinical trial. These issues involved testing, formulation and delivery of compounds that act in the vagina, or cervix, to inhibit transport and/or fertilization by sperm, and transport and infection by pathogens. Similar topics are also addressed in Chapters 10, 11, and 16 on screening of spermicides, animal model systems for testing spermicides, and a primate model for HIV transmission. The majority of time was devoted to contraception, in part because of the broader base of knowledge concerning chemical and barrier methods of contraception, and in part due to time constraints.

Many of the scientific and technical concepts and methods for contraceptive testing have direct relevance to assessment of compounds that are prophylactic against sexually transmitted diseases (STDs) when applied topically. This link is important, given the relatively primitive state of STD prophylaxis. Even though the pathogenic microorganisms are biologically distinct from sperm, and their mechanisms and routes of infection are different, they populate the same environment, and are subject to the same anatomical and physiological conditions of that environment. Some spermicidal compounds, such as nonoxynol-9 (N-9, a nonionic surfactant), are known to have virucidal activity *in vitro*, as well. It appears unlikely that an efficacious anti-sperm modality will have equal effectiveness against infection by any pathogen. However, the complementary, and possibly synergistic actions of contraceptive and prophylactic methods is a critical theme, which underlied virtually all of the specific issues addressed by this Working Group.

Barrier Contraceptives: Current Status and Future Prospects, pages 295–310

Opportunities to inhibit sperm are potential opportunities to inhibit pathogens, and vice versa.

A distinction was made between compounds that directly kill spermatozoa and compounds that alter sperm so as to render them dysfunctional. Most spermicides in use today act by non-specific destruction of spermatozoan membranes (i.e., surfactants). The most common spermicide in use worldwide is N-9, and this is the only approved spermicide in the U.S. Many other compounds have been brought to various stages of development, or are marketed in other countries. However, the vaginal spermicide market is essentially reliant upon a single class of compound. This class of compound has significant limitations, e.g., irritation of the vaginal epithelium, especially with chronic use. Compounds that are efficacious at lower doses would likely be less irritating, and would remain bioactive against sperm after dilution with the fluids of the vagina. The biological specificity of sperm function inhibiting compounds (SFIs) is attractive in this regard, since it might result in lower effective dose concentrations.

The cervix, and its complement of mucus, was considered as an important target for contraceptive action. This site has not been fully exploited for application of compounds. Such compounds could have direct bioactivity against spermatozoa and/or act to increase the mucus resistance to sperm penetration.

Salient physicochemical properties of compounds must be considered, such as solubility, ease of formulation and incorporation into the delivery systems. The resulting modalities must be assessed for irritation and toxicity. Discrepancies persist between results of *in vitro* testing and studies *in vivo*, and reduction of these differences could significantly impact product development. Finally, the costs of development must be anticipated as early as possible in this process.

Portals of entry and infection differ substantially among bacterial, viral and other STD pathogens. These different routes and mechanisms of infection are often not understood, e.g., for HIV. Distinctions exist between chemical action against isolated pathogenic microorganisms suspended *in vitro*, and inhibition of infection *in vivo*. No single test may be sufficient for screening of compounds active against some pathogens, including HIV.

IN VITRO SCREENING

Successful development of an eventual product requires that the goals and end-points of spermicide screening programs be clearly defined. When optimizing *in vitro* screening, it is important to recognize those conditions that apply *in vivo* and to simulate them as much as possible when designing an *in vitro* test system. These conditions include:

- The acidic pH in vagina, which becomes neutral after liquefication of the ejaculate;
- The chemical environments of vaginal/cervical fluids and seminal plasma, e.g., ionic composition and glycoprotein content;
- Varying concentrations of spermatozoa with differing susceptibility to spermicides;

- The abundance of cervical mucus at mid-cycle, extending from the cervical os into the floor of the vagina; and
- The potentially rapid movement of spermatozoa to regions of low spermicide concentration.

Details of current *in vitro* tests for measuring the potential of compounds to inhibit sperm motility (e.g., Sander-Cramer tests, computer-assisted semen analysis, analyses of mucus penetration, sperm enzymes, egg binding and penetration) are provided in other chapters (see Ch. 10). For many of the *in vitro* tests, N-9 is used as a positive control, although cut-off values are not yet standardized. Ejaculates from different donors should be tested against each potential spermicide, utilizing a range of concentrations for each specimen. A procedure that has been recently described [Chantler, et al., 1992] is the generation of dose-response curves, using these conditions, followed by calculation of the ED_{50} for the agent. This method allows a numerical comparison of spermicides or SFIs under defined conditions, including concentrations that are not totally spermicidal.

INHIBITORS OF SPERM FUNCTION

Current *in vitro* testing focuses almost exclusively upon sperm immobilization as its end-point, yet it is not necessary to kill spermatozoa in order to render them dysfunctional. Development of compounds that actively inhibit vital sperm functions is an underdeveloped area. During the past ten years, the biochemical control of sperm function has become much better understood, and it should be a target of future contraceptive development. Although this is a biologically complex area, it does provide an important feature of chemical, topical contraceptive action - biological specificity of action. Such specificity can, in principle, be exploited to increase contraceptive efficiency without increasing undesirable side-effects such as tissue irritation. However, development of SFIs necessitates more diversified testing protocols, specific to the particular actions of compounds. For example, compounds that might block sperm binding to the zona pellucida must be tested using the zona pellucida. The action of SFIs may very well not occur within the vagina. Therefore, such compounds must contact and remain associated with the spermatozoa until their biological activity against the sperm is irreversibly manifest. Cervical mucus acts to remove many moieties from the sperm surface, and this "scrubbing" action of the mucus must be taken into account.

The majority of SFIs that have been studied during the past ten years are inhibitors of platelet-activating factor, acrosin, hyaluronidase and cyclooxygenase. Other compounds have been shown to prevent fertilization when placed vaginally in an animal model, including zinc salts, a component of neem oil, and gossypol, among others. Significant progress has been made with the acrosin inhibitors, e.g., aryl 4-guanidinobenzoates. One of these compounds has been selected for clinical study, the ethylparaben (EGB) derivative. A number of *in vitro* studies with these compounds show their ability to prevent human sperm function at nonspermicidal concentrations (including zona penetration).

As noted above, testing strategies will not be identical for all SFIs, depending upon their targets in sperm function. An *in vitro* testing sequence for some of the sperm enzyme-directed SFIs could proceed in the following steps:

1. Irreversible inhibition/inactivation of the enzyme isolated from the spermatozoon;
2. Inhibition of the enzyme after treatment of intact spermatozoa;
3. Testing of inhibition of the functional activity of spermatozoa by focusing upon one or more of the following:
   a. preventing zona binding and/or penetration,
   b. inhibiting sperm penetration into zona-free hamster oocytes,
   c. inhibiting induction of the acrosome reaction by external agents, such as a calcium ionophore,
   d. preventing sperm penetration into and through cervical mucus,
   e. inducing sperm agglutination,
   f. altering certain sperm movement characteristics without necessarily blocking motility altogether, e.g., inhibiting motility hyperactivation,
   g. reducing basal sperm metabolic activity.

ANIMAL MODELS

The effectiveness of compounds when testing has progressed from *in vitro* analysis to animal models has been generally unsatisfactory. Surfactants such as N-9 have not functioned to kill sperm nearly so well *in vivo* as *in vitro*. Such failures could arise from a number of factors, including inadequate delivery and dispersion, dilution, chemical or physical inactivation, etc. It is clear that the bioactive dose delivered *in vivo* may be quite different from its apparent counterpart dose *in vitro*.

As the understanding of reproductive physiology has increased, the ability to select the most appropriate animal models for testing vaginal contraceptives has improved. In Chapter 11, the advantages and limitations of relevant animals was discussed in depth. For most pilot studies, the rabbit remains the animal of choice, although artificial insemination is usually needed. Fertilization of oocytes or implantation can be monitored. A basic physiological limitation is that the rabbit does not have a mucus-laden cervix, and the biology of sperm entry and transport through the rabbit cervix is different from that in the human. Nonhuman primates such as monkeys are more physiologically relevant to the human, but also more expensive. The stump-tailed macaque (*M. arctoide*) has been used in testing of chemical contraceptive compounds. This monkey mates readily and is behaviorally cooperative. Its cervix is relatively straight, enabling easy collection of cervical mucus.

In principle, the baboon could be used for testing of barrier contraceptive devices since in this primate, as compared to macaques, the female will not attempt to remove the object placed into the vagina. However, the devices would have to be scaled to the vaginal size in this, or any, animal. Postcoital analysis of the cervical mucus after barrier device insertion would be biologically meaningful, but has rarely been undertaken, and is not required by the USFDA.

CERVICAL MUCUS

The cervix appears to offer great potential as a target for anti-sperm (and/or anti-pathogen) actions. A clear means of inhibiting sperm function is to block passage into and through the cervical mucus. The mucus has a multifaceted role in sperm transport. For example, it functions to remove seminal plasma moieties from the sperm plasma membrane. Some evidence suggests that the cervical crypts function as a sperm reservoir, storing spermatozoa after insemination, before releasing them to continue migrating to the egg. The close association between the mucus macromolecular backbone and the sperm surface may act to remove anti-sperm compounds delivered in the vagina, just as it "scrubs" the sperm surface free of seminal plasma moieties. Some spermatozoa enter the mucus within seconds of ejaculation. Thus, deployment within the mucus of antisperm compounds may be essential in the absence of a complete mechanical barrier to sperm transport into the cervix. There are two basic strategies for application of contraceptive compounds to the cervical mucus: 1) alteration of mucus properties such that it physically resists sperm entry, and 2) permeation into mucus of compounds that are bioactive against the sperm.

There has been little work which addresses either of these targets for contraceptive action. Success of either strategy requires adequate permeation into mucus of the active compound. Molecular characteristics that govern such permeation need to be identified more thoroughly than current knowledge provides. Those properties that lead to strong attachment (mucoadhesion), enhanced permeability and maximal bioactivity should be vigorously investigated. Several spermicidal agents appear to thicken mucus, but the mechanism of action is poorly understood and quantitation is difficult. Permeation of a compound into mucus depends not just upon the properties of the compound but also upon those of its carrier. This is due, in part, to a partitioning phenomenon that occurs at the interface between the mucus and the carrier. The opportunity exists to utilize carriers that actually enhance mucus uptake of the compound.

Testing of the sperm-mucus-spermicide/SFI interaction requires a supply of ovulatory cervical mucus, or surrogates for it, adequate for replicate assays. The human mucus is available in small volumes, and requires careful monitoring of the menstrual cycle of donators to determine optimal mucus collection times. Use of surrogate mucus may have some value in such testing. Reconstituted ovulatory bovine cervical mucus is available commercially (Penetrak™) and has been used. However, this gel is not identical to ovulatory human mucus, and is more permeable to human sperm. No synthetic gel has yet been utilized which is as close to human mucus as the bovine material. This should be a goal for future development.

FORMULATIONS

Formulation considerations should be addressed as soon as antisperm activity is confirmed in *in vitro* testing, and when animal experiments are begun. Preliminary analysis of formulation problems can proceed by careful evaluation of the structure, and physical and chemical properties of the agent. It must be possible to measure the concentration of the active agent in any formulation. Development cost issues, and suitability of a compound for large scale industrial production must be assessed. Such studies should be performed by scientists with specific expertise in formulations. Although appropriate formulation is critical, there currently exist very few people with such expertise and interest in formulation of contraceptive compounds. It is of note that work on several very potentially useful vaginal contraceptive agents has been halted because of formulation problems.

DELIVERY SYSTEMS

Delivery vehicles for compounds are almost as critical as the compounds themselves. These vehicles may have independent bioactivity against sperm, and/or be protective of the vaginal epithelium. There are a number of desired features of the final vehicle including that it be easy to apply, non-messy, well dispersing and long acting.

Better assays are needed to objectively measure the dispersion and muco-adherence of a compound from a test delivery vehicle. Investigators at the R.W. Johnson Pharmaceutical Research Institute have employed a semiquantitative system for measuring vaginal spreading using a sheep model [DW Hahn, personal communication]. In this model, access is easy, the vagina is of similar length to the human and gravity will not influence the flow. Using a human vaginal speculum, spreading can be directly visualized and photographed. In the testing of various carrier agents, e.g., jelly, oils, detergents, cyclodextrins, etc., multiple comparisons should be performed, including an aqueous control medium. Several such tests can be performed each day using the same individual ewe, which would reduce inter-animal variation. Interpretation of other methods for measuring dispersion, such as objective assessment of spreading over tissue explants, is uncertain.

Uniform coating of the vaginal epithelium may be most relevant to protection against STD transmission, since the effective concentration of spermicidal agents at the cervical os, or even within the mucus and cervical crypts, may be more important for contraceptive efficacy. The amount delivered and its dispersion within the vagina may vary substantially. A delivery vehicle, such as a dental adhesive, that adheres to the epithelium around the cervical os, and gradually releases effective concentrations of spermicide over a period of hours, would be ideal. From the limited studies with assay systems for intravaginal spreading, it appears that foam formulations produce the fastest and most thorough spreading. However, it is unknown how long the delivered dose of compound remains in place, as compared to other formulations. It should be appreciated that extended retention of the active agent by the vagina (or cervix) may impose tradeoffs. As the retention time increases, so, too, could the absorption of a compound; this might or might not be deleterious. Alternatively, slow release over an extended period may reduce absorption, since lower concentrations would be absorbed per unit time.

STD PATHOGENS

In order to produce a topical contraceptive that also provides STD prophylaxis, understanding the portals of entry of different pathogens is required. At present, however, the mechanisms of transmission and infection by many STD pathogens are not understood. The entire vaginal epithelium must be covered for protection against many pathogens, a feature that may not be obligatory for contraceptive efficacy. Development of bifunctional agents - potent against both sperm and pathogens - may be difficult and time consuming, but should not be abandoned altogether. It is entirely possible that agents which are highly potent against one of the targets may exhibit at least secondary potency against the other. For example, N-9 is known, at least *in vitro*, to have such bifunctional capability. In the context of barrier development, the strategy of developing sperm-active agents, which may or may not provide chemical STD protection, should be pursued. Barrier contraceptives that protect against STDs are also discussed in Chapters 7, 14, 11 and 30.

It should be emphasized that the mechanism of sexual transmission of the most harmful pathogen, HIV, is not understood. This limits our interpretation of *in vitro* tests which could be used to assess compounds that inhibit such sexual transmission. Although lymphocytes are the primary cell type infected with the HIV virus, other cell types can be infected including phagocytes, non-phagocytic antigen-presenting cells (e.g., Langerhans cells and dendritic cells), and epithelial cells.

The *in vitro* tests for HIV inhibition/inactivation have typically relied upon three steps:

1 Incubation of a compound with cell-free and/or cell-associated virus;
2 Elimination of the compound from the medium; and
3 Detection and quantitation of infection of target cells.

Various measures of infectivity have included syncytia formation, HIV-specific antigen or enzyme production, electron microscopy and PCR of HIV nucleic acids. Primary cervical cell cultures, cultures using immortalized cells, e.g., Hela 229, and genital tract secretions, have already been used as test systems for assessing the antiviral and antichlamydial activity of different agents. It may be necessary to utilize a panel of tests, including some that incorporate spermatozoa. In the context of HIV, such tests would utilize cell-free and/or cell-associated virus.

The involvement of vaginal epithelial cells in sexual transmission of HIV is unclear, but one *in vitro* assay which may prove useful in identifying compounds which block the attachment of HIV-infected cells to vaginal epithelium and their subsequent infection has been developed [Phillips and Bourinbaiar, 1992]. This assay has identified potential inhibiting compounds, but is limited by cytotoxicity effects on the tissue culture cells used in the preparation.

Relevant animal models for testing vaginal contraceptives against many STDs are desperately needed. The pig-tailed macaque (*M. nemestrina*) has been used in study of infection by chlamydia and could be of value for combined testing of contraceptives.

Animal models for testing activity against HPV and gonorrhea are virtually non-existent. The rhesus macaque model described in Ch. 15 has the advantage of knowing exactly when SIV infection occurs. This model has been used to conduct studies which suggest that N-9-containing spermicidal products inhibit HIV transmission following vaginal inoculation.

Immunostimulatory agents from plants, which activate the cellular immune system non-specifically and attract macrophages, have been proposed as a mechanism to block pathogen infection. This is an intriguing concept, but concerns of actually enhancing HIV infection have led other groups not to consider this approach further. Combinations of antibacterial/antiviral agents with the known spermicide, N-9, are being considered. To the extent that time-to-market is critical, this approach may be the easiest to pursue.

Due to the lack of knowledge about the exact mechanism of heterosexual HIV transmission, it may be desirable to target "co-factors" of HIV transmission rather than the virus directly. Other STDs or lesions may increase infectivity. Thus, remediation of these co-factors may contribute to a reduction of HIV transmission, in addition to that afforded by spermicidal agents alone. Clearly, continued interdisciplinary collaboration is strongly needed to integrate topical contraception with STD prophylaxis.

REGULATORY AND SAFETY ISSUES

After a compound has been identified for continued development, additional tests are required before introduction into humans. In the US, the FDA has established guidelines for determination of the safety of barrier contraceptives. Depending upon the regulatory history and non-contraceptive experience with new spermicidal compounds and formulations, preclinical testing of an agent's safety may entail animal testing of toxicity and irritation, vaginal absorption, metabolic studies, assessment of fertility and teratogenicity, and laboratory testing of mutagenicity and carcinogenicity. These issues are discussed in more detail elsewhere in these proceedings (see Chapter 9).

Irritability considerations have become increasingly important with regard to possible enhancement of transmission of STDs. Initial irritation studies can be made less expensive by focusing on macroscopic observation of frank irritation of the vagina in a limited series using an animal model (e.g. rabbit), or the development of a skin test that results in reasonable correlations with results obtained from vaginal mucosal surfaces. A "pocket" technique is being studied in which tissue from the reproductive tract of the pig-tailed macaque is grafted onto the abdomen [Patton, et al., 1987]. These methods offer promise as a means of large scale assessment of subsequent irritation by contraceptive or prophylactic agents and could serve as early screening models. They would not replace subsequent *in situ* testing in animals with more complete histopathological analysis which, in turn, would precede clinical testing. However, they could identify significant potential for irritation and as such, would not be prone to false positive results. Testing for penile irritation in an animal model is not easy. Monkeys would, in principle, be the most relevant model, but the cost of testing in primates is prohibitive for use in routine screening.

There do exist *in vitro* tests of inflammation (cultures of vaginal epithelial and foreskin fibroblasts) which may be appropriate for some immune responses. However, it has not yet been shown that the surrogate markers followed actually correspond to *in vivo* irritation rather than to frank inflammation, and as such, they are not adequate for contraceptive applications. Other model systems have been employed, e.g., rabbit skin, but they do not correlate well with vaginal irritation since they do not utilize mucosal epithelium. Some of these systems work well at high and low concentrations, but do not discriminate well over the range of intermediate values. Moreover, they are not amenable to testing effects of chronic exposures which may cause an irritation not induced in short exposures. Successful development of such systems into viable tests for spermicides is not considered likely.

Apart from intravaginal application of steroids, absorption of compounds from the vagina has generally been a neglected field of study. Ten years ago, Benziger and Edelson reviewed the information available on absorption of compounds [Benziger and Edelson, 1983], but information is scarce for known or potential spermicides. The degree to which absorption occurs must be assessed for any new spermicide, with or without antibacterial or antiviral activity.

SUMMARY AND RECOMMENDATIONS

There are a great many challenges and constraints on the development of new compounds and modalities for topical contraception and STD prophylaxis. Existing knowledge must be more thoroughly integrated, and many new areas of investigation and testing are indicated. Such work must understand and reconcile the perspectives of the user, the clinician and the manufacturer. These perspectives include the following concerns:

User Concerns	Manufacturer Concerns	Clinician Concerns
High efficacy	Long-term stability	Sperm-specific activity, if possible
Taste and odor free	Easy synthesis	No systemic absorption
Instantly active	Easy formulation	No adverse toxicity
Non-irritating	Latex compatible	Not detrimental to vaginal flora
At least 12-hour duration	Low cost	Anti-STD activity
Easily applied	Patent protectable	Rapid entry into semen and cervical mucus
Aesthetically acceptable	Limited regulatory requirements	No attenuation by genital tract secretions
		pH independent
		Easy patient instructions
		Application possible before intercourse

In considering the challenge of addressing these concerns, the Working Group emphasized the necessity of distinguishing short-term and long-term goals. In the short term, four areas were defined as requiring the highest attention.

1) Better formulations and delivery systems are needed for compounds about which a large amount of safety and efficacy information is available. Examples of such agents are N-9 and benzalkonium chloride. Improved formulation technology would aid in the development of future candidate agents.
2) Screening should continue of compounds that exhibit potential topical contraceptive action. Use of mixtures should be included in such testing. Synergy of action might enhance efficacy while reducing bioactive dose levels.
3) Validation of *in vitro* and animal models for STD transmission is critically needed. This will require better understanding of the basic biology and physiology of such transmission.
4) Integrated efforts must replace the relatively disjoint history of scientific development of many aspects of topical contraception. Interdisciplinary meetings are needed to improve our understanding of the science, technology, and needs of topical contraception. Such meetings must include acceptability considerations and the perspectives of women's

health advocates. An important component of this new meeting structure would be focused symposia on specific aspects of topical contraception. These could generate new insights in development, which would then be incorporated into a more broadly based, integrative exchange.

The recommendations for long-term goals build upon the short-term goals. Better identification and understanding of those critical reproductive processes that can serve as targets, or yield new approaches for contraceptive action, must be achieved. The focus should be on opportune and potential "vulnerable sites" for such action, such as the sperm membranes and the cervix. The body of knowledge on the causes of human infertility should be exploited. Aspects of sperm function required for fertility should be targeted, rather than cell death alone, recognizing that such action might not affect STD pathogens. Mixtures of compounds with specific actions could overcome this limitation, and/or target multiple sperm functional sites for increased contraceptive efficacy. The final recommendation is to increase attention to development of long-acting delivery systems. Here, the technical expertise and experience gained with other biological delivery systems and implants should be utilized.

REFERENCES

Chantler E, Fisher H, Solanki S, Elstein M (1992): Quantification of the *in vitro* activity of some spermicidal compounds. Contraception 46:527-536.

Phillips D, Bourinbaiar A (1992): Mechanism of HIV spread from lymphocytes to epithelia. Virology, 186:261-73.

Patton DL, Kuo CC, Wang SP, Brenner RM, Sternfield MD, Morse SA, Barnes RC (1987): Chlamydial infection in subcutaneous fimbrial transplants in cynomolgus and rhesus monkeys. J Infect Dis 155:229-235.

Benziger DP, Edelson J (1983): Absorption from the vagina. Drug Metab Rev 14:137-168.

Katz DF, Dunmire EN (1993): Cervical mucus: problems and opportunities for drug delivery via the vagina and cervix. Adv Drug Deliv Rev, In Press.

DISCUSSION

Felicia Stewart: I'm wondering if there is any hope that we may be able to think about systemically active agents. In other words, agents that do not have to be applied locally, but instead could be taken by mouth or by patch.

David Katz: There is active work, as I mentioned, on contraceptive vaccines that inhibit sperm function, for example, compounds that block the ability of the sperm to bind to the egg.

Gabriel Bialy: I think what Felicia was addressing is compounds like ketoconazole, but without its toxicity, that get into the reproductive tract of the male and act as a systemic spermicide as opposed to local. I know that CONRAD is supporting research in this area. The issue that I would like to address to the group as a whole, is this: it is one thing to set up screening programs, but each screening program must have specific selection criteria which would identify a compound or a group of compounds with which one should like to proceed with further development. There is no point screening unless one is willing to go to the next step, which most likely involves formulation technology, testing for irritability, shelf stability, and all of those things that deal with pharmaceutical product development. As a final step, one has to have selection criteria for which type of compounds and formulations one is willing to proceed to clinical studies. There is no point with any one of these steps unless we have the means and understanding of the successive steps and go to the ultimate model which, obviously, is the human being.

David Katz: We were very mindful of the iterative process in which these many factors that influence the success of a compound, the influence that they play throughout the development process. For example, ease of formulation and the economic feasibility of scale-up for industrial production.

Lourens Zaneveld: I totally agree with these points. If you start the process you've got to finish it. One of the questions that came up yesterday and the day before is do we really have sufficient money to do all this work. I don't know that CONRAD by itself can do so. We are hoping that the agencies will get together, ultimately, and get the industry interested, particularly if the compound has some anti-STD activity. Then through group effort we can develop it further.

Gabriel Bialy: This is where the issue of setting program priorities comes in. With a limited amount of funding which is characteristic of any and all of the programs that I am aware of that sponsor research, it seems to me that if you have identified one compound or two compounds that fit all of the criteria that you set up, I am a firm believer that necessary funding to do the formulation and all the remaining product development will be found.

David Katz: One thing that I sense is that meetings like this reflect the fact that now, as opposed to say ten years ago, there are more people with more expertise available and contributing earlier on in the development process. I think we are sensitized to this issue now. I think we have to organize ourselves better so that we don't pursue a magic bullet that could never be manufactured.

Henry Gabelnick: I was intrigued about your suggestion that instead of doing the expensive vaginal irritation studies which are done traditionally in rabbits and sometimes in rats, that a skin test be looked at as a way to screen compounds, and I wondered if you could elaborate on this a little more?

Donald Waller: The primary purpose is to determine whether or not a material might cause some irritation. Although mucus membrane tissue is preferred -- and there are other models available besides the rabbit vaginal models such as the cheek pouch in the hamster -- the cost of those other models make it just about as expensive. So the question became do we have other methods we use to determine irritation. Skin irritation is a fairly simple methodology, but would need to be validated. There are a variety of skin models that could be looked at that might be more sensitive or more closely aligned to the vagina as far as sensitivity. Now we recognize that the skin is not a mucus membrane, but we would hope that maybe there could be some validation studies done to see whether or not that could used as an alternative model.

Gabriel Bialy: We were equally frustrated with rabbit vaginal irritation assay, which essentially is a good assay, but it's time consuming and expensive. So we did some work on the rat ear assay and indeed you can induce irritations, but we were not able to get a good correlation between the rabbit vaginal irritation assay and the assay based on the skin of the rat ear. Indeed, there may be other parts of the anatomy. Maybe there is a way of sensitizing skin with some other agent and then applying something else to it, meaning a spermicide or other vaginal sperm-active compound, but at least in this particular instance the rat ear is not a good model.

Penelope Hitchcock: I think there are sort of two perspectives to take in this. One is under ideal circumstances; how would we put together a series of recommendations that will ultimately give us the greatest degree of comfort with these compounds? The other side of that is in a resource-limited setting, what's the minimum we can do before we go into people? I think in this issue that we better err on the side of being very confident about the safety of these compounds. In a sense I think I would take issue with the extent to which this research agenda has been supported in the past as it relates to prevention of STDs. I think we do need more money. I think we need to develop very good preclinical parameters so that even the safety testing, when we get to people, is done with some confidence. I do not think we can take keratinized external epithelium and know perhaps what's going to happen in a mucosal surface like the vagina.

Henry Gabelnick: I interpreted the idea presented by David Katz that this is to be used as a screen. When you have hundreds of compounds one can't go the rabbit vaginal irritation study which costs $10,000 to $20,000 a compound, but one would like to be able to screen first and then ultimately those that look the best would obviously be taken through the more standard irritation studies before beginning human studies. At least that's how I interpreted it.

David Katz: Yes, that's correct.

Robert Staab: I am a toxicologist. I have done a lot of work in rabbit vagina for irritation for about 15 years or so. I don't know what your protocol is that you're running $15,000 or $20,000 per compound. I get a lot of very, very good predictive data from a very simple animal model with the rabbit or the rat. The rat is less sensitive than the rabbit, but in some ways the rat responds more like the women's vaginal tissue. But for $700 or so, one can buy six animals, and for screening, good predictive data can be obtained from three animals. If fluorescein is used to see whether or not the vaginal epithelium has been stripped, you can also increase the sensitivity fairly easily. That's one comment. Another comment is that toxicologists have been extremely frustrated in trying to correlate mucosal irritation with any part of the skin, so it has not been an easy task. I know what we're saying here is let's see if we can develop something and then come up with a correlation and a validated protocol. That is fine, but I think if you were in a toxicology meeting right now you would find the toxicologists just as frustrated with that question. So it's not an easy one for this group to take on.

David Katz: Let me explain. The reason it mounts up to that type of price is at $500 a rabbit with histopathology in most commercial laboratories, and if you eliminate the histopathology, then according to some of the people we've spoken with, the degree of irritation is not adequately assessed. So then histopathology is included with controls and sham operations, etc., and it gets pretty expensive. The other thing is, with respect to fluorescein, a lot of the compounds that we're working with -- or the base compounds -- take up fluorescein dyes, so you run into a problem with attempting to ameliorate the requirements for histopathology.

Robert Staab: I see what you're doing now. I still think there is probably room for streamlining the process in the short-run and getting more bang for your buck, so to speak, by getting around it. You don't have to use fluorescein, that is just one option. As far as histopathology goes, on a first-time screen, I still think you can eliminate the histopathology and get a pretty good indication of vaginal irritation. Those tall columnar cells strip right out with anything that is irritating. If you find something that's not stripping out those cells, even if it's not the most sensitive model, you have useful information, then you can start spending your money in more elaborate protocols.

Richard Cone: We have no animal models that represent the major problems in the human vagina, that is, with bacterial vaginosis (BV) or yeast infections and a lot of the irritation and toxicity problems that can occur in those conditions. I was thinking that although one can focus on trying to come up with a very cheap screening test, say on the skin, it's not a cost effective process if the results really don't correlate very well. I was thinking, in terms of the call for interdisciplinary meetings, this meeting didn't have a session devoted to creating the best possible animal models for toxicity testing, not to mention efficacy testing. We have so few models for STD research and, as far as I know, we don't have any models that are really functional for BV or yeast. I would just like to go back to the notion that we really need to focus on even better vaginal models.

David Katz: One of the themes in our working group was this need to improve animal models in this type of testing. A reevaluation of the trade off between *in vitro* testing and *in vivo* testing is important. Again, just to reiterate the point that our group was trying to make, these skin tests were not suggested as replacing any of the mucosal testing that involves histopathology, but rather in a situation where you may have a great many compounds that you are screening -- this is merely a first pass and by no means a substitute.

Penelope Hitchcock: Let me deal with the scientific issue here, because I think we face this in a lot of situations. That is, determining whether or not a "pretest" is really a good surrogate for a "real" test. A perfect example is going into the field with HIV vaccines in populations that have a number of factors that make them extremely difficult to protect, including other STDs. I think we're going to flush a lot of good vaccines down the toilet because we cannot protect a high risk population even though they would protect most of us. If you are going to screen using skin testing, we think of the adverse outcomes: if it inflames the skin, then it's surely going to inflame the vagina, and I am not convinced that either direction is well-understood. Something that might not be inflammatory for the skin could be for the vagina and the converse. I don't know how you develop a protocol that would speak to all of the various compounds that you're going to deal with, without doing both of those tests.

Michael Burnhill: I would like to second that. I spent most of my life working with chronic vulvovaginitis and I now have a list of at least 30 classes of substances that cause vulvitis with distinct colposcopic or vulvoscopic changes. They include substances that women are using every day on their entire body: bath soap, shampoo, hair conditioner, etc. These women have no changes anywhere on their body except in the introitus and possibly an inch into the vagina where you can see a series of toxic changes to the mucosa using that wonderful new invention for checking the vagina called a colposcope. Between the colposcope and vaginal smears which show cytolysis you can begin to pick up degenerative and irritative changes, and my guess is that the skin is too insensitive and does not mirror the microecology of the vagina or its defense system.

Donald Waller: Just one comment. The perfect model is, of course, the human vagina. The question is how much money do you have and what is ethically possible to evaluate for toxicity? Obviously the skin is not going to be the vagina and it's not going to give you any information regarding the vagina directly, but we have to decide at what level we're going to assess effects. Toxicologists always have this question: which model should be used? How close to the human do you need to be? We could use the primate and get much more information, but I doubt if anybody is going to support looking at toxicity in the primate vagina before we do all these tests. We have to look at the most reasonable, most efficient, cost-efficient manner in which we can make an initial assessment of toxicity. That's not an easy question. It's easy for any body to get up and say we need to look at a vagina and do all these very sophisticated tests, but the practicality of the situation is we can't afford to do that.

Preclinical Study Design Working Group Contributors

Mariano Busso
Mt. Sinai Medical Center

Eric Chantler
University of Manchester

Douglas Colvard (Rapporteur)
Contraceptive Research and Development Program

Gustavo Doncel
Contraceptive Research and Development Program

Erik Dunmire
University of California, Davis

Sanjay Garg
National Institute of Immunology
New Delhi, India

DoWon Hahn
RWJ Pharmaceutical Research Institute

Michael J.K. Harper
Baylor College of Medicine

David Katz (Moderator)
University of California, Davis

Christopher Miller
California Regional Primate Research Center

David Phillips
Population Council

Shakti Upadhyay
National Institute of Immunology
New Delhi, India

Donald P. Waller
The University of Illinois at Chicago

Lourens Zaneveld
Rush-Presbyterian-St. Luke's Medical Center

27

CLINICAL STUDY DESIGN WORKING GROUP REPORT

Milton Cordero (Moderator)
Lisa Rarick (Rapporteur)

Profamilia
Calle Socorro Sanchez #64, Gazcue
Apartado Postal 1053
Santo Domingo, Dominican Republic (MC)
Food and Drug Administration
Parklawn Bldg., 14B04
5600 Fishers Lane
Rockville, MD 20857 (LR)

The working group entitled "Clinical Trial Design" consisted of participants from multiple disciplines including health care providers, social scientists, policy administrators, and clinical researchers. The group met for two afternoon sessions to review and discuss multiple questions including the proposed Family Health International guidelines for clinical studies of barrier contraceptives, both drugs and devices. They also discussed recommendations for future achievements in the area of trial design.

Several specific issues concerning clinical trial design were discussed at length. The major topics are presented here.

LENGTH OF STUDY REQUIREMENTS FOR CONTRACEPTIVE STUDIES

There was general consensus that six months was an adequate study length to assess contraceptive efficacy of barrier methods, both drugs and devices. The very nature of barrier method use precludes the easy recruitment and maintenance of subjects in a lengthy trial protocol. It was felt that issues of safety, although evaluated in the six month efficacy trial plan, should also be explored through other means. For example, colposcopic and bacteriologic assessments might be done under standardized conditions--after a pre-specified dose and duration of use.

Barrier Contraceptives: Current Status and Future Prospects, pages 311–321

While it is not always possible and at times may be unnecessary, the "ideal" clinical trial design would involve a randomized, controlled trial performed in the appropriate target population.

DISCONTINUATION OF SUBJECTS FROM TRIALS

The group agreed with the usual criteria for individual subject discontinuation from studies--such as adverse events, pregnancy, personal reasons, etc.. Much discussion surrounded the concepts of lack of monogamy, decreased coital frequency, and use of other methods, either instead of or concomitantly with the study method, as grounds for discontinuation.

Mutual monogamy has been a requirement for subject enrollment in some clinical studies, for purposes of ensuring that partners continue to meet entry criteria, or for avoiding exposure to STDs in trials in which such exposure could confound evaluation of inflammation or other adverse experiences, or in studies evaluating seroconversion in HIV discordant couples. The group agreed that, although monogamy may be a desirable feature in some specific efficacy trial settings, it is impossible to assure or enforce. In most settings, monogamy should not be a prerequisite for enrollment. Individualization to specific study design should be considered.

On the subject of coital frequency, it was felt that for admission to a study, subjects should report coital frequency of at least twice weekly. The participants concluded that subjects should not be dropped from the study as long as they remained sexually active (even as infrequently as once per month). If a certain coital frequency were required of subjects to remain enrolled it could conceivably encourage either dishonest reporting or possible coercion--for example, forcing a subject to locate a sexual partner in order to continue with the perceived benefits of the clinical trial (health care, financial compensation, etc.).

On the subject of concomitant use of other contraceptive methods, the panel recognized that in almost every setting it would be appropriate to counsel specifically regarding condom use. Whenever possible, the population chosen should be at low risk for sexually transmitted diseases (STDs) thus avoiding a high prevalence of condom use.

If a subject or couple does find it necessary to use condoms or another non-study method during a trial, information should be sought regarding frequency of use, reason for use, and intended future use. Subjects whose intent is to remain in the study and to use the test method should be maintained in the study and the data collected and analyzed appropriately. Subjects who intend to use condoms, for example, on a regular basis should not be encouraged to discontinue condom use, but should be encouraged to terminate the study.

In the case of studies done to evaluate the ability of a barrier to prevent STDs, the population enrolled will of necessity be at high risk for STDs. In this setting subjects must be counselled regarding the recommendation for condom use. Although this introduces increased complexity and the need for a larger sample size to the clinical study design, to withhold this recommendation would be ethically irresponsible.

END-OF-STUDY REQUIREMENTS AND MANAGEMENT OF SUBJECTS LOST TO FOLLOW-UP

It was agreed that in most circumstances a standard final follow-up visit, including pregnancy testing, should be included in contraceptive clinical trial design. After the study method has been discontinued (due to study completion or subject termination) a follow-up visit at 2 to 4 weeks is desirable. Reporting of adverse events and pregnancy testing to evaluate the last cycle of study method use would be done at this time.

Several participants pointed out that an actual visit and pregnancy testing may be impractical in some settings and could be replaced by ascertainment of adverse events and return of menses information by telephone or mail-in information.

The data from subjects "lost to follow-up" should be included in analysis up to either 2 weeks before the last study encounter or one week before if the encounter included a negative pregnancy test.

MEASUREMENTS OF COMPLIANCE AND CORRECT USE

Assuring correct and consistent use of a study method has always presented difficulties in clinical trial design, data collection, and analysis. Several methods of evaluating these issues were presented.

Separate small studies of various subgroups of subjects could be performed to intensively evaluate effectiveness of educational and counselling materials regarding instructions for consistent and correct use. Ability to correctly place a method could be validated by observation of individual use.

Another method of measuring compliance could be included in the larger study design. Here a random subset of subjects could undergo periodic assessments (for example, initially and at some later point) such as questionnaire completion and physical observation in order to evaluate subjects' understanding and retention of information given regarding instructions and ability to comply.

A further suggestion for assessing compliance with method use would involve partner interviews and comparison to study subject information obtained.

FAILURE RATE INFORMATION

The participants agreed with the concept of failure rates as defined in the presentation by James Trussell in the plenary sessions. There was no agreement as to which types of failure rates are the most important or useful to report. The merits of the various forms of reporting failure rates ("method", "user", "total", or "trial" failure rates, and failure rates occurring during "perfect" and "imperfect" use) were discussed and each was considered potentially useful. As the ability to accurately define these rates depends on the quality of data obtained, the discussion turned to issues surrounding data collection.

Aside from continuous subject surveillance, many submitted that data collection by monthly encounters and by several methods (such as interview techniques, prospective diary information and retrospective questionnaires) would be the most appropriate way to ascertain compliance and sexual activity information in a contraceptive study setting. Monthly encounter information could potentially improve data analysis and collection of similar information by different methods would be useful for comparing results and methods of obtaining information.

Many of the working group participants, although they appreciated the philosophy behind the idea of monthly information gathering, felt this would be impractical at many clinical sites, especially in developing countries.

NON-TRADITIONAL CLINICAL TRIAL DESIGNS

Although the randomized clinical trial has become a mainstay for evaluating contraceptive efficacy, some participants felt that other methods of evaluation of contraceptives may be more appropriate. The classical randomized trial is probably most useful when one wants to compare the efficacy of two similar methods (for example a plastic versus latex male condom). The randomized trial study design may have significant drawbacks and may, in fact, be impossible to design in less traditional settings.

First, it may not be realistic or possible to randomize between two methods, especially where one represents a highly novel method and is difficult to compare to a "standard" method. Second, some effects may be rare or obscure enough that it is not practical to assess them within the usual sample size and design of the standard clinical trial. Lastly, randomization, selection criteria, quality of care differences, and other situational factors necessarily cause the clinical trial environment to be significantly different from "real world" conditions.

Perhaps most importantly, random allocation in a clinical trial setting can eliminate or markedly alter a major attribute of barrier methods: in actuality, users generally select the method(s) they feel most comfortable with and which best suits their personal needs. This is especially important for highly behaviorally dependent methods such as barrier methods.

Accordingly, it seems reasonable, under certain circumstances, to pursue techniques to assess efficacy and acceptability of methods outside of the traditional clinical trial setting. Techniques might include service delivery and population based approaches.

RECOMMENDATIONS FOR CLINICAL STUDY DESIGN

- Standardize
    - Develop and make available standardized guidelines for clinical study of both contraceptive and STD preventive barrier methods. These guidelines should address the duration of studies, grounds for discontinuation, and final follow-up visits.
    - Present and maintain standard definitions across studies (adverse experience and failure rate reporting, etc.).

  - Standardize data collection methods and analyses, including calculation of failure rates.
  - Standardize counseling content.
- Correct existing shortcomings
  - Develop improved informed consent procedures including novel approaches to communication techniques and validation of subject comprehension and agreement.
  - Modify protocols, including counseling techniques, to meet appropriate cultural and educational needs. This would include early involvement of population-sensitive advocates in the design process.
  - Make available raw data for analyses for all contraceptive studies (past, present and future) within an appropriate time frame (one to two years from end of study)--especially for those studies funded by public sector funds.
  - Develop approaches and study designs to supplement the "traditional" randomized clinical trial.
- Find needed funds and optimize resource use.
- Develop interdisciplinary approaches to maximize use of resources.
  - This would include communication between agencies/companies regarding study goals, possible sharing of funds, cooperation on study design and conduct, and sharing of data. In this way, one study could serve the needs of more than one group and duplication of effort would be reduced.

DISCUSSION

James Shelton: The randomized clinical trial has always been considered the ideal way to study these methods, but in fact, it may not be the best way. One of the major properties of these methods is that they are selected and controlled by the user. A randomized clinical trial may create an artificial environment in which you are not really studying the method in the way it would actually be used. I recognize that your charge was "clinical trials," but was there any talk about the alternative of studying efficacy in a service delivery environment? There is the alternative of making a method such as this available in some fairly well-controlled service delivery environment where it is freely chosen and then observing through some follow-up procedure or some concurrent process what, in fact, the efficacy is *vis a vis* other methods that people might select. I would like to enter a plea that we not be so Pavlovian in the way that we think that a randomized trial is the best way to conduct science.

Lisa Rarick: Absolutely. But we need to have some kind of comparison group so that we don't get in too much hot water by having, for example, a high failure rate. If you had an arm that also had an approved product that also had a high failure rate you'd be in a lot better position. I think as studies get completed, and there are some comparison groups in studies, it will be easier to use at least a more recent historical control.

Rosalie Dominik: I think that we should really consider the randomized controls more. I think we have to pick appropriate controls.

James Shelton: I don't want to sound like I don't think there is a role for randomized trials at all. There certainly is. But what we've got to recognize is that they're still a very artificial situation and whatever pregnancy rates you might find might not apply to the general population.

Roberto Rivera: In our discussion we recognized that there was a need for efficacy information for different purposes or for different intentions. The problematic efficacy was certainly important. We were somehow limited because we were supposed to address clinical trials. But we didn't mention the so called introductory research, something that has to be pursued for the evaluation of matters and more population-based type of studies addressing efficacy, issues of acceptability, issues in general that are more problematic.

Clark Taylor: I didn't hear a very realistic approach to clinical trials in STD prevention. For example, how do you assess the risk in a population? How do you look at serial monogamy when you're looking at something like chronic hepatitis? How do you establish what a sexual cohort is? What is the risk of exposure to an STD in the first place? What are the various techniques that people are using in a sexual encounter? For example, one may be testing the product for pregnancy and that is very important, but if you are also concerned about STD prevention, you also have to look at whether subjects are engaging in other forms of sexual intercourse or play that are not related to pregnancy. We know herpetic Whitlow, for example, is a serious concern with digital stimulation. What do fingernails and the like cause in terms of irritation, cutting, and so forth in vaginal tissues? All these issues seem to be ground level issues. It seems to me that we ought to get those into the discussion. If you did discuss them and just didn't bring them up, would you tell us what your other thoughts were?

Kamran Moghissi: We felt we should separate those studies which are targeted to establish efficacy of contraceptives, and studies which are targeted primarily at prevention of STDS, because the target population for these two groups are quite different. Therefore, we should not try, at least at the outset, to mix up these studies, although we recognize there are devices or techniques that can prevent both pregnancy and STDs.

Clark Taylor: I'm confused as to how the target group for prostitutes could be different when you look at conception and STDs.

Kamran Moghissi: But everybody is not a prostitute. We don't devise contraceptives just for prostitutes.

Clark Taylor: As a sexologist, I have to point out that very few people are also monogamous.

Kamran Moghissi: No, but I think the point is that if we establish that a certain device is efficacious in preventing pregnancy, we can extend that to see whether it also prevents STDs. We can select monogamous couples to study contraceptive efficacy, but this may not be a good target group for studying STDs.

Jeffrey Spieler: I want to ask about the subject of compliance. You told us that you couldn't insist that people use only one method or that they use your method for every act of intercourse but that we had to collect that kind of data. How can you establish rules in a multicenter study so all investigators apply them relatively similarly? How much noncompliance can you accept in terms of either using a method other than the method that is being evaluated or not using any method at all? This can't come out through some novel data analysis. We have to figure out quantitatively, if at all possible, when to discontinue or recommend to a client not to stay in the study because they are not compliant enough. We really need some quantitative help in this area. It just cannot be qualitative only.

James Trussell: I believe that those issues can be adequately dealt with in the analysis stage. What we were trying to avoid was dropping women from a trial when, in fact, they may get important health benefits from a trial. We don't want to present a woman with the choice of having to go out and pick up a partner to keep her coital frequency up or being dropped from the study. We don't want people to not use an adjunct method of contraceptive if they find themselves in a position where they need to do so. What you do want, it seems to me, is to encourage people to be honest about what they're doing during the trial. If you set up rules that say you do x, y or z or you're going to be dropped, it seems to me that you're inviting people to not tell you the truth. That makes the analysis of the data much harder. I would rather have it be clear that subjects are not going to be punished or dropped from the trial for any reason, but when one analyzes the data, one analyzes it separately for those cycles in which only the method under study was used, cycles in which other methods were used, or cycles in which no sex took place, etc. I think you're kidding yourself if you set up rigid rules and think that people are then going to tell you the truth.

Jeffrey Spieler: I'm not interested in setting up rigid rules. I think we need to reach a balance between being too lax and therefore not really evaluating our method, and being so rigid so that people will not be honest in their reporting. The other issue is, as James Shelton was saying, that there isn't always a need for a clinical trial as we define clinical trials. Maybe for existing methods a different approach ought to be looked at where we can collect some data on acceptability, use effectiveness, and continuation. But in a much less rigid, much more realistic manner, and get out of the clinical trial mode.

James Trussell: I would rather have one good clinical trial than 40 pieces of garbage. I think that when you look at most of the studies we have, service delivery kinds of trials for spermicides and for condoms, they aren't worth a doodle.

Penelope Hitchcock: I would like to hear the recommendation with respect to testing barrier methods for protection against sexually transmitted infections. The standard is to counsel condom use, but there are clearly some methods that we would like to look at that are incompatible with condom use. Perhaps some of the compounds that we heard about yesterday may be oil-based therefore, destructive to latex for example. What is your recommendation in terms of how to handle those problems?

Lisa Rarick: We didn't discuss very much specifically related to how you would do a trial for STDs.

Penelope Hitchcock: I would like to encourage a microbiological assessment at the beginning of these studies -- not so much so to show efficacy with respect to sexually transmitted diseases. Especially if you're testing contraceptive efficacy. But so that we have some way of interpreting irritation and inflammatory changes, in terms of what the mediators of those inflammatory changes might actually be.

Alfred Shihata: We all realize that compliance with barriers is always a problem and will always remain a problem. It is incumbent upon all of us to try everything we can to enhance compliance. For example, with the second phase for the Femcap, each woman will be given two Femcaps. One would be in her purse all the time with instructions to never leave home without it, and the other one would remain home.

Elizabeth Ngugi: I would like to go back to Clark's point about control of STDs. I think that if we are serious about controlling STDs, as we do clinical studies I would suggest that we go from the simple, meaning couples and then to the complex, meaning people with multiple sexual partners. If we don't do that, I'm afraid we cannot control STDs.

James Shelton: I have a request that James Trussell open his mind and heart a little bit to the concept of collecting useful information on efficacy outside of a clinical trial. If studies that have been done are not worth a doodle, then I think that maybe that has to do with the methodology. I would point out that he and others have done an excellent job of analyzing efficacy data from the National Survey of Family Growth, which does not come at all from clinical trials, and does seem to be reflective of what really does go on in people's lives and the ways that they use methods. Those kinds of techniques can be applied to contraceptive technology outside of a clinical trial and give us meaningful information on how what methods really work in actual use.

Irv Sivin: I guess I very seldom agree with James, but in this I agree with James. The problem is that most of our introductory studies are not well-designed and are not well-executed. The problem with Jim Shelton's suggestion is that we have to wait for years before we can have national probability samples such as the Survey of Family Growth with a sufficient number of users that we can look at. We need good study design at the beginning when we first want to evaluate something for the safety and the effectiveness of the new method.

Jacqueline Forrest: I think the difficulty is deciding what we're trying to measure. Are we trying to get at the perfect use pregnancy rate of a method? Or are we trying to get some indication of what this is going to be like in a general population of users? Those are very different goals. We can get the first through clinical trials. The second is going to be almost impossible because of the population sub-group characteristics and the characteristics of who is going to go through the trouble of being in a clinical trial. Who needs the money? Who needs the services? Who doesn't need it? So, I think we need to figure out what our goal is and then we design the study to meet those goals.

Lillian Yin: I do not believe that for OTC products we must have randomization because it just doesn't make sense. Women tend to look at the product and self-select. So, if we are going to stress randomization, sometimes the study won't get done. I do believe that there are good parameters for a good study. Comparability is important even if it's a historical control. I agree with Jim Trussell that a good study is better than 40 little who-knows-what. A good study does not have to be randomized but there are many, many basic parameters you must control. Sometimes you may want to do a short, study, like Jim Shelton did. You could do that as long as you can stratify all the important parameters that you're looking for. The most important thing is that we must give our women something we can say is half-way decent. You can't just wait for ten years and Then say that's really no good. You need certain assurances and then follow-up. Have a small well-controlled study and then have the nerve to cut it and go and follow-up later.

Lisa Rarick: I think that your statistician would consider well-controlled, differently than you might be presenting right now.

Lillian Yin: I disagree. Our statistician believes that we are not going to go for the drug model and even our new center director suggested to us that he doesn't believe that we need 1,000 women or things like that. We are trying to be very innovative knowing that the barrier method may be the only thing that we may have for women at this point. Let's see what we can do. Our statistician is very willing and we do work with Dr. Bert Peterson from CDC. We are looking for many innovative studies that we can do.

Lisa Rarick: Let me just make sure that I understand. You are happy to accept historical controls if they are reasonably appropriate?

Lillian Yin: You can have historical controls, but there are some very basic parameters that you must meet. If you cannot meet those basic parameters there is no such thing as good historical control.

Pamela Stratton: Part of my concern is that when we use STDs as surrogates for HIV transmission, we may not understand sexual behaviors among people who know they are at risk and try other ways of having intimacy that may also cause HIV transmission. As I listen to this very feminist group, I get the sense that we're worried most about male to female transmission because it's so much more efficient than the other way. I'd like the group to remember we know very little about sexual transmission of HIV, or transmission of it from mother to child. We need to work to be able to identify and quantify HIV in semen, in cervical vaginal secretions, and understand just as much work has been done regarding routes of infection for other STDs, characteristics of the life-cycle of HIV and be able to, in the context *in vivo* studies, quantify if new chemical agents actually inhibit the amount of HIV that is there that is infectious. I believe very strongly in clinical trial designs and doing things *in vivo*. But we must make sure that we stay knowledgeable about what's happening in HIV research and provide guidance about how to think about the genital tract. Although a lot of HIV researchers do have background in STDs, I have from my perspective as one of the few OB/GYNs involved in this in my past, recognized that they know very little about the female reproductive tract.

Christopher Elias: Just a couple of quick comments about STD efficacy testing. I agree that we're going to have to do different studies for efficacy for contraception and STDs. I think that the STD efficacy test is going to be much more difficult. I think that we need randomized controlled trials that involve enough people at significant risk of

STD infection who are going to be infected after we counsel them and provide condoms for them. That is going to mean a large number of subjects. Particularly because we know so little about the sexual transmission of HIV that we can accept a surrogate yet. Maybe that's one area to look for: reasonable surrogates for HIV infection. We are talking about a many centered, many country study that, because of the confounding effect of IV drug use in industrialized countries, is primarily going to have to take place in the developing world. That is where we could focus, some of our public sector resources. One of the major obstacles to private sector development of effective barrier methods for STD prevention is the issue of clinical efficacy testing because companies don't have access to large multicenter, multicountry testing infrastructures. If we, in the public sector, were to focus on the development of a clinical efficacy infrastructure for evaluating barrier methods for STD prevention, we would see a lot more private sector investment in this area. I think people recognize its importance and its potential profitability.

Penelope Hitchcock: I need to deal with something that Chris said because it's really important. I agree that other STDs are not surrogates for understanding whether or not we've interfered with HIV transmission. But in the context of a clinical study and the variable incubation period, do you think we're going to be able to measure the very event that you've pointed out as the issue in the context of a timely result?

Chris Elias: I think it's going to be very hard. I think we should study the other STDs. I think before we get to issues of actually recommending products as HIV prevention technologies we are going to need data on HIV. We need to look at opportunities for efficacy testing, whether that's working with the vaccine efficacy testing structures and thinking about sort of a structure for testing HIV prevention technology, whether it's a vaccine or a vaginal product. I think we need to start now to establish that clinical testing infrastructure. It's going to take a long time to establish. The studies are going to take a long time. In our paper we make some estimates of sample size calculations and they are large -- even under the conditions that we assume. Given that we're going to study populations with a high risk of STDs we're going to have of lots of lost to follow-up, lots of design problems. So, I think yes, we should look at other STDs first but, ultimately, we have to look at HIV efficacy.

Lisa Rarick: I want to thank again all of the members of our working group for a very thoughtful and exciting discussion.

Clinical Study Design Working Group Contributors

David Archer
Contraceptive Research and Development Program

Susana Bassol
Department of Reproductive Biology, Medical School, Mexico

Gerald Bernstein
University of Southern California

Carmela Cordero
Association for Voluntary Surgical Contraception, Inc., Dominican Republic

Milton Cordero (Moderator)
Profamilia, Dominican Republic

Rosalie Dominik
Family Health International

Marc Karam
World Health Organization, Switzerland

Suporn Koetsawang
Siriraj Family Health Research Center, Bangkok

Oladapo Ladipo
South to South Corporation in Reproductive Health, Brazil

Kamran Moghissi
Wayne State University

Lisa Rarick (Rapporteur)
Food and Drug Administration

Arturo Remes
Veracruz General Hospital, Mexico

Roberto Rivera
Family Health International

Carol Sampson-Landers
Ortho Pharmaceutical Corporation

Irv Sivin
The Population Council

Felicia Stewart
Valley Center for Women's Health

James Trussell
Princeton University

28

ISSUES OF ACCEPTABILITY AND BEHAVIOR WORKING GROUP REPORT

Virginia S. Cain (Moderator)
Nancy E. Williamson (Rapporteur)

National Institutes of Health
6100 Executive Blvd., Room 8B13
Bethesda, MD 20892 (VSC)
U.S. Agency for International Development
SA-18, Room 1200
Washington, D.C. 20523-1817 (NEW)

Acceptability is one of the three attributes we seek to achieve in the development and promotion of new barrier methods, whether for prevention of unwanted pregnancy or STD/HIV infection. The others are safety and effectiveness. The three are closely related since products are unlikely to be acceptable if they are not safe and effective. However, acceptability can encompass a great deal more and the importance of different characteristics of the methods can vary among individuals and groups. It is important to try to address acceptability issues early in the development process. To this end, the acceptability and behavior working group contributed the following observations.

ASSESSING ACCEPTABILITY

For barrier methods, acceptability is often a function of the type of partner as well as the method itself. Thus, condoms may be considered acceptable for use with a temporary or new partner but not for use with a lover or stable partner. Similarly, condoms may be acceptable for prevention of infection but not for family planning (or vice versa). For barrier methods, more than for other methods, the acceptability of the method to the partner is often critical, especially for sustained use. It is likely to be difficult for one partner to use male or female condoms for a long period of time if the other objects.

Barrier Contraceptives: Current Status and Future Prospects, pages 323–337

As noted in the Forrest paper in this edited collection, barrier methods such as condoms and spermicides are not widely used in most countries. Yet they may meet the needs of certain types of users (e.g., young people with infrequent sex) or under certain circumstances (e.g., while waiting for another method). This raises the question of whether we need to achieve mass appeal for new barrier methods or if it will suffice if we identify new methods which attract a small but devoted following or are used at certain life stages.

The answer to this may differ for family planning and STD/HIV prevention. Since family planning is fairly well established and has always relied on a cafeteria of methods, we might be able to justify adding several more to the repertoire to expand choice, even if they do not attract large numbers of users. But to stop the spread of STDs/HIV, it will be necessary to find methods which appeal to huge numbers of people and, especially in the case of HIV-infected persons, will be used consistently for long periods of time.

In clinical trials, acceptability is often inferred from observed continuation rates. This is logical since individuals who are unhappy with the method are likely to discontinue it. However, this is only a crude measure of acceptability or user satisfaction. Some continuers may be quite dissatisfied with a method but feel there are no alternatives. With methods requiring provider removal (Norplant and IUDs), continued use may sometimes be due to lack of access to removal. Conversely, some discontinuers may have intended to use the method for only a short time or as a bridge to another method. The method may have been very satisfactory during the time that it was used.

Outside the clinical trial setting, using continuation rates to measure acceptability and satisfaction are even more problematic, since users, in developing countries at least, often discontinue due to supply problems. If continuation rates are used to assess acceptability, it is important that investigators thoroughly explore the "personal reasons" that users discontinue rather than aggregating them.

Clinical trials typically pay both providers and users. It is important that these payments are carefully calibrated. If providers are paid too much, they may steer users toward the method which they are being paid extra to provide, thus distorting the method mix. If users are paid too much, they may be interested in participating in the trial mainly because of the money, thus compromising the information they can provide on the method. Yet if users are paid too little, they may be lax in complying with the considerable demands of being in a clinical trial.

The stance of providers toward a method being tested in a clinical trial is crucial. They should be objective, positive and thorough in their explanations of the method. However, if providers are too enthusiastic, users may be unwilling to bring up problems or admit to poor compliance. On the other hand, if providers are negative, this may discourage use.

Potential barrier method users may differ significantly from the kinds of people typically eligible for clinical trials. For example, typical barrier method users may be single, have sex only irregularly, and be non-monogamous. Clinical trials frequently require that participants be in a stable relationship and have a certain sexual frequency (e.g., 2+ per week). In order to get typical use efficacy and continuation rates, more naturalistic field trials would be desirable to supplement the results of clinical trials.

These could be non-clinic based and have very few exclusion criteria. They could collect follow up information through home visits by study staff rather than requiring clinic visits for a non-clinical method.

PROVIDER BIAS AGAINST BARRIER METHODS

It has been argued that one of the reasons that barrier methods are not more popular (see Chapter 1) is that family planning providers have not promoted them enthusiastically. Family planning providers may be concerned that barrier methods are less effective than other family planning methods and may require more time for instruction. Consequently, they may not make the effort to ensure that the methods are stocked, made available, and taught. This applies particularly to methods like the diaphragm which is, at this point, a provider-dependent method. In support of this position, there is U.S. evidence that if a woman goes to a nurse practitioner, she is more likely to get a barrier method than if she goes to an obstetrician/gynecologist.

However, there is some evidence that providers have only limited opportunity to steer clients to or from particular methods because many potential users, at least in the U.S., decide first which method they want and then follow what they see as the quickest and cheapest course to get the method. Thus, if they want to use condoms or spermicides, they will go to a shop or pharmacy and not to a health care facility.

For devices requiring fitting (diaphragm, cervical cap), there is no question that providers are crucial in increasing access to these methods. More research could be done on the feasibility of improving providers' knowledge, attitudes, and skills in providing barrier methods and seeing what impact this might have on correct, consistent, and sustained barrier use.

Barrier methods and natural family planning (NFP) methods have roughly similar use effectiveness rates; both are highly dependent on user behavior, user motivation, and good instruction. Women's health advocates and promoters of natural family planning have something in common. Both have despaired that providers are an obstacle to use of their preferred methods by having a bias against them. Both groups complain that providers put an overemphasis on just one factor (effectiveness), ignore other aspects of acceptability (such as lack of side effects), and do not acknowledge that there are some people who really want another option (barrier or NFP method).

LOCAL PRACTICES AFFECTING USE OF BARRIER METHODS

Basic scientists may have a rational model of how human beings behave. Thus they may seek the best animal model for studying the human vagina or studying its normal and pathological states but not even consider that traditional practices like putting substances into the vagina can change the vaginal environment. In Zaire and Haiti, for example, it is not unusual for people to put certain kinds of leaves or substances into the vagina to enhance "dry and tight" sex [Brown et al., 1993].

Another widespread practice is douching after sex. This is done with a variety of liquids. This practice may directly conflict with the requirements of some spermicides that douching be postponed for 6-8 hours.

Still another traditional practice is to re-use anything that can be re-used. This is especially common in low income countries. In some cases, latex condoms are re-used and this practice is likely to become even more common with stronger products such as plastic male or female condoms. Given the huge commodity requirements worldwide for barrier methods, the enormous expense, and the very difficult logistical problems, providing a re-usable device is an attractive idea but only if the devices could be washed adequately. Otherwise, they could become a public health menace rather than a prophylactic. The HIV virus is relatively fragile and short lived outside blood or other bodily fluids and could probably be destroyed with simple washing in water. But what about other micro-organisms? Clearly, more research is required on this before any public health position can be established.

A final common practice is to add lubricants to barrier products. The use of oil-based lubricants with latex condoms can lead to the rapid deterioration of the latex resulting in breakage. Often users are unaware of the need to use water-based lubricants or the water-based lubricants are unavailable.

In Thailand, getting female prostitutes to use condoms is a crucial component of the Thai government's AIDS prevention strategy. Yet many women complained of vaginal irritation with the condoms being provided. There were also complaints about breakage. Recently, Thailand has changed the specifications for the amount of lubrication to be added to condoms purchased by the Ministry of Health. The increased amount of lubricant is one step taken to solve women's complaints. Another is the recommendation that women purchase a tube of KY jelly and use just a few drops of it when they have sex. Finally, a condom quality assurance program has been initiated. These responses of the AIDS program in Thailand are impressive examples of public policies which are sensitive to users' preferences and problems.

In a similar vein, it might be possible to change manufacturing specifications for condoms to increase their performance and acceptability, a point made in the plenary discussion of this working group's findings.

PARTNER INVOLVEMENT AND BARRIER METHODS

From an egalitarian point of view, the "duet method" might be ideal: one partner uses a spermicide and the other, a condom. Indeed, this approach seems to be accepted among some populations, for example, discordant couples (one HIV+, the other HIV -) who are highly motivated. This approach allows each partner to express care and responsibility. But whether the "duet" approach will ever be practiced on a widespread basis, we do not know. Of course, the costs and logistics would be more burdensome if large numbers of couples begin using two methods on a regular basis.

Although some women do not like barrier methods, there is certainly a widespread perception that men are more opposed to them than women. Among heterosexual couples (who were the main focus of this working group), the cooperation of the male partner is usually necessary for male condoms and probably also for female condoms as well as for methods like foam. Methods such as the contraceptive film and the diaphragm may be less obvious to the male partner and thus require little or no cooperation from him.

Women may fear economic loss, physical violence, or disapproval if they insist on barrier method use against the wishes of their partner. Introducing a barrier method often adds an unwelcome element of distrust into a relationship. Methods which can be used surreptitiously by women may be popular in many locales including in Africa where men often object to using condoms. (Referring to the difficulty of getting joint responsibility for contraception in Africa, an African participant in the working group observed: "In the West it may take two to tango; in Africa, it involves only one.")

DUAL PURPOSE DEVICES

Both condoms and spermicides have been shown to reduce the risk of pregnancy and some sexually transmitted diseases although there are still unresolved questions about the effectiveness of spermicides against HIV. This dual nature of condoms and spermicides has advantages and disadvantages.

On the positive side, it is sometimes possible to promote the device for one purpose but not for another. Thus in some settings, the Catholic Church has opposed mass media campaigns for condoms for family planning but not for disease prevention. At the couple level, users are sometimes able to reduce partner opposition by citing one purpose or the other. Thus, a married women who is concerned about getting an STD from her husband can introduce the condom as a family planning device. ("My doctor told me today I should no longer use the pill and suggested we try condoms.")

On the negative side, the image of condoms is often a poor one. Since the advent of effective medical methods of contraception, many people associate condoms with commercial or extra-marital sex and with disease prevention. There are concerns that it would be insulting to one's spouse or steady partner and introduce an element of distrust to propose using condoms in a stable relationship. However, there is some evidence from Mexico that promoting condoms for AIDS prevention does not harm the reputation of condoms for family planning [Stover and Bravo, 1988].

SUSTAINED USE OF BARRIER METHODS

In the family planning context, use of barrier methods is typically fairly short term. Individuals may use barrier methods early in their sexual careers, early in a relationship, between steady partners, between other methods, and during breastfeeding. When women and men settle into a steady relationship or complete their childbearing, they may gravitate toward longer term and eventually, permanent methods.

For the prevention of STDs/HIV, sustained use is an essential and new element to barrier method use. We need to know more about how to encourage people to use barrier methods on a long-term basis. "Recidivism" has been observed among gay men in San Francisco. In some settings, people may tire of condom use and seek out young people who are thought to be uninfected. This underscores the importance of understanding strategies and obstacles to sustained use.

How can we study sustained use? One means would be to find individuals or couples who have used barrier methods throughout their reproductive lives. It would be interesting to see what strategies they have employed. Are they bothered by boredom or lack of variety? Do they stick with only one type of barrier method or do they switch among barrier methods? Do they eroticize barrier use or consider it more of a habit and not think much about it?

MESSAGES WE NEED TO CONVEY ABOUT CONDOMS

Although efficacy data on condoms and spermicides are incomplete (see Chapter 2), it is clear that consistent and correct use of barrier methods is essential to their effectiveness. It is also very likely that correct and consistent use, at least for condoms, yields high efficacy. (The Centers for Disease Control is planning a new social marketing campaign with the message: "condoms work!"). One important message is, therefore, that people should use a barrier method correctly every time they are exposed to the risk of infection or unwanted pregnancy (assuming they are not using another method).

At the same time, using a barrier method in a less than perfect fashion can also be beneficial in many situations. The probability of becoming pregnant from a single act of unprotected intercourse is quite low. The likelihood of becoming infected with an STD from a single act of unprotected intercourse varies with the micro-organism involved, but is not 100 percent. While consistent use is a goal, public health professionals should not give the impression that barrier method use is a waste of time if it is not done perfectly.

DOUBLE METHOD PROBLEM, RISK PERCEPTIONS AND RISK ASSESSMENT

Barrier methods are not the most effective family planning methods; on the other hand, most family planning methods do not protect against STDs/HIV. This raises the question of whether family planning programs should recommend that people at risk of STDs/HIV use two methods: one for family planning and the other for STD protection. The two method solution could involve the use of two barrier methods (spermicide and condom) to increase their effectiveness but also could include both barrier and non-barrier methods such as condoms and oral contraceptives.

When a person is at risk of STDs/HIV, a method like condoms should be used no matter what the life cycle stage or other method used. Thus, in reference to heterosexual women, condoms are potentially needed during pregnancy, postpartum amenorrhea, after menopause, before menarche, and during use of pills, IUDs, implants, and sterilization. But the limited U.S. research available suggests that if a woman is using a method like pills or sterilization, she is less likely to use condoms even when she is at risk for infection. The situation in developing countries has not yet been assessed except among discordant couples who are often willing to use two methods.

A key factor in barrier method use (with or without another method) is whether an individual perceives he/she is at risk. How accurate are people's self assessments? Even more problematical, how accurate are their assessments of their partners' risks? How can health care providers and the mass media help people to be more accurate in their risk assessments? Is it worthwhile for family planning facilities to do risk assessment to help people decide whether they need to use two methods or if they want to use only one, to decide which one? These risk assessments may not be completely accurate but would they be better than nothing?

Another approach is to advise people to assume all partners are potentially infected, the analogy of "universal precautions". But is this practical advice and will people do this or will they consider it totally unrealistic and perhaps discount all the advice given? Clearly, the most cautious position of protecting oneself against STDs from all partners is not feasible for women who want to become pregnant.

An additional variant of double method use is to combine the use of a barrier method with fertility awareness. There are two ways this could be done. A barrier method may be used in the times of the menstrual cycle which are less risky for pregnancy while abstinence is practiced during the riskiest days. More commonly, no method is used during the less risky days while a barrier method is used during the risky period. This second approach appears to be a common practice in developing countries, although couples may not know enough to correctly define the risky period.

The potential advantages of this approach are that fewer commodities are required and people may be willing to use this method longer since they do not always need to use a barrier method. Use of the barrier method in the risky period may also be done with unusual vigilance since the couple knows that they are in the risky time of the woman's cycle. The main disadvantage is that this timed use of a barrier method does not fully protect against STDs/HIV.

IMPORTANT QUESTIONS RELATED TO ACCEPTABILITY AND BEHAVIOR THAT SHOULD BE ADDRESSED IN THE COMING DECADE

1 Provider Attitudes: We recommend that operations research be conducted on the following topic: Is it possible to change providers' attitudes toward barriers and their sensitivity to gender and power issues? If so, what would be the impact of these changes on users' behaviors, including correct and consistent barrier use?

2 How can we increase initial acceptability (i.e., user willingness to try a method) of barrier methods?

- What are the best ways of changing community norms?

- Reaching men: How can men be convinced that using barrier methods is in their best interest?
- How can we best convince people that condoms can enhance sex?
- What are the best ways of promoting condoms for which groups?
- How can we give condoms a positive image (condoms are clean, condoms work)?

3 How can we <u>increase experiential acceptability</u> (i.e., willingness to continue use of barrier methods)?

- How can sustained use be promoted?
- How can community norms be changed regarding use of barrier methods?
- What are the "strategies" of successful sustained barrier method users?
- How acceptable is the "duet approach"?
- Is it more effective to encourage people to use a mix of barrier methods or only one?
- Is it better to eroticize barrier method use or encourage people to just make it a habit?
- How safe and acceptable is re-use of barrier methods such as the plastic male and female condoms?
- Who should be advised to use double methods? Who will comply?
- What should be recommended if people at risk of STD and unplanned pregnancy will not use two methods?

4 Basic research on <u>sexual behavior and gender power</u>

- Decision making regarding use of barrier methods
- Gender/power issues
- Baseline data on sexual behavior against which to measure changes
- Local practices which may affect acceptability of barrier methods

IMPORTANT ACHIEVEMENTS THAT COULD BE MADE IN THE COMING DECADE

1 Given that we will never get a perfect method, i.e. a single method that meets the diversity of needs of individuals and couples, our goal should be to increase the range of choice of methods. We need methods which are simple, easy to use, highly effective for both pregnancy and STDs, female controlled, able to be concealed from one's partner, quicker acting and longer lasting (i.e., 8 hrs? 24 hrs?).

One idea for meeting women's needs for STD and pregnancy prevention is to have, for example, two vaginal rings: one for STD prevention and one for pregnancy prevention. They could be used together or separately.

2 We would like to see an increase in the number of countries which provide barrier methods as a part of their national health/FP programs. We recommend that contraceptive service policies and providers attempt to integrate STD/HIV prevention (and in some settings, STD treatment) with FP. This will require that providers broaden their skills, knowledge, and attitudes toward reproductive health. If STD services are begun, it will also be desirable to initiate treatment of male partners so women will not get reinfected. An important component of this recommendation is to support programs which will increase male involvement. An example is a recent project in Zimbabwe [Piotrow, et al., 1992].

3 We would like to see more widespread promotion of barrier methods through the mass media including messages directed toward men. One example of a successful program is condom social marketing which uses modern advertising and commercial marketing to promote condoms to men. But we would also like programs directed toward men which promote a variety of barrier methods including condoms.

REFERENCES

Abdool Karim Q, Preston-Whyte E, and Abdool Karim SS (1992): Teenagers seeking condoms at family planning services, part I: a user's perspective. South African Medical Journal 82:356-59. (There is also a companion piece on providers' attitudes).

Beckman LJ, Harvey SM, and Murray J (1992): Perceived contraceptive attributes of current and former users of the vaginal sponge. J Sex Res 29:31-42.

Beckman LJ, Harvey SM, and Murray J (1992): Dimensions of the contraceptive attributes questionnaire. Psychol Women Q 16:243-259.

Boulos ML, Boulos R, Nichols DJ (1991): Perceptions and practices relating to condom use among urban men in Haiti. Stud Fam Plan 22:318-325.

Brown JE, Ayowa OB, and Brown RC (1993): Dry and tight: sexual practices and potential AIDS risk in Zaire" Soc Sci Med, in press.

Bruyn M (1992): Women and AIDS in developing countries," Soc Sci Med 34:249-262.

Cates W and Stone KM (1992): Family planning, sexually transmitted diseases and contraceptive choices: a literature update---part I. Fam Plann Perspec (Mar/Apr) 24:74-84. Part II (1992): Fam Plann Perspec (May/June) 24:122-128.

Coleman S (1981): The cultural context of condom use in Japan. Stud Fam Plan 12:28-39.

Germain A, Holmes KK, Piot P, Wasserheit JN (1992): "Reproductive tract infections: global impact and priorities for women's reproductive health" New York and London:Plenum Press.

Hardon AP (1992): The needs of women versus the interests of family planning personnel, policy-makers and researchers: conflicting views on safety and acceptability of contraceptives. Soc Sci Med 35:753-766.

Jivasak-Apimas S (1991): Acceptability of the vaginal sheath (Femshield) in Thai couples. Contraception.

Joanis C (with Steiner M, Spruyt A, Piedrahita C and others) Series of papers on acceptability of latex condoms, plastic male condoms, female condom, and spermicides. Available from Family Health International, Research Triangle Park, N.C.

Jaost K and Forrest JD (1992): American women's sexual behavior and exposure to risk of sexually transmitted diseases. Fam Plann Perspec (Nov/Dec) 24:244-254.

Liskin L, et al. (1990): Condoms: now more than ever. Population Reports. Johns Hopkins University.

Maticka-Tyndale E (1991): Sexual scripts and AIDS prevention: variations in adherence to safer-sex guidelines by heterosexual adolescents. J Sex Res 28:45-66.

Piotrow PT, Kincaid DL, Hindin MJ, Lettenmaier CL, et al. (1992): Changing men's attitudes and behavior: the Zimbabwe male motivation project. Stud Fam Plann (Nov-Dec) 23:365-75.

Plichta SB, Weisman CS, Nathanson CA, Ensminger ME, Robinson JC (1992): Partner-specific condom use among adolescent women clients of a family planning clinic. J Adolescent Health 13:506-511.

Potts M and Short RV (1989): Condoms for the prevention of HIV transmission: cultural dimensions. AIDS 3 (suppl 1), Current Science Ltd., S259-S263.

Shain RN (1980): Acceptability of contraceptive methods and services: a cross-cultural perspective. In RN Shain and Pauerstein (eds): "Fertility Control" New York:Harper and Row, pp 299-312.

Shelton JD, Harris JR (1990): Role of the condom in combatting global AIDS: the application of sutton's law to public health. In Alexander, NJ et al., (eds): "Heterosexual Transmission of AIDS" New York: Wiley-Liss, Inc., pp 327-337.

Sonnex C, Hart GJ, Bounds W, Williams P, Guillebaud J, Adler MW (1989): Sexual behaviour and attitudes to barrier contraception among family planning clinic attenders. British J Fam Plann 15:71-75.

Stover J and Bravo M (1988): The impact of AIDS on the use of condoms for family planning in Mexico: a SOMARC special study. p 23. (Subsequently, published in International FP Perspectives (1991) 17:61-64).

Weisman CS, Plichta S, Nathanson CA, Ensminger M and Robinson JC (1991): Consistency of condom use for disease prevention among adolescent users of contraceptives. Fam Plann Perspect 23:71-74.

Williamson N (1991): Summary: international acceptability task force meeting. Family Health International, Research Triangle Park, NC.

Worth, D (1989): Sexual decision making and AIDS: why condom promotion among vulnerable women is likely to fail. Stud Fam Plann 20:297-307.

DISCUSSION

Carlos Conde-Glez: Have you considered the need to develop any special studies to deal with the issues of acceptability and behavior in adolescents?

Nancy Williamson: I think we really neglect the adolescent issue. We did not do that. We did point out that adolescents are the major consumer of barrier methods in some settings and that they are very important.

Penelope Hitchcock: I want to address two issues that I think are really the essence of issues that we face. The first has to do with clandestine use. Let me lay out some examples which I think absolutely mandate the development of some sort of barrier method for STDs and they are as follows.

One is in adolescent populations. It is very clear that we as a species begin physical intimacy often before we are able to be intellectually and emotionally intimate. That makes negotiation of barrier methods among young couples very difficult.

Secondly, we all engage in perhaps pseudo monogamy. If you look at the divorce rates in the United States, 50 percent of marriages end up in divorce. How many of those divorces, at some point, encompassed extramarital relationships? None of those would be dealt with in the context of negotiation.

Thirdly, the partners of IV drug users are at enormous risk for heterosexual transmission and may not be able to negotiate barrier method use.

And fourth, we have bisexuality, which we understand to be the acceptable expression of homosexual preferences in many cultures. I think that all of those things mandate that women have a method they can use without negotiation with their partner.

Finally let me expand on the issue of serial monogamy. In an editorial in the Washington Times last year, Jim Mason described a little mathematical scenario for us. He postulated that a 15-year-old would have a first sexual experience and would have two partners a year in a serial monogamous fashion thereafter. Of course each time imagining that this was their partner forever. The given in the situation is that each one of their partners entertained the same sort of pattern of partner experience. If you do the math, by the time these kids are 24 years of age, they have been exposed to over 500 partners. In the District of Columbia where we have close to 2 percent HIV prevalence in a cohort of 15 to 17-year-olds, that means that these kids are going to be exposed to HIV- infected partners. I think that we have to push for the development of barrier methods suitable for clandestine use. We have beautiful models that you have given us in the field of contraceptive development. We have the pill, we have the IUD, we have surgical methods, and now we have Norplant. None of these require partner cooperation, approval, or knowledge and women have demonstrated that they will use these methods and they will comply if we make them available.

The second thing I want to mention is risk assessment in partners. It is absolutely impossible to look at a partner or talk to a partner and know anything about whether or not he or she has a sexually transmitted disease. We know that 30 million Americans have herpes infections, many of which are asymptomatic. We know that probably 40 million people have human papilloma virus infection, many of whom have no symptoms.

Furthermore, there was a very disturbing letter in JAMA a couple of years ago polling sophomores in college. The questions were basically, have you ever been dishonest with a partner in order to get them into bed. Both men and women in alarming numbers said yes. I think the most disturbing thing was that when they were asked whether they would withhold information about their HIV status, if it were positive, in order to have a sexual relationship, many of them said yes. So, I do not think you can assess your partner's risk. We all have to act as if our partners are infected and that we are susceptible. That's the only way that we can protect one another.

Roberto Rivera: I have a comment because of what you said on point number two, consideration of clinical trials. At the end of a clinical trial there is a research methodology to assess safety and efficacy. Acceptability is not a normal or expected outcome of a clinical trial. There is information like discontinuation rates and reasons for discontinuation that are commonly taken as evidence of acceptability but they cannot be taken at face value because of the circumstances under which such information was obtained. Very commonly people from your field approach our clinical trials people and ask them to ask some questions regarding acceptability. This could be done, but again, that information has to be taken with a lot of caution. On the other hand, if one wants to expand on that type of information, one has to bear in mind that the conditions or the procedures used to obtain such information may have an effect on safety and efficacy outcomes and one wouldn't like that to happen.

Michael Free: I'd like to bring up a possible way that the acceptability of condoms could be improved very quickly, namely by simply changing public procurement specifications. There is mounting evidence, I think, that many, perhaps most, of the complaints coming from the field recently have had nothing to do with poor quality, *per se*, or deterioration of condoms; but have had to do with the fact that the design of the product was not appropriate for the target groups. In particular, I'm referring to something we call modulus, which is the amount of force required to stretch the condom to any given length. The condoms that have been procured by most public agencies throughout the world happen to be condoms that are on the very stiff side of the spectrum. It turns out that when they go into hot countries they continue to get stiffer - up to 20 percent stiffer - and we're learning now that stiff condoms break much more easily.

My point is that, if we can simply change specifications and provide condoms that don't break, and that aren't too small or too tight, we are going to make a big difference very quickly in the acceptability of these products.

Alfred Shihata: I'd like to address the negative bias of the health care provider against barriers. There are two main things that we can do to greatly improve on that bias: 1) improve the effectiveness of barrier methods; 2) save the time of the health care provider. He or she does not like to spend an hour, a half-hour, or even 15 minutes teaching women to use barriers. A gynecologist told me he can do a hysterectomy much faster than teaching one woman the Prentiff Cap. If we give good educational tools to the woman, with very clear illustrated brochures and low literacy video tapes, it will cut down greatly on the time needed from the health care provider to teach women.

Eric Miller: I want to address the issue of male involvement. In the clinical trials group, the issue of males having to sign a consent form was raised. Of course it drew a lot of fire, but in light of STD/HIV transmission and the male's risk of infection through sexual intercourse, and also the increasing need for males to take responsibility for their children -- for example, there is increasing legislation cracking down on dead-beat dads -- I think that there will be an increasing need to involve men in our clinical trials. In addition, targeting couples instead of individual women as subjects in a clinical trial may encourage sexual communication.

Nancy Williamson: I'll try to respond to all comments real quickly. The issue of clandestine use is related to the products themselves. For example, if a product is very short-lived like some of the spermicides, then the woman can't put it in long in advance. She needs to put it in pretty close to the sex act which affects the clandestine use. The vaginal film is an example of a product which could probably be used quite well clandestinely. But we don't know about its effectiveness.

Regarding the question of risk assessment in partners, I think the points that Dr. Hitchcock made about risk assessment and partners are very well taken. The question is whether laboratory risk assessment is better than verbal risk assessment. We don't have the resources to test people for herpes, we don't have the resources to test people for AIDS, and so on. Verbal risk assessment is a relatively low cost thing, if you add some questions on a family planning visit, for example. It's not a question of whether it is perfect. Is it better than nothing?

Regarding the point about research methodology, I agree with Roberto's comments that clinical trials are not the ideal vehicle for looking at acceptability. I thought it was an interesting point that if you added too much work on acceptability it might actually distort the clinical trial.

Regarding Michael's comments about changing the specifications for condoms, I think that is a very important way to go. Honestly, I think it is so much easier to change the methods than it is to change the people, and we should make every effort to meet peoples's need to the extent that we can. The condom industry has been fairly complacent. They have had a product for 100 years and they really haven't been aggressive in trying to improve it. So, I really applaud his remark. I thought the suggestions regarding negative bias of providers were important. Certainly one of the reasons providers and family planning programs don't promote barriers is they just feel they're ineffective. If there is anything that a provider hates, it is a person coming in with an accidental pregnancy who has been using the method you told them to use. It's really a very depressing experience. Regarding the comment about couple involvement in clinical trials, I think it's very interesting in trying to get consent from men. In the family planning field we've never tried to make a requirement of male consent, but it's certainly very, very desirable.

Issues of Acceptability and Behavior Working Group Contributors

Gloria Alvarado
Institute of Scientific Research, Mexico

Marie Jose Araujo
Colectivo Feminista Sexualidade e Salude Bartolomeo

Virginia Cain (Moderator)
National Institutes of Health

Jane Cottingham
HRP\World Health Organization

Grace Delano
Association for Reproductive and Family Health, Nigeria

Jacqueline Forrest
The Alan Guttmacher Institute

Carol Joanis
Family Health International

Judy Norsigian
Boston Women's Health Book Collective

Wiwat Rojanapithayakorn
Ministry of Public Health, Bangkok

Pramilla Senanayake
International Planned Parenthood Federation

Rochelle Shain
University of Texas, San Antonio, Health Science Center

Nancy Williamson (Rapporteur)
United States Agency for International Development

29

PRODUCT DEVELOPMENT WORKING GROUP REPORT

Gabriel Bialy (Moderator)
Jeffrey Spieler (Rapporteur)

National Institutes of Health
6100 Executive Blvd., 8B07
Bethesda, MD 20892 (GB)
Office of Population
U.S. Agency for International Development
Washington, DC 20523 (JS)

The Product Development Working Group discussed the management aspects of product development as opposed to the scientific aspects of the R&D process. There were 35 members of the Working Group representing multi-national companies and large industrial concerns, very small entrepreneurial companies as well as individuals from not-for-profit R&D organizations and representatives of funding agencies such as WHO, A.I.D., and NIH. Although much of the discussion concerned development of products in the USA, the responsibility to discuss worldwide needs with special emphasis on the developing countries was accepted.

It was recognized that there are various approaches to product development, such as:

1 The traditional pharmaceutical companies, some of whom are already in the condom, spermicide, and/or diaphragm market. In their decision-making process for new product development, these companies would assess the nature of their capabilities in terms of available scientific and managerial staff, fiscal resources, marketing niche(s), and potential for profit generation. If the company considered that any one of the strategic elements were missing for a new product, it is very unlikely that the company would pursue the development of the product.

Barrier Contraceptives: Current Status and Future Prospects, pages 339–351

2 Companies that examine the availability of products in certain markets outside their country and then try to bring those products into their own or other markets. Frequently, this may involve the need for additional toxicology and clinical trials. At the same time, the uncertainty of whether the product will be effective is of less concern since it has probably already undergone considerable testing.

3 Innovative entrepreneurs who seek venture capital, frequently mortgaging their own homes or taking on extra jobs in order to secure sufficient funding, to develop prototypes of their products.

4 Public sector agencies such as CONRAD, FHI, Population Council, WHO, A.I.D., and NIH which try to define the nature of the products needed and are willing to provide funding for their development.

The operational styles and motives of these approaches may be different, but the sought-after end result is similar.

There was considerable discussion about the size of the market for barrier methods, especially in terms of dollars and numbers of "consumers." It was estimated that the U.S. yearly sales of condoms amount to \$200 million while spermicides bring in \$150 million. Trailing far behind are the sales of diaphragms at about \$15 million. Compared to the \$1 billion market for oral contraceptives in the U.S., barrier methods are not very profitable. However, with the AIDS pandemic and renewed concerns about other STDs, the barrier methods market has tremendous potential for expansion. It was recognized during the course of the Workshop, and during deliberations of the Working Groups, that what is required is the development of several new products. It was further recognized that such products, whether physical or chemical barriers, face a number of obstacles. Barrier methods historically have been shown to have lower efficacy than oral contraceptives, implants, injectables and IUDs, due largely to a lower acceptability and compliance with their proper use. Additional obstacles to development include regulatory requirements, lack of funding, and lower return on investment. The issue of product liability is invariably a potential impediment to the development of new products and cannot be dismissed even though it may represent a threat at varying levels to different companies.

Another important subject discussed related to the process that must be followed to expand the barrier contraceptive market. Does one emphasize the contraceptive purpose, or the STD purpose, or a combination product with sufficient information permitting indications for both? The companies represented in the Working Group expressed the view that products to prevent STDs appear to them to have more appeal than contraceptives. However, the issue of developing specific vaginal microbicides, without potential contraceptive action, may be outside the scope of work of organizations that are primarily charged with the development of new contraceptives.

Expansion of the barrier contraceptive market will, to a very large extent, depend on how the products can be promoted. Without a change in the regulations controlling fertility regulation; this change must be explored to its fullest potential.

Finally, the private sector participants expressed the view that the public sector agencies must have a clear goal in terms of their support for product development. An agency such as NIH sponsors development of products primarily for the market in the U.S.. While U.S.-based funding agencies may sponsor research on products such as the Reality female condom and on polyurethane condoms, and help to bring these products to the market, they recognize that these products may be too expensive for developing countries. The primary interest of agencies like A.I.D. and WHO are to develop methods suitable for use and supply in developing countries. In return for their contribution towards the development of such products, public sector agencies expect and attempt to negotiate a preferential price for the public sector.

As the final outcome of the Working Group deliberations, the five documents listed below were prepared and they are included as Appendices to this Chapter:

- A Recommendations for the Next Decade
- B Recommendations for Improving the Device and Drug Regulatory Approval Process
- C Public and Private Sector Partnerships for Development of Barrier Methods
- D Changing Behavior
- E An Urgent Manifesto

APPENDIX A: RECOMMENDATIONS FOR THE NEXT DECADE

- Develop several new barrier methods (physical and chemical) to prevent HIV and other STD transmission, and to prevent unplanned pregnancy.

  There is an urgent need to develop improved and new methods to prevent the sexual transmission of HIV, other STDs and unplanned pregnancy. Both physical barriers and chemical methods are required but it was recommended that, regardless of the approach, priority be placed on efficacious methods to help control the spread of AIDS. It was recognized that developing methods that people will not use, or that have requirements with which people cannot comply, will do little good in stemming STDs or unplanned pregnancy. Thus, special attention has to be paid to issues related to cost to the consumer, those related to the ideal characteristics of barrier methods, and to product initial and experiential acceptability (see chapter 3). The experimental plan should include the development of methods that represent an improvement on the currently available ones as well as methods that represent a breakthrough technology.

- Streamline the device/drug regulatory process and/or establish a new Reproductive Health Products division at the FDA and within the national regulatory agencies of other countries.

  The FDA should, in collaboration with both public sector agencies and the private/commercial sector, develop new guidelines for evaluating and approving vaginal products aimed at preventing the transmission of STDs. Without such guidelines, the private sector will be reluctant to develop new prophylactic products. In view of the urgency created by the AIDS pandemic, the FDA should consider organizational changes to ensure expedited and speedy review of all products (chemical and physical) that not only prevent life threatening diseases like AIDS, but that also prevent other STDs that act as cofactors for HIV transmission; these products must be placed in the "fast track" system of review. (See Appendix B for a fuller discussion of recommendations for improving the device and drug approval process.)

- Establish a forum for better communication between the Public and Private sectors, and the drug and device regulatory agencies (e.g., the FDA).

  The pharmaceutical industry has been relatively inactive in the area of developing new spermicidal entities or female physical barrier methods. Public sector agencies have tried to fill the void with some success. The funding levels and the industrial know-how are limited in the public sector. It is recognized that only private industry with its production and marketing expertise can bring new products to the marketplace. Consequently the public and private sectors and the regulatory agencies must establish a mechanism to promote a productive dialogue that will result in getting new, safe and effective products on the market in the shortest possible time. Very early in the research and development process, the regulatory agencies should provide clear guidance on what specific safety and efficacy studies must be performed in order to speed up the approval process.

- Develop ways to improve compliance with the use of currently available methods to prevent pregnancy and the transmission of AIDS and other STDs.

  The poor compliance with the utilization of barrier contraceptives is strongly suggestive that the educational messages do not always have the intended affect on the target populations, and that education alone is not enough to change behavior. Careful examination of the differences that exist in use and compliance rates among the different groups, regions or countries is suggestive of the fact that access has to be improved. It would appear that use and, perhaps, compliance would improve under the circumstances where there exists unrestricted access to the mass media, especially television, both for educational programming and advertising. Restrictive advertising policies with respect to contraception and prevention of STDs coupled with the explosion of sexually explicit programming has a negative effect on compliance and this practice should be corrected. (See Appendix D on changing behavior).

- Create a new way of doing business among public and private sector groups and drug regulatory agencies in product development. (See above for discussion of this item).

- Public sector agencies and organizations need to establish a mechanism to liaise on an ongoing basis for research and development in the area of HIV/STD prevention, including discussion of joint public and private funding of research.

  Limited fiscal, scientific, and managerial resources have had a negative impact on the development of improved and new HIV/STD prevention methods. There are few if any organizations in the public sector that have the means to address alone the complex issues involved in product development. Some kind of system or mechanism needs to be established on an ongoing basis to promote, coordinate and facilitate the necessary fiscal, technical and human resources of the public and private sector organizations involved in a product development in this field. The earliest possible commitment from the industrial sector must be sought in order maximize public-private sector collaboration and bring new products to the market as quickly as possible. The participants in the Product Development Working Group recognized that the dialogue established at this Workshop needs to be continued on a more formal and long-lasting basis.

- Tort reform/liability and risk sharing.

  Product liability has been frequently mentioned as an important deterrent to the development of new contraceptive products. This is especially true in the area of barrier methods where sales and profit are considerably lower than for hormonal contraceptives. Legislative initiatives aimed at tort reform have not been successful in the past, but that should not be taken as a deterrent for trying to formulate new legislative strategies. Funding agencies (public and private) may consider establishing a liability insurance fund which would be accessible to manufacturers. Access could be on restricted basis and products would be judged on their potential and, in the future, real impact on STD/HIV and pregnancy prevention.

- Conclusive studies need to be conducted on the effectiveness of existing barrier methods (physical and chemical) to prevent HIV transmission.

  While the literature is replete with numerous studies concerning the efficacy of barrier methods for the prevention of STDs, the majority of these studies suffer from a variety of shortcomings which leave us with uncertainty about the results. Furthermore, the interpretation of the results of these studies for the purposes of product labeling is difficult, if not impossible. Consequently, well-designed clinical studies must be initiated in order to ascertain the effectiveness of the various existing barrier methods for STD and HIV prevention. Recognizing the difficulty that implementation of such studies entails, it is mandatory that protocol development involve the participation of several disciplines and the various agencies that conduct and support research in this area.

Other recommendations discussed by the Working Group included: developing methods to prevent STDs but not pregnancy; developing improved *in vitro* assays for testing the safety and efficacy of STD prevention agents and contraceptives; considering the necessity of American companies going off-shore for product development and marketing; educating the medical community on over-the-counter and prescription medication for contraception and STD prevention; expanding the FDA monograph on Class I spermicides; developing non-latex condoms; developing methods to prevent non-curable STDs other than AIDS; and, creating tax incentives for R&D companies for the development of methods to prevent STD transmission.

APPENDIX B: RECOMMENDATIONS FOR IMPROVING THE DEVICE AND DRUG REGULATORY APPROVAL PROCESS

The regulatory approval process, despite the desires of all involved, often degenerates into an adversarial, and sometimes even antagonistic, process. It has become obvious that this is inefficient and counterproductive, and as such, requires immediate improvement. The Product Development Working Group has, in the very limited time available, come up with several suggestions for ways of improving communication, understanding and cooperation between officials of regulatory agencies and the private sector. In view of the fact that the "industrial" participants on the Product Development Working Group were predominantly representatives of U.S. companies, the statements in this section reflect on interaction between the product development sector and the U.S. Food and Drug Administration (FDA). However, it must be recognized that suggestions for the implementation of more positive interactions may very well apply to the drug regulatory agencies of other developed and developing countries.

In the United States, industry and the FDA need a better mutual appreciation of each other's respective goals and needs during the approval process. Therefore, we suggest the establishment of a Panel consisting of representatives of the FDA, industry, and public sector agencies to identify, review, and recommend actions for removing the troublesome aspects of the regulatory review process, such as slow review mechanisms and changing expectations and standards. As one example, precisely how the FDA reviews products - its mechanisms, chain of command, legal concerns, whether submissions are reviewed in a linear or parallel manner - is often a mystery or "black box" to industry. With a better understanding of the FDA's perspective and processes, perhaps industry will be able to better design their studies and submissions to facilitate the approval process - or mutually explore ways to address frequently perceived FDA constraints in available staffing or expertise. In turn, this Panel might allow the FDA to gain greater understanding and confidence in industry. For example, there is every incentive for industry to understand and comply fully and precisely with all FDA regulations and requests. Funding, time, and personnel considerations as well as liability concerns make this a fact of survival for industry. Therefore, we believe the impression that the FDA must be adversarial to meet its regulatory responsibilities could be lessened to a great degree. It might be replaced with the understanding that the FDA can offer specific guidance and collaboration without compromising its objectivity or responsibilities to protect the public.

There is a strong perception in industry that the regulatory process is like "trying to hit a moving target." In other words, requirements, methodologies, expectations, goals and regulatory interpretations are not clearly defined at the outset. This leads to misunderstanding and changes, both real and perceived, from previous positions or commitments. This, in turn, adds tremendously to the time, expense, and communication barriers already inherent in the process. We, therefore, recommend that a Working Group be established for each new product entered in the review process. The Working Group would:

- Consist of appropriate FDA representatives from the appropriate internal divisions involved (e.g., statistician, toxicologist, etc.) and appropriate Company representatives.

- Appoint a team leader or facilitator to act as the funnel through which all information and communications would be channeled. The facilitator would be responsible for coordinating and keeping the process on track.

- Agree at the outset upon expectations of the steps to be followed and the analyses to be performed, and record the commitments of both parties in writing. It would also indicate and monitor adherence to performance time frames.

- Provide a mechanism for answering questions or clarifying regulations or other official guidelines or responses.

- Provide a mechanism and consistent platform for addressing ongoing test requirements (e.g., IDE or stability studies).

- Provide a mechanism for handling complaints, disagreements and misunderstandings internally, to get away from public squabbles, unwarranted political intervention and other counterproductive measures that can interfere in the process.

The approval of new products should not be delayed by holding them to standards inordinately different from those applied to similar products already on the market, particularly when such standards are beyond state-of-the-art knowledge or science, or when they are required for specific products to answer more universal or general basic science questions.

Perhaps the most efficient and effective way to implement these recommendations would be through the establishment of a new unit within the FDA to be responsible for evaluating all products, both drugs and devices, related to reproductive health.

APPENDIX C: PUBLIC AND PRIVATE SECTOR PARTNERSHIPS FOR DEVELOPMENT OF BARRIER METHODS

Public agencies have played a key role in the development of all major methods of contraception now available. Public-private partnerships for contraceptive development were among the earliest examples of cooperation in product development not linked directly to public procurement. Thus, there is precedence established for collaboration in the expedited development of new and improved barriers for contraception and the prevention of STDs, including AIDS.

Interested U.S. agencies are able to adopt flexible approaches to joint development, scale-up, introduction and promotion of products that are affordable and appropriate for use in developing and developed countries. Special target groups include resource poor and/or high risk groups worldwide.

These agencies apply resources to research and development frequently leading to patentable new or improved products. They can also assist in the advanced development, evaluation, and introduction of proprietary products in the commercial and non-profit sectors.

A variety of agreements have been negotiated between agencies and private sector partners. In general, these agreements recognize the essential role of the commercial sector in the developing, manufacturing, marketing, and distributing of technologies. The predominant role of entrepreneurs, small companies and venture capitalists in the development of new barrier methods is also recognized. Public-private sector agreements include memoranda of understanding, joint development agreements, licenses, creative financing arrangements and performance contracts that may or may not be linked to procurement. The terms of these agreements are negotiated, recognizing that circumstances and goals differ for each product and partner. Confidentiality and other intellectual property issues are understood and accommodated if they are not in conflict with public sector rights.

In general, the public sector agencies and donors seek to protect the rights of the public sector and to meet the goals of appropriateness, availability and affordability for their target groups in exchange for the value brought to the product by public investment. The agencies may also require royalties or other considerations on private sector sales depending on the contribution of the public sector agencies.

Recommendations for expediting development of barrier methods in the pipeline and exploring new technologies through public-private sector partnerships include:

1 Ad hoc joint public-private working groups should be formed to better define needs, problems and markets, and to develop performance-based criteria for product categories.

2 Entrepreneurial approaches by public sector agencies should be encouraged.

3 Communication between the collaborating sectors for the purpose of information exchange and resource sharing on new product opportunities should be improved through appropriate forums, clearing houses and networks.

4 Creative means by which the public sector can share liability risks (and profits) should be devised.

5 Joint strategies should be devised to secure the sources of essential and non-substitutable new materials.

6 Joint planning should be done to avoid or mitigate competing interests between the sectors, e.g., public distribution of free goods versus development of local enterprise.

7 Research on issues of industry-wide interest should be emphasized.

APPENDIX D: CHANGING BEHAVIOR

While considering the urgent need to develop new and more acceptable devices and chemical agents to prevent pregnancy and the transmission of STDs (particularly the viral diseases), it is important to recognize that massive changes in behavior will be needed to maximize the benefit from any new products.

Recent reviews of behavior - changing strategies and programs such as the IOM study on "AIDS - Reducing Risk Taking" and Joy Dryfoos's book Adolescent Pregnancy have sharpened our awareness that it is simple to increase knowledge but very difficult to change behavior. It has been shown that in a population of human beings there will always be some (at one extreme of a normal distribution curve) who will take risks no matter what programs are directed towards them. At the other end of the curve are the "no risk" takers; they do not take risks and need practically no programmatic attention.

Near the center of the curve, where most of us are located, are people who may take a risk sometimes and who comply with recommended preventive behaviors at other times, but not all of the time. The "centrists" need to be constantly reminded of the need to protect themselves and others.

To move more towards a population of perfect compliers, it will be necessary to change the entire process of moral and behavioral education. This must begin practically at birth and involve parents, religious leaders, and the educational and health services system.

Fundamental behavioral change rather than the miraculous discovery of a "magic bullet" (that would remove problems of unintended pregnancy and sexually transmitted disease), is necessary as a long term solution to the current problems. These changes will require a massive and continuous effort to redirect moral, psychological and sexual behavior into a more healthy pattern.

Instructive, prudent, and low risk behavior must be marketed to a population that has, in recent years, been largely exposed through mass media to dangerous and destructive behavioral patterning. The first and easiest step in this process is to remove all marketing and mass media constraints to public information on contraception and STD prevention products.

APPENDIX E: AN URGENT MANIFESTO

The worldwide epidemic of serious and/or fatal STDs, particularly AIDS, is of such critical importance that it mandates as a first priority a maximum effort to coordinate the discovery, testing, and worldwide distribution of barrier methods and anti-viral (microbicidal) products.

This endeavor must be well coordinated and combine the efforts of the private and public sectors (including funding, regulatory, and distribution/marketing agencies) worldwide. Such a collaborative effort should accelerate the entire process resulting in shorter time span from discovery to use of new products.

As mentioned by Dr. Malcolm Potts in his keynote address to the Workshop, HIV alone could result in the deaths of up to 40 million human beings; a death toll equivalent to that of World War II. A catastrophe of such magnitude requires an effort comparable to the "Manhattan Project" that led to the development of the atomic bomb and shortened the war.

The second priority is the need to reduce further the morbidity and mortality that result from unwanted pregnancy. To this end, we recommend a major commitment of resources and personnel to develop new, improved, and less expensive contraceptives, including those that are under the control of the woman.

Every day that we delay committing ourselves fully to these ends will only magnify the devastation now being wrought by sexually transmitted disease and unwanted pregnancy.

DISCUSSION

Douglas Colvard: To what degree are the U.S. liability issues also a similar type of constraint in other countries?

Gabriel Bialy: I don't have the kind of experience to answer the question directly. But when I look at clinical trials being conducted by the World Health Organization on a world-wide basis on products that were developed by public sector agencies, including WHO, while they've been thinking about liability issues, WHO did not come up with any plan. They continue to do those studies they feel are necessary. The fact that there is good informed consent, that there is care for the participants in the trials, and because the folks in other parts of the world are not looking to sue just for the sake of obtaining money, reduces the urgency in the developing countries *vis a vis* liability. This was not something that the Product Development Working Group specifically addressed but, I would respectfully submit that as long as we take proper precautions in product development we should not let that liability issue perpetually act as a deterrent to new product development.

Bradley Pugh: First of all, I would like to say that it was a great pleasure being part of the Product Development Working Group. As a manufacturer of barrier products, it gave me new insight on what we are doing and where we need to be going. But there is one issue that I would like to raise again. We have to improve compliance with existing methods. While it is certainly important that we develop new methods of barrier contraception especially for STD prevention, we must remember that we do have effective methods out there already. The compliance with these is terrible and we need to be thinking about improving compliance, whether by promotion through the media or other ways.

Erica Gollub: I think the whole idea of a new, across-the-board FDA office looking at women's reproductive health methods which would include prevention of STDs, contraception and various other parts of women's reproductive lives is a very good idea and one that deserves some very serious further attention. It sort of goes against FDA dogma, or gospel. So, I think that the feasibility of it needs to be looked at. It potentially solves lots of problems that are dogging us right now in terms of what to do with combination methods, ones that fit both into drugs and devices review sections. Also, another problem I think that has been a issue for women is the kind of backlash created by the FDA considering all devices the same. I think in many ways we have suffered from issues related to the Dalkon Shield and, probably, breast implants. The FDA must distinguish between devices that actually have very low safety risks and health risks from ones that have potentially very high safety risks. I think that as a group, if there are people interested, we have to continue talking about this and to try to create a national working group to further the idea and to work with FDA personnel to determine its feasibility.

Gabriel Bialy: Not long ago, you may remember you expressed some great concerns about proposed action of the FDA on labeling of contraceptive products. We found that there was great responsiveness on the part of the FDA. They did not wish to harm the population by placing inappropriate labels on contraceptive products. The FDA asked for the participation of NIH and CDC to bring together all available information and, if need be, develop new information that could provide for adequate product labeling. Working groups have been established to address a variety of issues. I think that we are moving in that direction. I also think that the public pressure is indeed eliciting a response and I believe this response will be greater as time progresses.

Malcolm Potts: Following up on Erica's remarks and in the belief that the FDA Commissioner and staff are very professional and rational people, I have talked to a number of the manufacturers and to people in CONRAD and FHI, and I have suggested informally, after this meeting has completed that everybody interested in the subject should get together. As an ad hoc group of concerned manufacturers and people from not for profit institutions we should send a letter to the FDA Commissioner covering some of the things that we have discussed here.

Robert Staab: Essentially, I would like someone to comment on how we are going to take this statement of urgency up to where it might have an effect and what we might expect. Malcolm Potts certainly touched on that, but maybe others have some comments on what we intend to do.

Gabriel Bialy: Well, we have people in the audience representing the Food and Drug Administration, Penny Hitchcock representing NIAID, Pam Stratton from the Contraceptive Development Branch of NICHD; we shall bring the message of this meeting back to our organizations. I am sure Dr. Yin and Dr. Rarick will communicate the spirit of this meeting to their superiors and there certainly should be no problem in having follow-up meetings.

Michael Burnhill: I want to make a statement about something we discussed in the working group that didn't get presented. That is, we have to remove free marketing constraints to these products. We are at war. We are fighting deadly diseases and we are fighting unintended pregnancies. We can advertise detergents and hemorrhoidal suppositories, but the products that finally squeak through the FDA approval process at great cost are forced to be marketed as if we lived in the stone age. If we don't free ourselves from the media suppression on advertising these products we're going to have great difficulty even if we get great products out there. We need some way to lean on those responsible for constraining the written, visual and radio media to let us market these products openly.

Penelope Hitchcock: I think it is very clear that the public sector agencies on an international scale have a very clear mandate. But the interface of science and politics is, in fact, going to be the most fertile ground for progress here. Although each of you are experts in your field, and you contribute that expertise to the solution of the problem, you are also citizens with loud voices both in your neighborhoods and your states and your countries. What happens when scientists talk about more funding for science, is that we are seen as a group who has a vested interest in sustaining ourselves. The advocacy groups that are attending this Workshop have, in the AIDS epidemic in general and in women's health issues quite specifically, provided a lesson for us. That is, to be apolitical is to be unfunded and to be a political is to be ineffective. We in the institutes cannot do this. We need your help, we need you to mobilize.

Product Development Working Group Contributors

Phyllis Barber
London International U.S. Holdings, Inc., United Kingdom

Gabriel Bialy (Moderator)
National Institutes of Health

Lance Bronnenkant
Finishing Enterprises

Barbara Brummer
Ortho Pharmaceuticals

Anne-Marie Corner
BIOSYN, INC.

Luis de la Macorra
Grupo Profamilia, Dominican Republic

Fred Dietsch
Ansell, Inc.

María Louisa Eluf
SEMINA Industria e Comercio Ltda., Brazil

Michael Free
Program for Appropriate Technology in Health

Shlomo Gabbay
YAMA Inc.

Curt Hamann
SmartPractice

Mary Ann Leeper
Wisconsin Pharmacal Company

Donald Nicholson, II
PROFIT

Howard Price
Family Health International

Bradley Pugh
ALADAN Corp.

Patrick J. Rowe
World Health Organization, Switzerland

James Shelton
United States Agency for International Development

Alfred Shihata
Scripps Institute

Jeffrey Spieler (Rapporteur)
United States Agency for International Development

Robert Staab
Allendale Labs

Lillian Yin
Food and Drug Administration

30

HIV AND OTHER STDs WORKING GROUP REPORT

Penelope Hitchcock (Moderator)
Lee Claypool (Rapporteur)

National Institutes of Health
9000 Rockville Pike
Bethesda, MD 20892 (PH)
Contraceptive Research and Development
(CONRAD) Program
1611 N. Kent St., Suite 806
Arlington, VA 22209 (LC)

Blocking transmission is the ideal way to prevent and control sexually transmitted diseased (STDs), and interest in developing more effective mechanical and chemical barrier methods for this purpose is intense. The barrier methods that are currently available, however, have limitations. Most require male partner consent, if not active cooperation, and therefore cannot be independently implemented at the discretion of the female partner. Although epidemiological studies have demonstrated that the use of condoms or spermicides can reduce the risk of acquiring some STDs, there are many situations in which personal, social, or cultural barriers interfere with a woman's ability to successfully negotiate and implement their use. Specifically, the need for a method which can be implemented by women is grounded in the high prevalence of: 1) non-consensual sex; 2) sex without condom use; and 3) risky behaviors that occur without partner knowledge.

Therefore, just as oral contraceptives dramatically enhanced the ability of women to avoid unwanted pregnancy, effective female-controlled mechanical and chemical barriers are urgently needed to enhance the ability of women to avoid sexually transmitted infections. Ideally, topical microbicides could be developed with or without contraceptive efficacy; these would be responsive to women who wish to become pregnant while avoiding disease, who use contraceptive methods which either have no effect on or increase risk of STD transmission, or who wish to prevent STDs during their non-reproductive years. Furthermore, chemical barriers that inactivate pathogens in

Barrier Contraceptives: Current Status and Future Prospects, pages 353–362

vaginal/cervical secretions as well as in the ejaculate could reduce female-to-male as well as male-to-female transmission.

In order to develop products and the necessary behavioral interventions, many questions regarding the biological, biochemical, anatomical, microbiological and behavioral aspects of the problem must be addressed. Among the relevant issues are: 1) the acceptability of barrier methods that prevent infection alone or also prevent pregnancy; 2) the early steps in pathogenesis; 3) the mechanisms of action; and 4) the associated inflammation, irritation, and toxicity.

ACCEPTABILITY OF BARRIER METHODS FOR STD PREVENTION

Important acceptability factors should be addressed for methods that are only antimicrobial as well as for those that are both antimicrobial and contraceptive. The product should be invisible (i.e., colorless, odorless, and tasteless); stable, with a long shelf life; convenient to carry; easy to use; available over the counter; inexpensive; rapidly effective for the appropriate duration; and safe for daily single use (however, daily multiple use is desirable).

It is recognized that potential users of topical microbicides are socially, economically and culturally heterogenous, therefore, product characteristics should be modifiable based on specific user needs. For example, the invisible nature of the product might not meet the needs of some users - visible assurance of product use, medication-like qualities, or erotic features might be preferred. Unique packaging and distribution systems may also be needed to reach particular user groups such as adolescents, post-menopausal women, and sex workers, as well as non-monogamous and monogamous women. Behavioral research on acceptance and compliance issues must be undertaken. This agenda should include basic as well as intervention research.

MECHANISM OF ACTION FOR BARRIER METHODS

Identification/characterization of *new* microbicidal compounds is likely to be essential to develop chemical barrier methods for STD prevention. A number of potential compounds have been suggested including sulfated polysaccharides such as the dextran sulfates and carageenins [Ueno and Kuno, 1987; Busso and Resnick, 1990; Pearce-Pratt and Phillips, 1993] or plant products such as gossypol [Polsky et al., 1989] and neem tree derivatives [Riar et al., 1990]. After efficacy has been demonstrated in screening tests, the effects of active ingredients on normal flora and on the female and male reproductive tract should be thoroughly investigated. The effect of microbicides on yeast infections and bacterial vaginosis should also be characterized. It is presumed that microbicides will need to be fast acting in order to block infection, although our understanding of the early steps and chronology of establishing infection is still very limited. Mechanism of action studies should be undertaken on promising compounds.

The use of agents that might prevent infection by maintaining the acidic pH (4.0-

5.0) [Moller and Kasperson, 1991] of the vagina certainly merits investigation. Mononuclear cells and free HIV have been shown to be very labile at pH 4.0-4.5, [Ongradi et al., 1990; Voeller and Anderson, 1992]. Whether acidic pH could be maintained in the vagina given the effects of the alkaline ejaculate [Fox et al., 1973] is unknown. The dynamics and distribution of pH changes relative to tissue surfaces and the vaginal lumen also needs to be further clarified.

The role of H_2O_2-producing lactobacilli, and the possibility of developing better strains for explant requires further study. Whether the lactobacilli contribute to disease resistance merits additional attention, as well. Relevant research is already underway in the rhesus monkey to determine the effects of seminal plasma and pH on infectivity of simian immunodeficiency virus (SIV). It may also be possible to consider systemic approaches to acidify the ejaculate by systemic therapy such that infected mononuclear cells and pathogens would not survive. Associated effects on spermatozoa and fertility should be characterized.

The microbial spectrum of activity within the genital environment must be established. Bacterial STDs, such as chlamydial infection and gonorrhea occur at very high incidence rates and are typically established within the columnar epithelium of the cervix; these organisms may ascend to the upper tract causing pelvic inflammatory disease (PID). Chronic sequelae include tubal scarring, which leads to tubal infertility, ectopic pregnancy, and pelvic pain [Westrom, 1980; Wasserheit, 1989]. Furthermore, human papilloma virus infection of the cervix may lead to cervical dysplasia or cervical neoplasia. It may therefore be prudent to prioritize sites in the lower tract which should be protected, placing cervix first, vagina second, and vulva third. This likely represents the order in which protection is most feasible through currently available barrier contraceptives [Cates and Stone, 1992].

The role of spermatozoa in transporting pathogens or otherwise facilitating pathogen movement up the female reproductive tract should be clarified by further research [for HIV, see Anderson, 1992; Scofield, 1992]. Whether microbicides could target pathogens on the surface of, or within, spermatozoa without being spermicidal is unknown.

Once efficacy has been determined, the potential impact of all products on public health should be assessed. From the perspective of the individual, the ideal method is 100% efficacious. However, from a public health perspective, there is merit in considering the usefulness of approaches that are less effective. These may, with correct and consistent use, significantly reduce disease transmission and prevalence of infections. The problems associated with developing and licensing such partially or modestly effective products are formidable, however.

The efficacy and safety of barrier methods that are used by men to prevent sexually transmitted diseases should be viewed in the context of a variety of sexual behaviors. This includes condoms made of latex or of other materials. The development of behavioral interventions that target increased and correct use of condoms are needed, as well as new and better condom designs. Additional research to clearly establish the effects of vasectomy on the transmission of HIV and other pathogens is also warranted.

PATHOGENESIS

A more complete understanding of the early steps in infection, pathogen survival, and the kinetics of infection is critical. These features of infection are microbe-specific and relevant to the feasibility of both precoital and postcoital microbicides. Currently, the only postcoital practices that are purported to influence infection rates are urination in males and douching in females; neither have been demonstrated to reduce infection rates, however. In fact, douching has been associated with increased risk of PID [Wolner-Hanssen et al., 1990; Phillips et al., 1992]. Interestingly, douching is very widely practiced by large numbers of women in different cultures, who use a variety of techniques in doing so. The effects of different douching techniques on infection rates is not known, but interventions based on microbicide use may need to take into account the fact that women may be very reluctant to abandon this practice. This is especially likely where it is a traditional custom, or when associated with hygienic standards, e.g., when intercourse with different partners occurs within a relatively short interval of time.

The feasibility of targeted microbicidal agents is also related to the specific pathogens. Among the approaches that may be useful are: 1) immunoglobulins directed against conserved antigens on pathogens; 2) compounds that block receptor function; or 3) antibiotic peptides such as magainins or defensins. Other biological agents, such as plant products, may be useful, although much more knowledge regarding efficacy, safety, and their mechanisms of action is necessary. Specificity of action, however, is a very important property of effective, safe products, especially if deleterious effects on the natural flora and on the genital epithelium can be avoided. Achieving a high degree of specificity may also be a key element in developing products that are active against pathogens, especially HIV, without being spermicidal.

IRRITATION, INFLAMMATION, AND TOXICITY

An ideal microbicide must be *safe* as well as effective. In its most concentrated form, it must be non-inflammatory but potent enough to retain sufficient efficacy after being diluted with semen and vaginal secretions. The extent and effect of this dilution may vary greatly between individuals or within the same individual over time. With active agents that are more specific than the detergent compounds currently used in spermicides, the window between safe, efficacious concentrations and toxic concentrations may be much wider. In any case, more dose-response data is needed for current products as well as for any new agents and formulations. For dose-response data, it will also be necessary to identify parameters to be used in preclinical animal studies and to define their relevance as predictors of safety and efficacy in human clinical trials.

It is also desirable that the vehicles be neutral or synergistic with regard to antimicrobial activity, and neutral or ameliorative with regard to inflammatory effects. Better vehicles might promote more efficient coating of the vaginal, ectocervical, and penile tissues, and thereby contribute to preventing infection. Vehicles might also reduce the inflammatory effects of active ingredients.

For existing products and future candidates, it is necessary to define the frequency of use parameters for which the microbicide is safe. For example, if nonoxynol-9-containing spermicides cause genital ulceration with high-frequency use [Kreiss et al., 1992; Roddy et al., 1993], it is important to determine whether there is a frequency of use at which they are completely safe. In addition to frequency, there may be other parameters of use that would contribute to decreased safety. These include pre-existing inflammation or traumatic injury. If so, these need to be identified and safety limits defined.

The vaginal mucosal surface might be altered by microbicide-induced inflammation, thereby increasing infection risk by decreasing the infectious dose. The microbicide might affect tissue integrity (increased friability), vaginal pH, presence of target cells, mucus coat, endogenous normal flora and other more subtle properties of the normal reproductive tract. In other words, it is desirable that a normal healthy vaginal environment be maintained - changes induced by the use of microbicides should be characterized. Qualitative and quantitative effects and their duration should be evaluated.

In order to address these questions, the development of new methods/models as well as the adaptation of those that have been developed for other purposes may be necessary. Preclinically appropriate standardized methods for quantifying and documenting effects need to be defined and implemented. Parameters observed in preclinical animal studies, such as cytokine release, changes in tissue permeability, and cell mobilization, may differ greatly from those in human clinical studies, so the correlates of toxicity in humans will have to be established.

Standardized and improved clinical methods are also necessary for evaluating inflammatory responses and associated reports of irritation. In this regard, characteristics of the normal or typical reproductive tract tissues should be better defined at the level of both the individual and the population. Tissue changes that occur due to the hormonal variations of the menstrual cycle, hormonal contraceptive use, and many other factors should be investigated. Clinically, the correlation between symptoms and signs of inflammation is poorly defined. Furthermore, it is unclear what degree of inflammation is significant with respect to alteration of infection risk or comfort. Inflammatory responses may also be very site specific; the variability and significance of this response should be investigated.

Appropriate regulatory guidelines must be defined to expedite the development of safe and effective microbicides or mechanical barrier methods. Adequate regulatory guidelines already exist for estimating the toxicity and teratogenicity of new drugs (e.g., United States Department of Health and Human Services, Federal Register 45:82014-82049, 1980). Unfortunately, most existing barrier methods were marketed prior to the development of these FDA regulations, consequently safety and efficacy data for most barrier methods have never been gathered. Ideally, compounds will not be absorbed systemically and will be safe for use during pregnancy. Products which may be highly efficacious in preventing disease, but damage spermatozoa without being completely spermicidal, will present special regulatory problems.

The potential for antagonistic or synergistic interactions with combination products using both chemical and barrier approaches needs to be examined. Use of methods combining a chemical barrier with a mechanical barrier, or a chemical barrier with a delivery device, may be associated with increased or decreased inflammatory effects, and should be addressed experimentally. If mechanical barriers are used, it is desirable that sustained pressure on tissues be minimized. Ease of insertion is also relevant for reducing trauma and irritation during product use.

MAJOR QUESTIONS TO ADDRESS

In order to facilitate development of effective barrier methods, three major questions should be addressed:

1. What is the safety and efficacy of currently available barrier methods for preventing sexually transmitted infections?

2. What new strategies and new methods are feasible to use as protective barriers against sexually transmitted infections without partner knowledge or cooperation?

3. What are the general and culturally specific factors that influence compliance with barrier method use? Based on knowledge of these factors, what interventions will result in sustained, correct and consistent use of barrier methods to prevent sexually transmitted infections?

MAJOR ACHIEVEMENTS FOR THE COMING DECADE

Progress in this critical area will require the establishment of international, multidisciplinary, and coordinated efforts that include:

1. A significant increase in funding for research on barrier methods for STD prevention, with special emphasis on topical microbicides.

2. The establishment of a global clinical trials network to test the safety and efficacy of barrier methods for STD prevention.

3. The establishment of cooperative working relationships with regulatory agencies to define and implement an expeditious and responsible approval process for barrier methods for STD prevention.

4. The establishment of a creative and rational resolution of associated liability issues, perhaps including the establishment of a "no-fault" international compensation program based on pooled resources from manufacturers.

REFERENCES

Anderson DJ (1992): Mechanisms of HIV-1 transmission via semen. Journal of NIH Research 4:104-108.

Busso ME, Resnick L (1990): Anti-human immunodeficiency virus effects of dextran sulfate are strain dependent and synergistic or antagonistic when dextran sulfate is given in combination with dideoxynucleosides. Antimicrob Agents Chemother 34:1991-1995.

Cates W Jr., Stone KM (1992): Family planning, sexually transmitted diseases and contraceptive choice: a literature update - Part I. Family Planning Perspectives 24:75-84.

Fox CA, Meldrum SJ, Watson BW (1973): Continuous measurement by radio-telemetry of vaginal pH during human coitus. J Reprod Fertil 33:69-75.

Kreiss J, Ngugi E, Holmes K, Ndinya-Achola J, Waiyaki P, Roberts PL, Ruminjo I, Sajabi R, Kimata J, Fleming TR, Anzala A, Holton D, Plummer F (1992): Efficacy of nonoxynol-9 contraceptive sponge use in preventing heterosexual acquisition of HIV in Nairobi prostitutes. JAMA 268:477-482.

Moller BR, Kasperson P (1991): The acidity of the vagina. In Horowitz B, Marc PA (eds): "Vaginitis and Vaginosis." New York: Wiley-Liss, Inc., pp 63-68.

Ongradi J, Ceccherini-Nelli L, Pistell M, Spector S, Bendinelli M (1990): Acid sensitivity of cell-free and cell-associated HIV-1: clinical implications. AIDS Res Hum Retroviruses 6:1433-1436.

Pearce-Pratt R, Phillips DM (1993): Sexual Transmission of HIV. Biol Reprod 48:431-445.

Phillips RS, Tuomala RE, Feldblum PJ, Schachter J (1992): The effect of cigarette smoking, *Chlamydia trachomatis* infection, and vaginal douching in ectopic pregnancy. Obstet Gynecol 79:85-90.

Polsky B, Segal SJ, Baron PA, Gold JWM, Ueno H, Armstrong D (1992): Inactivation of human immunodeficiency virus *in vitro* by gossypol. Contraception 39:579-589.

Riar SS, Devakumar C, Ilavazhagan G, Bardhan J, Kain AK, Thomas P, Singh R, Singh B (1990): Volatile fraction of neem oil as a spermicide. Contraception 42:479-487.

Roddy RE, Cordero M, Cordero C, Fortney JA (1993): A dosing study of nonoxynol-9 and genital irritation. AIDS, in press.

Scofield VL (1992): Sperm as vectors and cofactors for HIV-1 transmission. Journal of NIH Research 4:104-111.

Ueno R, Kuno S (1987): Dextran sulfate, a potent anti-HIV agent *in vitro* having synergism with zidovudine. Lancet 1987i:1379.

Voeller B, Anderson DJ (1992): Heterosexual transmission of HIV. JAMA 267:1917-1918.

Westrom L (1980): Incidence, prevalence, and trends of acute pelvic inflammatory disease and its consequences in industrialized countries. Am J Obstet Gynecol 138:880-892.

Wasserheit JN (1989): The significance and scope of reproductive tract infections among third world women. Int J Gynaecol Obstet, Supplement 3:145-168.

Wolner-Hanssen P, Eschenbach DA, Paavonen AA, Stevens CE, Kiviat NB, Critchlow C, Derouen T, Koutsky L, Holmes KK (1990): Association between vaginal douching and pelvic inflammatory disease. JAMA 263:1936-1937.

DISCUSSION

Gabriel Bialy: The assumptions are correct that there is indeed a crisis situation and that in some way NIH will have to get away is from its rigid and parochial method of reviewing applications. I think the system as it currently exists was fine for basic research, but we need to develop strategies with room for at least some initial investment in innovative approaches that are not extensively developed with a lot of data, so that the investigator can establish applicability, effectiveness, etc. Our current approach is stifling. While I do not necessarily take issue with the great value of the peer review system, I think we need to alter our criteria. How we do this remains a question, but we need to work together and create a new forum for the review and implementation of critical projects.

Clark Taylor: It was immensely exciting to work with this particular committee. In addition to the ground we've covered here, we also need to develop a new paradigm for looking at what we're doing with barrier techniques. We need to clarify our understanding and language about sex. We mustn't confuse sexual orientation and psychological issues with sexual techniques. We mustn't confuse what causes conception with what transmits disease. This means thinking about things in a different way. Something we haven't talked about in this meeting is the immense suffering, such as the blindness, the pain, and the death, that are caused by sexually transmitted cofactors like cytomegalovirus, Epstein Barr virus, and herpes, for people who are already HIV-infected. This raises issues, such as kissing, oral sex, anal sex, etc., that we haven't thought about when we limit our focus to preventing conception with STD prevention as a secondary benefit. What we really have to do is re-think our goals as we incorporate STD prevention as a major focus of our activity.

Penelope Hitchcock: I think that all of us want to take action and we should recognize that this meeting represents a very substantial step in this process. This meeting has taken an interdisciplinary approach, looking at things from different perspectives, and talking about public and private sector interaction. It is not easy to develop a new discipline. It is not easy to address emerging problems. I think this meeting, certainly from the perspective of my own education and understanding of the problem, and the opportunity to network with all of you, has been very successful. I want to thank the organizers for having the insight to do this, and everybody in this room for their contribution and commitment. I think it's been a very productive and important meeting.

James Koch: Following on the comments that were just made, I'd like to know if anyone has plans for carrying on the tremendous impetus that was generated by this conference. If I were among those who planned and organized this I would be very proud, but I'm not close enough to these organizations to know what might be feasible to carry forward this impetus and not let it simply die after this conference is over.

Henry Gabelnick: As I said at the beginning of the Workshop, and repeated this morning, one of the primary objectives of the working groups was to develop action plans. Although not all of the items can implemented with specific actions, there are certainly aspects which can be, and we intend to do so, whether it's those things that are directly within our power, such as putting together interagency groups, or things that are indirectly within our power, such as establishing public sector/private sector collaborations. There are many other issues that will come forward. As you know, the proceedings will be published and the statements and recommendations from the working groups included. We intend to prioritize the most urgent recommendations and see that they are implemented.

HIV And Other STDs Working Group Contributors

Luis Acosta
Secretaria Distrital de Salud, Colombia

Michel Alary
Hôpital Du St-Sacrament, Canada

Lee Claypool (Rapporteur)
Contraceptive Research and Development Program

Carlos Conde-Glez
Instituto Nacional de Salud Publica, Mexico

Ann Duerr
Centers for Disease Control

Penelope Hitchcock (Moderator)
National Institutes of Health

Marie Laga
Institute of Tropical Medicine, Belgium

Alina Llop
Instituto Superior de Ciencas Medicas de la Habana, Cuba

Elizabeth Ngugi
Kenyatta National Hospital, Nairobi

Dorothy Patton
University of Washington-Seattle

Amy Pollack
Association for Voluntary Surgical Contraception

Lionel Resnick
Mount Sinai Medical Center

Ronald Roddy
Family Health International

David Soper
Medical College of Virginia Hospitals

Zena Stein
New York State Psychiatric Institute

Katherine Stone
Centers for Disease Control

PART VII: WORKSHOP PARTICIPANTS

Workshop Participants

Mary Ann Abeyta-Behnke
University Research Center
Federico Henriquez y Carvajal
#11 Sector Gascue
Santo Domingo, Dominican Republic

Dr. Luis Acosta
Secretaría Distrital de Salud
Cra. 24 No. 45-97
Santa Fé de Bogota
Colombia, South America

Claude Aguillaume, M.D.
The Population Council
1230 York Avenue
New York, NY USA 10021

Michel Alary, M.D, Ph.D.
Centre de Recherche
Hôpital Du St-Sacrament
1050 Chemin Ste-Foy
Quebec, Qc, G1S4L8 Canada

Dr. Gloria Alvarado
Instituto de Investigación Científica
Avenidad Universidad y Fanny Antinua
Apartado Postal B-7
3400 Durango, Dgo., Mexico

Frank Alvarez, M.D.
Profamilia
Calle Socorro Sanchez #64
Gazcue
Apartado Postal 1053
Santo Domingo, Dominican Republic

Lic. Rosa Rita Alvarez
Mujeres en Desarollo, Inc.
Edificio San Rafael , 6to Piso
Calle Leopoldo Navarro
Santo Domingo, Dominican Republic

Lydia Antolin
Contraceptive Research and
Development (CONRAD) Program
Eastern Virginia Medical School
601 Colley Avenue
Norfolk, VA USA 23507

Felice Apter, Ph.D.
Office of Population
U.S. Agency for International
Development
SA-18 Room 820
Washington, DC USA 20523

María José Araujo
Colectivo Feminista Sexualidade
e Salude Bartolomeo
Rua Bartulemeo Zunega, 44
cep 05426-020 Pinheiros
Sao Paulo-SP, Brazil

David F. Archer, M.D.
Contraceptive Research and
Development (CONRAD) Program
Eastern Virginia Medical School
601 Colley Avenue
Norfolk, VA USA 23507

Dr. Marcos Arévalo
International Planned Parenthood
Federation (IPPF)
902 Broadway 10th Floor
New York, NY USA 10010

Phyliss M. Barber
London International
U.S. Holdings, Inc.
1819 Main Street, Ste. 800
Sarasota, FL USA 34236

Mark Barone, D.V.M.
Office of Population
U.S. Agency for International
Development
Washington, DC USA 20523

Susana Bassol, M.D.
Departamento de Biología Reproductiva
Escuela de Medicina
Avenida Morales 900 Ote.,
ZP 27000
Torreon Coahula, Mexico

Gerald Bernstein, Ph.D., M.D.
Department of OB/GYN
University of Southern California
School of Medicine
Women's Hospital
1240 N. Mission Road, Rm 1028
Los Angeles, CA USA 90033

Gabriel Bialy, Ph.D.
National Institutes of Health
6100 Executive Blvd., 8B07
Bethesda, MD USA 20892

William J. Bologna
Columbia Research Laboratories, Inc.
501 East, 79 Street
New York, NY USA 10021

Lydia Bond, M.D.
Pan American Health Organization
525 23rd St., NW
Washington, DC USA 20037

Vivian Brache
Profamilia
Calle Socorro Sanchez #64
Gazcue
Apartado Postal 1053
Santo Domingo, Dominican Republic

John H. Bratt
Family Health International
P.O. Box 13950, RTP Branch
Durham, NC USA 27709

Lance Bronnenkant, Ph.D.
Finishing Enterprises, Inc.
825 Wurlitzer Dr.
N. Tonawanda, NY USA 14120-2089

Richard Coleman Brown, M.D., M.P.H.
Hôpital Sainte Croix-Leogane, Haiti
Box 407139
Ft. Lauderdale, FL USA 33340-7139

Barbara Brummer, Ph.D.
Advanced Care Products
Ortho Pharmaceutical Corp.
Route 1 South & Milltown Road
P.O. Box 6024
North Brunswick, NJ USA 08902

Michael S. Burnhill, M.D.
Robert Wood Johnson Medical School
Academy Health Science Center
CN19
New Brunswick, NJ USA 08903

Mariano Busso, M.D.
Mt. Sinai Medical Center
Pearlman Research Bldg.
Department of Research
4300 Alton Road
Miami, FL USA 33140

Martha Butler
AIDSCAP/Santo Domingo
Calle Moises Garcia
Sector Gazcue
Santo Domingo, Dominican Republic

Virginia Cain, Ph.D.
National Institutes of Health
6100 Executive Blvd., Room 8B13
Bethesda, MD USA 20892

Patricia Campbell, MBChD
Chartex International PLC
33 Cavendish Square
London WIM 9HF, United Kingdom

Lic. Magaly Caram
Profamilia
Calle Socorro Sanchez #64
Santo Domingo, Dominican Republic

Wes Cetnarowski, M.D.
Warner-Lambert Company
201 Tabor Road
Morris Plains, NJ USA 07950

Eric Chantler, Ph.D.
Department of OB/GYN
University Hospital of South Manchester
University of Manchester
Nell Lane, West Didsbury
Manchester M20 8LR, United Kingdom

Lee Claypool, Ph.D.
Contraceptive Research and
Development (CONRAD) Program
1611 N. Kent Street, Suite 806
Arlington, VA USA 22209

Douglas Colvard, Ph.D.
Contraceptive Research and
Development (CONRAD) Program
1611 N. Kent Street, Suite 806
Arlington, VA USA 22209

Carlos J. Conde-Glez, M.S.
Instituto Nacional de Salud Pública
Apartado Postal 222
Oficina de Correos No. 1
C.P. 62000, Cuernavaca, Mor., Mexico

Richard A. Cone, Ph.D.
The Johns Hopkins University
Department of Biophysics
Charles and 34th Streets
Baltimore, MD USA 21218

Carol Connell
Family Health International
P.O. Box 13950
Research Triangle Park, NC USA 27709

Carmela Cordero, M.D.
Association for Voluntary Surgical
Contraception, Inc. (AVSC)
Centro Médico Dr. Betances
Avenida Bolivar #754
Santo Domingo, Dominican Republic

Milton Cordero, M.D.
Profamilia
Calle Socorro Sanchez #64
Gazcue
Apartado Postal 1053
Santo Domingo, Dominican Republic

Anne-Marie Corner
BIOSYN, INC.
3401 Market Street
Philadelphia, PA USA 19104

Jane Cottingham
HRP/World Health Organization
1211 Geneva 27
Switzerland

Elizabeth W. Cravey
World Health Organization
1211 Geneva 27
Switzerland

Dr. Luis De La Macorra
GRUPO Profamilia
Apartado 407
Queretaro, Mexico

Grace E. Delano
Association for Reproductive and
Family Health
13 Bajayi Osungbekun Street
Nr Ikolaba High School Ikolaba Gra Ibadan
P.O. Box 30259 Secretariat
Ibadan, Nigeria

Juan Diaz, M.D.
Caixa Postal 6181
cep 13081
Campinas, SP, Brazil

Fred Dietsch
Ansell, Inc.
P.O. Box 1252
Dothan, AL USA 36302

Dr. Elias Dinzey
ADOPLAFAM
Calle Luis F. Thomen #309
Apto. 2-B
Ensanche Quisqueya
Santo Domingo, Dominican Republic

Rosalie Dominik
Family Health International
Box 13950
Research Triangle Park, NC USA 27709

Gustavo Doncel, M.D., Ph.D.
Contraceptive Research and
Development (CONRAD) Program
Eastern Virginia Medical School
601 Colley Avenue
Norfolk, VA USA 23507

Laneta Dorflinger, Ph.D.
Family Health International
Box 13950
Research Triangle Park, NC USA 27709

Ann Duerr, M.D., Ph.D.
Centers for Disease Control
1600 Clifton Road
Atlanta, GA USA 30333

Erik Dunmire, M.S.
Department of OB/GYN
Suber House
School of Medicine
University of California, Davis
Davis, CA USA 95616

Christopher Elias, Ph.D.
The Population Council
One Dag Hammarskjold Plaza
New York, NY USA 10017

María Luisa Eluf
SEMINA Industria e Comercio Ltda.
Rua Vergueira 2.994 - Cop 04102
Vila Marina, Sao Paulo, Brazil

Adria Estrada
Contraceptive Research and
Development (CONRAD) Program
1611 N. Kent Street, Suite 806
Arlington, VA USA 22209

Albarhina Espinal
Uruguay No. 16, Gazcue
Santo Domingo, Dominican Republic

Anibal Faúndes, M.D.
Department of OB/BYN
School of Medical Sciences
State Univeristy of Campinas
Caixa Postal 6181
cep 13081, Campinas SP, Brazil

Jacqueline D. Forrest, Ph.D.
The Alan Guttmacher Institute
111 5th Avenue
New York, NY USA 10003

Ian S. Fraser, M.D.
Deparrtment of OB/GYN
The University of Sydney
Sydney, NSW 2006, Australia

Michael Free, Ph.D.
Program for Appropriate Technology in
Health (PATH)
4 Nickerson St.
Seattle, WA USA 98109

Ron Frezieres
Los Angeles Regional Family Planning
Council, Inc.
3600 Wilshire, Suite 600
Los Angeles, CA USA 90010

Lea Gabbay
YAMA Inc.
P.O. Box 884
Millburn, NJ USA 07041

Shlomo Gabbay, M.D.
YAMA Inc.
P.O. Box 884
Millburn, NJ USA 07041

Henry Gabelnick, Ph.D.
Contraceptive Research and
Development (CONRAD) Program
1611 N. Kent Street, Suite 806
Arlington, VA USA 22209

Sanjay Garg, Ph.D.
National Institute of Immunology
Shaheed Jeet Singh Marg
New Delhi, 110 067, India

Sara George
USAID/ Santo Domingo 3470
Washington, DC USA 20521

Erica Gollub, Ph.D.
HIV Center for Clinical
and Behavioral Studies
New York Psychiatric Institute
722 West 168th Street, Unit 53
New York, NY USA 10032

Léon F. Gosselin
Axcan Ltée
597, boul. Laurier
Mont-Sainte-Hilaire
Quebec, Canada J3H 4X8

Ernesto Guerrero, M.D.
Union Latinoamericana Control de
Enfermedades de Transmisión Sexual
(ULACETS)
Apartado Postal 1139
Santo Domingo, Dominican Republic

DoWon Hahn, Ph.D.
RWJ Pharmaceutical Research Institute
1000 Route 202, Room B132
Raritan, NJ USA 08869

Curtis Hamann, M.D.
SmartPractice
3400 East McDowell
Phoenix, Arizona 85008-3846

Michael J.K. Harper, Ph.D.
Dept. of Obstetrics & Gynecology
Baylor College of Medicine
1 Baylor Plaza
Houston, Texas 77030

Dr. Michael Harrison
Carter-Wallace Inc.
Half Acre Road, PO Box 1001
Cranbury, NJ USA 08512

Nash Herndon
Family Health International
P.O. Box 13950
Research Triangle Park, NC USA 27709

Penelope J. Hitchcock, D.V.M., M.S.
Sexually Transmitted Disease Branch
Division of Microbiology & Infectious Disease
National Institutes of Health
9000 Rockville Pike
Bethesda, MD USA 20892

Supanee Jivasak-Apimas
Siriraj Family Health Research Center
Department of OB/GYN
Faculty of Medicine Siriraj Hospital
Mahidol University, Siriraj Hospital
Bangkok 10700 Thailand

Carol Joanis
Family Health International
Box 13950
Research Triangle Park, NC USA 27709

Marc Karam, M.D.
Global Programme on AIDS
World Health Organization
CH-1211 Geneva 27
Switzerland

David Katz, Ph.D.
Department of OB/GYN
Suber House
School of Medicine
University of California, Davis
Davis, CA USA 95616

James P. Koch, M.D.
Brigham and Women's Hospital
1037 Beacon Street
Brookline, MA USA 02146

Suporn Koetsawang, M.D.
Department of OB/GYN
Faculty of Medicine Siriraj Hospital
Siriraj Family Health Research Center
Bangkok 10700 Thailand

Stephanie Koontz
Population Action International
1120 19th Street, NW, Suite 550
Washington, DC USA 20036

Oladapo Ladipo, M.D.
South to South Cooperation in Reproductive Health
Rua Caetano Moura 35
Federacao - 40210 Salvador
Bahia, Brazil

Marie Laga, M.D.
Institute de Tropical Medicine
Nationalestraal 155
9000 Antwerp, Belgium

Katherine LaGuardia, M.D., M.P.H.
New York Hospital-Cornell Medical Ctr.
525 East 68th Street
New York, NY USA 10021

Pekka Lahteenmaki, M.D.
The Finnish Population and Family Welfare Federation
Kalevankatu 16 B
00100 Helsinki, Finland

Marlene LaMont
Contraceptive Research and Development (CONRAD) Program
1611 N. Kent Street, Suite 806
Arlington, VA USA 22209

Virginia Lamprecht, R.N., M.S.P.H.
Institute for Reproductive Health
Georgetown University
3800 Reservoir Road, NW
Washington, DC USA 20007

Mary Ann Leeper, Ph.D.
Wisconsin Pharmacal Company
919 N. Michigan Avenue, Suite 2208
Chicago, IL USA 60611

Howard Levine, Pharm.D.
Columbia Laboratories, Inc.
745 5th Avenue, Suite 1808
New York, NY USA 10151

Laurie Liskin, Sc.M.
The Johns Hopkins University
Center for Communications Program
527 St. Paul Place
Baltimore, MD USA 21202

Alina Llop, M.D.
Instituto de Medicina Tropical
"Pedro Kuri"
Instituto Superior de Ciencas Médicas
de la Habana
Avenida 15 y Calle 200
Siboney, Ciudad de la Havana, Cuba

Peter Mason, Ph.D.
Department Medical Microbiology
University of Zimbabwe Medical School
P.O. Box A178, Avondale
Harare, Zimbabwe

Christine Mauck, M.D., M.P.H.
Contraceptive Research and
Development (CONRAD) Program
1611 N. Kent Street, Suite 806
Arlington, VA USA 22209

Deborah Menius, MPA
Family Health International
P.O. Box 13950
Research Triangle Park Branch
Durham, NC USA 27709

Marilyn Metcalf-Whittaker, M.S.
Family Health International
P.O. Box 13950
Research Triangle Park Branch
Durham, NC USA 27709

Christopher Miller, D.V.M., Ph.D.
California Regional Primate Research Ctr.
University of California, Davis
Davis, CA USA 95616

Eric Miller, M.H.S.
Contraceptive Research and
Development (CONRAD) Program
1611 N. Kent Street, Suite 806
Arlington, VA USA 22209

Susan Mitchell
PROFIT
1925 N. Lynn Street, Suite 601
Arlington, VA USA 22209

Kamran Moghissi, M.D.
Department of OB/GYN
Hutzel Hospital
Wayne State University
4707 St. Antoine Blvd.
Detroit, Michigan 48201

John W. Moushall
Ansell, Inc.
P.O. Box 1252
Dothan, AL USA 36302

Robert M. Nakamura, Ph.D.
Livingston Research Center
1321 N. Mission Road
Los Angeles, CA USA 90033

Elizabeth Ngugi, M.D.
Department of Community Health
Kenyatta National Hospital
University of Nairobi
P.O. Box 19676
Nairobi, Kenya

Donald R. Nicholson, II
PROFIT
1925 N. Lynn Street, Suite 601
Arlington, VA USA 22209

Judy Norsigian
Boston Women's Health Book Collective
240a Elm Street
Somerville, MA USA 02144

Robin Ogilvie
Trimensa Pharmaceuticals
1050 Lawrence Dr.
Newbury Park, CA USA 91320-1301

Susan Palmore
Family Health International
P.O. Box 13950
Reaserch Triangle Park, NC USA 27709

Dolores J. Patanelli, Ph.D.
8324 Raymond Lane
Potomac, MD USA 20854

Dorothy Patton, Ph.D.
Department of OB/GYN, RH-20
University of Washington
Seattle, WA USA 98195

Herbert Peterson, M.D.
Centers for Disease Control
1600 Clifton Road
Atlanta, GA USA 30333

David Phillips, Ph.D.
The Population Council
1230 York Avenue
New York, NY USA 10021

Amy Pollack, M.D., M.P.H.
Association for Voluntary Surgical
Contraception, Inc. (AVSC)
79 Madison Avenue
New York, NY USA 10016

Malcolm Potts, M.B., B.Chair., Ph.D.
University of California, Berkeley
2408 Parker Street
Berkeley, CA USA 94704

Howard Price, Ph.D.
Family Health International
Box 13950
Research Triangle Park, NC USA 27709

Alexandre Psychoyos, M.D.
Hôpital Bicentre Batiment INSERM
80 Rue General Leclerc
94276 Kremlin-Bicetre, France

Bradley Pugh
ALADAN Corp.
Box 8308
Dothan, AL USA 36303

Luis Alberto Quiroz
Instituto Dominicano de
Seguro Social
Santo Domingo, Dominican Republic

Lisa Rarick, M.D.
Food and Drug Administration
Parklawn Bldg., 14BO4
5600 Fishers Lane
Rockville, MD USA 20857

Dr. Frances C. Rawle
Medical Research Council
20, Park Crescent
London, W1N 4AL, United Kingdom

P.K. Reddy
Ansell, Inc.
P.O. Box 1252
Dothan, AL USA 36302

Arturo Remes, M.D.
Hospital General de Veracruz
Avenida 20 de Noviembre
Veracruz, VER. Z.P. 91910
Mexico

Lionel Resnick, M.D.
Mount Sinai Medical Center
Pearlman Research Bldg.
Department of Research
4300 Alton Road
Miami, FL USA 33140

Roberto Rivera, M.D.
Family Health International
P.O. Box 13950
Research Triangle Park Branch
Durham, NC USA 27709

Ronald Roddy, PA-C, M.P.H.
Family Health International
P.O. Box 13950
Research Triangle Park, NC USA 27709

Debbie Rogow, M.P.H.
International Women's Health Coalition (IWHC)
24 E. 21st Street, 5th floor
New York, NY USA 10010

Wiwat Rojanapithayakorn, M.D., M.P.H.
Department of Communicable Disease Control
Ministry of Public Health
Devavesm Palace, Samsen Road
Bangkok 10200 Thailand

Gabrielle Ross
International Women's Health Coalition
24 East 21st Street
New York, NY USA 10010

Patrick J. Rowe, M.B., B.S., F.R.C.O.G
Special Programme of Research, Development and Research Training in Human Reproduction
World Health Organization
Avenue Appia 20
1211 Geneva 27, Switzerland

Subir Roy, M.D.
Women's Hospital
1240 N. Mission Road, Room L-1022
Los Angeles, CA USA 90033

Carol Sampson-Landers, M.D.
Advanced Care Products
Ortho Pharmaceutical Corporation
Route 1 South and Miltown Rd.
P.O. Box 6024
North Brunswick, NJ USA 08902-0724

Edward Scholl
USAID/Santo Domingo
Washington, DC USA 20521

Dr. Pramilla Senanayake
International Planned Parenthood Federation (IPPF)
Regent's College, Regent's Park
Inner Circle
London MWI 4NS, United Kingdom

Rochelle Shain, Ph.D.
Department of OB/GYN
University of Texas, San Antonio
Health Science Center
7703 Floyd Curl Drive
San Antonio, Texas 78284

James Shelton, M.D., M.P.H.
Office of Population
U.S. Agency for International Development
Washington, DC USA 20523

Alfred A. Shihata, M.D.
Scripps Institute
13565 Mira Montana Drive
Del Mar, CA USA 92014

Irving Sivin
The Population Council
1230 York Avenue
New York, NY USA 10021

Suren Solanki
London International Group
Unit 205, Cambridge Science Park
Milton Road
Cambridge CB4 4GZ, United Kingdom

David Soper, M.D.
Medical College of Virginia Hospitals
Main Hospital
8th Floor, Suite 220
12th and Marshall Street
Richmond, VA USA 23298

Joanne Spicehandler
World Health Organization
1211 Geneva 27
Switzerland

Jeffrey Spieler
Office of Population
U.S. Agency for International Development
Washington, DC USA 20523

Robert J. Staab, Ph.D.
Allendale Labs
Regulatory and Technical Assoc.
73 Franklin Turnpike
Allendale, NJ USA 07401

Zena Stein, M.A., M.B., B.Ch.
HIV Center for Clinical
and Behavioral Studies
New York State Psychiatric Institute
722 West 168th Street
New York, NY USA 10032

Felicia H. Stewart, M.D.
Valley Center for Women's Health
87 Scripps Drive, Suite 202
Sacramento, CA USA 95325

Katherine Stone, M.D.
Centers for Disease Control
1600 Clifton Road
Atlanta, GA USA 30333

Pamela Stratton, M.D.
National Institutes of Health
6100 Executive Blvd/Room 8B13
Bethesda, MD USA 20892

G.P. Talwar, Ph.D.
National Institute of Immunology
Shaheed Jeet Singh Marg
New Delhi, 110 067, India

Clark Taylor, Ph.D., Ed.D.
Sexologist's Sexual Health
10 Corwyn Street, #1
San Francisco, CA USA 94114

Rosemarie Thau, Ph.D.
The Population Council
1230 York Avenue
New York, NY USA 10016

Jack Thomas
USAID/Santo Domingo 3470
Washington, DC USA 20521

Ioanna Trilivas, Ph.D.
RD/H/HIV-AIDS
U.S. Agency for International
Development
SA-18, Room 1200
Washington, DC USA 20523-1817

Tim Truitt
USAID/Santo Domingo 3470
Washington, DC USA 20521

James Trussell, Ph.D.
Office of Population Research
21 Prospect Avenue
Princeton, NJ USA 08544

Yun Yen Tsong, Ph.D.
The Population Council
Center for Biomedical Research
1230 York Avenue
New York, NY USA 10021

Shakati Upadhyay, Ph.D., D.Sc.
National Institute of Immunology
Shahid Jeet Singh Marg
New Delhi, India 110067

Lisa Usher
USAID/Santo Domingo 3470
Washington, DC USA

Donald P. Waller, Ph.D., D.A.B.T.
Department of Pharmacodynamics
The University of Illinois at Chicago
P.O. Box 6998
Chicago, Ilinois 60680

Terry Walsh
Los Angeles Regional Family Planning
Council, Inc.
3600 Wilshire Blvd., Suite 600
Los Angeles, CA USA 90010

Martha Watson
Axcan Ltée
597 boul. Laurier
Mont-Sainte-Hilaire
Quebec, Canada J3H 4X8

Kevin Whaley, Ph.D.
Department of Biophysics
The Johns Hopkins University
Charles and 34th Streets
Baltimore, MD USA 21218

Nancy Williamson, Ph.D.
R&D/H/AIDS
U.S. Agency for International Development
SA-18, Room 1200
Washington, DC USA 20523-1817

Celia Woodfill, Ph.D.
Office of Population
U.S. Agency for International
Development
SA-18, Room 820
Washington, DC USA 20523

Lillian Yin, Ph.D.
Food and Drug Administration
1390 Piccard Drive
Rockville, MD USA 20850

Lourens Zaneveld, D.V.M., Ph.D.
Department of OB/GYN
Rush Medical College
Rush-Presbyterian-St. Luke's Medical
Center
1653 West Congress Parkway
Chicago, IL USA 60612-3864

INDEX

Abortions
 contraceptive failure and, 18, 22, 74
 spontaneous, 227
 survey reliability and, 18
Abstinence, 18, 240, 329
 HIV and, 113
 periodic, and ovulation method, 20
 vaginal trauma and, 83
Adenosis, DES daughters and, 111
Adolescents, sexually active
 HIV and, 229
 preventative behavior and, 13, 347–348
 puberty and, 94, 227
 and use of barrier methods, 9–10, 15
Age
 condom use and, 31
 contraception demographics and, 13–14, 188, 277–278
 preventative behavior and, 10–11, 13, 14–15
 spermicide use and, 32–36, 278
 use of vaginal barrier methods and, 32–43, 188
AIDS, 213, 215, 340, 348
 prostitutes and, 326
 vaginal microflora and, 97
Allergic reactions
 condoms and, 127
 spermicides and, 110, 153, 268
Anal sphincter, 77
Anatomy
 familiarity with, 71
 vaginal, 77–86, 214
Animal testing
 baboon, 168–169, 298–299
 and HIV, 214–219
 penile/vaginal irritation and, 145, 274–275
 rabbit, 164–166, 275, 307
 spermicide trials and, 154, 157, 163–169, 217–219, 274–275, 297–298
 stump-tailed macaque, 166–168, 213–219, 222, 274–275, 298–299, 302
Anterior vaginal suburethra (G-spot), 83
Antibiotics
 disease resistance to, 240
 sexually transmitted diseases and, 204, 240
 vaginal flora and, 95, 102, 111
Arcuate tendon, 79
Arousal, sexual
 bulbocavernosus muscles and, 77
 orgasm and, 83
 secretions and, 83, 111
 and stages of sexual response, 83
 vaginal engorgement and, 83
Arthritis-dermatitis syndrome, 228
Autonomic nervous system, 83
Azithromycin, 240

Bacteroides, 92, 94, 96, 111
Barrier contraception
 and acceptability evaluations, 58–63, 113–114, 145–157, 163–169, 323–331, 354, 356–358
 access to, 5, 9–10, 15, 54–63, 255–259
 adolescents and, 9–10, 15
 and age of user, 9–10, 14–15, 188
 coital positions and, 85
 cultural precedents and, 15, 66, 71, 112, 209, 255–259
 defined, 3, 10–11
 demographic use of, 4, 6–10, 11, 13–14, 123, 186–195, 255–259, 261–269, 275, 277–278

development issues for, 114–115, 137–144, 147–157, 163–169, 185–196, 197–200, 323–331, 339–350, 354–358
efficacy of, 17–22, 28–29, 53, 59–60, 83, 113, 124–126, 181, 183, 185–196, 197–200, 208, 232, 245–249, 261–269, 281, 283–285, 335, 339–350, 354–358
failure [rate of], 22–28, 30–43, 124–126, 195, 257
feminist perspective on, 69–74
first intercourse and, 9–10
infection and, 105–116, 208–210, 212, 245–249, 249–253, 261–269, 281, 283–285, 353–356, 360–361
and married couples, 8–9, 10, 15, 187
motivations for using, 53–64
unmarried individuals and, 9–10, 15, 261–269
user-controlled, 73–74, 353
See also individual barrier method by type

Bartholin's glands, 77, 91
Bladder, 79
Bulbocavernosus muscles, 77

Candida albicans, 96
and HIV, 204
microbicides and, 107, 275, 277, 282, 284
polyherbal cream and, 275, 277, 279, 281, 282, 284
spermicide use and, 115, 277

Candidiasis
barrier contraceptives and, 105, 106, 109
Depo-Provera and, 111
and diabetes, 111, 122
and HIV, 204

Cervical cap
demographic use of, 5–10, 11
failure [rate of], 18, 23–26, 28–29, 38–39, 50
fittings, 50–51
infection risk and, 109, 208
and spermicides, 50
and user age, 38–39
See also Vaginal methods

Cervicitis, 226, 227
health care costs of, 229–230
nonoxynol-9 and, 247

Cervix, 81
animal testing and, 164, 166, 167–168, 171, 222, 298–299
disease/infection and, 89, 105, 106, 207, 209, 214, 227–228, 355
ectopy of, 227
embryologic development of, 77–78
immunoglobulin concentrations and, 111, 115
microbiology of, 91–93, 97, 101, 214
mucus secretions, 82, 83, 89, 91–92, 110, 111, 150–153, 154–159, 173–181, 183, 298–299
and postcoital tests, 173–181
spermicides and, 50, 150–153, 154–159, 167, 297, 298–299, 355
sperm/semen [prostaglandins] and, 107, 173–181, 183

Chancroid, 203, 207
Chemical vaginal contraceptives. *See* Microbicides; Spermicides

Chlamydia, 203, 207, 212, 229
barrier contraceptives and, 105, 106, 107, 108, 232, 245–248, 249–253
condoms and, 113, 245–248, 249–253
contraceptive failure and, 49
health care costs of, 229–230
and HIV, 204, 229, 240
men and, 226–227
microbicides and, 107, 284
spermicides and, 153, 245–249, 268
vaginal microflora and, 96, 225, 226–227, 230–233

Chorioamnionitis, 227
Clitoris, 83
Clostridium, 92, 94
Coccyx, 77
Complement deficiencies, 228
Condom, 3, 124, 126–130, 249, 340
acceptability issues and, 57–58, 62, 63–64, 66–67, 123–124, 127–130, 323–331

availability of, 3–5, 6–7, 123–124, 255–259, 340
demographic use of, 5–10, 11, 14, 61–62, 123
efficacy of, 18, 19, 22, 30–31, 53, 59–60, 113, 123, 124–126, 185, 232, 245–248, 257–259, 312, 328–329, 335, 355
failure [rate of], 19, 22, 30–31, 124, 257, 261
first intercourse and, 10
infection risk and, 95, 108, 109, 113, 123, 208, 232, 245–248, 353, 355
married couples and, 5, 8–9, 31
motivation for using, 13, 61, 66–67, 257–259
prevalence of, 14, 123–124
reliability/manufacturing of, 124–130, 132, 257–259, 335, 339–350, 355
spermicidally lubricated, 22, 31, 62, 63–64, 132, 288
stages of sexual response and, 66–67
unmarried individuals and, 9–10
and user age, 31
vaginal trauma and, 83
See also Female condom
Contraception
acceptability evaluations and, 54–63, 113–114, 147–157, 163–169, 323–331, 354, 356–358
age and, 9–10, 14–15, 188
cervical mucus and, 83, 150–153, 154–159, 173–181, 183
condom and, 3, 4–10, 11, 123–130, 249–253, 255–259, 312, 339
correct use of, 19–22, 194–195
cultural precedents and, 15, 66, 71, 112, 209, 255–259
demographics of, 4, 5–10, 11, 13–14, 123, 186–195, 255–259, 261–269, 275, 277–278
development issues, 339–350, 354–358
efficacy of, 17–22, 28–29, 53, 59–60, 83, 124–126, 181, 185–196, 197–200, 208, 232, 245–249, 261–269, 335, 339–350, 354–358
evaluation/trials, 137–144, 147–157, 163–169, 194–195, 198–200, 245–249, 261–269, 281, 283–285, 295–297, 300, 302–305, 311–312, 344–347, 354–358
failure [rate of], 22–28, 30–43, 124–126, 195, 257
feminist perspective on, 69–74
merchandising and, 53–63, 339–350
and role of providers, 66, 191, 325–326, 329, 339–350
and safety, 53, 55, 67, 114–115, 126–128, 138, 139–144, 354–358
user control/motivation, 9–10, 13, 53–64, 353
vaginal barrier methods, 3, 5–10, 11, 14, 15, 54–63, 245–249, 323–331, 339–350, 353–356, 354–358, 360–361
Corynebacterium, 92
Creams, spermicidal, 3
Cystitis, 109
Cytomegalovirus, 96, 203, 360
female condom and, 249–253

Dalkon Shield, 71
Diaphragm
acceptability issues and, 58, 64
and availability, 15
demographic use of, 5–10, 11
failure [rate of], 18, 23–26, 28–29, 40–42, 50
infection risk and, 108, 109, 110, 208
during menstruation, 101
prevalence of, 15
and provider biases, 67, 74
provider profitability and, 340
and spermicides, 50, 66
and user age, 40–42
vaginal flora and, 102
See also Vaginal methods
Disease prevention, 231–232, 255–259, 353–356, 360–361
animal testing and, 213–219, 298–299, 302

barrier methods and, 10–11, 113–114, 137–144, 146, 185, 203, 208–210, 245–248, 249–253, 323–331, 341–350, 353–356
health care costs/profits and, 229–230, 340
menses and, 89
microbicides and, 106–107, 225, 226–227, 228, 230–233, 268, 271–275, 277–282, 283–285, 354–358
spermicides and, 95, 106–107, 108, 110, 114, 137, 141, 143–144, 146, 147, 153, 245–248, 261–269, 354–358
vaginal flora and, 95–97, 153, 225, 226–227, 230–233, 232, 277–285, 355, 356
Doderlein's bacilli, 97
Douching
animal testing and, 167
cultural practice of, 112
pelvic inflammatory disease and, 121
product brands, 115
vaginal flora and, 101, 102, 122
Dysplasia, cervical, 89, 355
See also Neoplasia, cervical

Ectopic pregnancy
barrier contraceptives and, 105, 106
pelvic inflammatory disease and, 227–228
See also Pregnancy
Enterobacterium, 92, 94, 95, 96
Epstein-Barr virus, 360
Escherichia coli, 92, 109
microbicides and, 107, 279, 281, 282
Estrogens
Lactobacillus and, 111
menopause and, 95
mucus-secreting cells and, 184
neonate and, 93–94
vaginal wall and, 81
Eubacterium, 92
Excitement phase, sexual response and, 83

Fallopian tubes, 227
gonococci infection and, 231
Family planning programs
access to contraceptives and, 4, 15, 328
and condom use, 22
demographics and, 11, 186–195
Female condom
acceptability issues and, 58, 64, 73
clinical trials and, 51
failure [rate of], 18, 26–27, 28, 29
sexually transmitted diseases and, 232, 249–253
Fem Cap, 72, 144, 318
Fertility
animal testing and, 163–169
barrier contraceptives and, 187–188, 355
disease and, 227
spermicides and, 147, 149, 163–169, 355
Fimbriae, 227
Flora, bacterial
aerobic/anaerobic, 82, 92, 94, 97
antibiotics/drug effects and, 95, 102, 111
HIV/AIDS and, 97, 204
menopause and, 95
nonoxynol-9 virucide and, 106, 153
pregnancy and, 94
spermicides and, 106–107, 108, 110, 114, 153, 355
systemic diseases and, 111–112
vaginal pH and, 82–83, 91–92, 94, 97, 106, 110, 114–115, 156
See also Infections
Foams, spermicidal, 3
acceptability issues and, 59
brands, 32–35
cervical mucus and, 83
demographic use of, 14
efficacy of, 23, 50
and user age, 32–36
Fusobacterium, 92

Gardnerella, 92, 94, 96
microbicides and, 107, 279, 281
polyherbal cream and, 277, 279, 281

Genitalia, 79
  and sexual response, 83
  ulcerous, 106, 203
Gonad, undifferentiated, 77
Gonorrhea, 71, 207, 212, 229
  and antibiotic resistance, 240
  barrier contraceptives and, 106, 108, 232, 245–249
  condoms and, 113
  health care costs of, 229–230
  and HIV, 204, 229
  microbicides and, 107, 225, 226–227, 230–233, 282, 284
  Neisseria bacterium, 93, 96, 101, 107, 203, 226, 230, 282, 284
  spermicides and, 153, 245–249
G-spot, 83

Haemophilus ducreyi, 203
Hawthorne effect, 18
Hepatitis B virus, 203, 212
Herpesvirus, 96, 203, 334, 360
  female condom and, 249–253
  microbicides and, 107
HIV
  animal testing and, 213–219
  barrier methods and, 11, 53, 106, 203, 208–210, 232, 247, 249–253, 261–269, 323–331, 341–350, 353–356
  condom usage and, 13, 113, 127, 247, 261
  contraceptive failure and, 49, 261
  epidemiology of, 228–230
  female susceptibility and, 209
  menstrual blood and, 85
  and microbicide developments, 271–275, 277–282, 283–285, 355
  prevention strategies, 113–114, 127, 247, 261–269, 271–275, 277–282, 283–285, 319, 341–350, 353–356
  semen and, 85, 156, 209
  STD potentiation of, 204, 207, 209, 225, 226–227, 228, 230–233, 240, 247
  vaginal microflora and, 95, 97, 102, 204
Hormones, reproductive, 110–112
  cervical mucus and, 173–181, 184
  systemic disease and, 111–112
  See also Estrogens
Hysterectomy, 222

Infections
  animal testing and, 213–219
  barrier contraceptives and, 10–11, 53, 105–116, 123, 185, 203, 208–210, 232, 245–248, 249–253, 255–259, 323–331, 341–344, 353–356
  health care costs/profits and, 229–230, 340
  hygiene and, 112, 115, 121, 122, 277
  microbicides and, 106–107, 225, 226–227, 228, 230–233, 268, 281, 283–285, 354–358
  spermicides and, 95, 106–107, 108, 110, 114, 137, 141, 143–144, 146, 147, 153, 225, 226–227, 230–233, 261–269, 271–275, 277–282, 283–285, 354–358
  vaginal flora and, 82–83, 92–93, 96–97, 106, 109, 153, 204, 225, 226–227, 230–233, 284, 354–356
  See also Disease prevention; HIV; Sexually transmitted disease
Intercourse
  age and, 10–11, 13, 188
  cervical mucus and, 173–181
  feminist perspective on, 69–74
  first, adolescents and, 9–10
  frequency, 189, 198, 205, 263
  last active, 13
  preventative behavior and, 13, 341, 347–348
  stages of sexual response and, 66–67, 83
  trauma and, 83
  vaginal flora and, 103, 204
Intrauterine device
  availability of, 4, 340
  bleeding/pain and, 20
  demographics and, 11
  failure [rate of], 18
  infection risk and, 108, 109, 328
  and side effects, 71
  vaginal flora and, 95

IUD. *See* Intrauterine device

Jellies, spermicidal, 3

Labia, 83
Lactobacilli, 92, 93
  estrogen and, 111
  menopause and, 95
  microbicides and, 107, 355
  pregnancy and, 94
  spermicides and, 153
  vaginal pH and, 114
Lactoferrin, 97
Langerhans cells, 214, 222, 223
Lea's Shield, 72, 144, 146
Levator ani muscle, 79
Literature, contraceptive, 69–70
Lymphoid cells, 110

Males, sexually active
  experimental animals, 214–219, 222
  and use of barrier methods, 9–10
Married couples
  barrier contraception and, 4–5, 8, 10, 11, 15
  and condom usage, 14, 22, 31
  contraception demographics and, 13–14, 187
  See also Unmarried individuals, sexually active
Mary Lane effect, 23
Mast cells, 107
Meningococcal infections, 228
Menopause, 328
  vaginal flora and, 95
Menstruation
  animal testing and, 166
  diaphragm and, 101
  puberty and, 94
  sexual activity during, 85, 89, 263
  toxic shock syndrome and, 121
  vaginal flora and, 94
  and viral sensitivity to pH, 217–218
Merchandising, contraception, 53–63, 255, 295–297
  condoms and, 123–126, 127–130, 133, 142, 339
  and new product development, 339–350
  spermicides and, 141–144, 148, 152, 339
Microbicides, 3, 72
  acceptability of, 281–282, 354–358
  barrier contraceptives and, 105, 353
  development of, 271–275, 277–282, 283–285, 354–358
  polyherbal, 273–275, 277–285
  vaginal infections and, 106–107, 225, 226–227, 228, 230–233, 268, 277–280, 281, 283–285, 354–358
Micrococcus organism, 279
Mobiluncus sp. bacteria, 96
Müllerian ducts, 77, 78
Mycoplasmas, 93, 94, 96

National Surveys of Family Health (NSFG), 18
Neem seed extract, 273
Neonate
  Chlamydia and, 227, 229
Neoplasia, cervical, 105
  barrier contraceptives and, 106, 108, 109
Nonoxynol-9, 106, 138, 147, 152, 161–162
  efficacy of, 245–248, 304
  and HIV, 216–217, 223, 247, 261–269
  See also Spermicides
Norplant, 66, 73

Oral contraceptives
  access to, 4, 15
  and failure rates, 18, 49
  infection risk and, 108, 111, 227
  and side effects, 71
  vaginal flora and, 95
  See also Pill, contraceptive
Orgasm, excitement cycles and, 83
Orgasmic platform, 83
Otitis media, 227
Ovulation method, 20
Papillomavirus, 96, 203, 334, 355
  barrier contraceptives and, 105, 232

Pelvic diaphragm, 79, 80
  sexual response and, 83
Pelvic inflammatory disease (PID), 71, 229
  barrier contraceptives and, 105, 106, 108
  complications of, 227
  douching and, 121, 122
  health care costs and, 229–230
Peptococcus, 92
Peptostreptococcus, 92, 94
Perineal muscles, transverse, 77
Pill, contraceptive, 4
  demographics and, 11
  and failure rates, 18, 49
  first intercourse and, 10
  infection risk and, 108, 111, 227
  and side effects, 71
Polyherbal microbicide/spermicide, 273–275, 277–285
  clinical response to, 281, 283–285
Postcoital test, 173–181
Potassium sorbate, 115
Praneem polyherbal cream, 273–275, 277–285
  clinical response to, 281, 283–285
  See also Microbicides
Pregnancy
  barrier contraceptives and, 187, 195, 327–331, 341, 353
  condom use and, 9–10, 22, 327
  contraceptive failure and, 18, 50, 195
  diagnosis of, 138
  and disease, 72, 146, 203, 204–205, 227–228, 343, 353
  protection from, 10–11, 353
  teenage, 229
  vaginal microflora and, 93, 94
  See also Ectopic pregnancy
Progesterone, 184
Prostaglandins
  semen and, 107
  vaginal immune response and, 111
Protozoa, 96, 203, 228
  and vaginal discharge, 277
Pseudomonas aeruginosa, 92
  microbicides and, 279
Puberty
  estrogen levels during, 94
  pre-, 110–111

Quinine hydrochloride, 273, 277

Rape, and vaginal trauma , 83
Reality condom, 72
Rectum, 79

Sanitary napkins, vaginal flora and, 95
Saponins, 273
Schistosoma haematobium, 95
Semen
  prostaglandins, 107
  STD transmission and, 207, 209, 226–227, 232, 355
  vaginal pH and, 82, 103, 156
  See also Sperm
Septicemia, 228
Sex education
  condom usage and, 123–124, 255–256
  providers of, 69–71
Sexually transmitted disease (STD), 89, 203–205, 229, 340, 348
  animal testing and, 213–219, 298–299
  barrier methods and, 10–11, 53, 105–116, 185, 203, 208–210, 232, 245–248, 249–253, 261–269, 323–331, 341–350, 353–356
  condom usage and, 13, 123, 245–248, 249–253, 255–259, 340
  contraceptive failure and, 49, 256–259, 312, 313–314
  douching and, 122
  epidemiology of, 228–230
  health care costs/profits and, 229–230, 340
  menstrual blood and, 85, 252, 263
  microbicides and, 271–275, 277–282, 283–285, 354–358
  pathogen transmission, 206–208, 209, 214–219, 226–227, 228, 232, 249–253, 265, 301–302, 341–344, 353–356

prevention, 9–10, 11, 13, 113–114, 137, 203, 208–210, 232, 245–248, 249–253, 255–259, 261–269, 273–275, 277–285, 287–292, 323–331, 341–344, 347–348, 353–356, 360–361
spermicides and, 106–107, 108, 110, 114, 146, 147, 153, 209, 219, 245–249, 302, 354–358
vaginal microflora and, 97, 102, 106, 225, 226–227, 230–233, 355
See also HIV
Simian immunodeficiency virus (SIV), 214–218, 223, 355
Sims-Huhner Test, 173–181
Single individuals, sexually active
and use of barrier methods, 9–10
Skene's glands, 91
Sperm
and animal testing, 163–169
cervical mucus and, 83, 154, 173–181, 183
and disease transmission, 207, 208, 212, 226–227, 232, 355
fertilizing ability of, 148, 150, 153–154, 157, 162, 173
spermicides and, 147–157, 355
transport, barrier methods and, 83–84, 147–157, 183, 209, 232, 297–298
See also Semen
Spermicides, 3, 138–139, 340
acceptability issues and, 58, 64, 138–139, 147–157, 163–169, 281–282, 323–337, 354–358
and allergies/irritations, 110, 138–141, 153, 268–269, 274, 303, 354, 356–357
animal testing and, 163–169, 217–219, 302
brands, 32–36, 50, 106, 138, 147, 152, 162, 261, 265–269
cervical mucus and, 83, 150, 153–154, 159
and condoms, 22, 31, 62, 63–64, 245–249
demographic use of, 14, 275, 277–278
and diaphragms, 50, 66, 340
evaluation/regulation, 137–144, 147–158, 161–162, 163–169, 245–249, 295–297, 300–305, 323–357, 354–358
failure [rate of], 18, 22–23, 28, 31, 32–36
infections and, 95, 105, 106–107, 108, 110, 114, 144–145, 153, 208, 209, 216–217, 218, 225, 226–227, 230–233, 232, 245–249, 261–269
polyherbal, 273–275, 277–285
toxicity of, 114–115, 138, 274–275, 309, 354, 356–357
and user age, 32–36
vaginal flora and, 95, 102, 110, 111, 153, 217–218, 225, 226–227, 230–233
Sponge, contraceptive
brands, 51, 265–269
demographic use of, 5–10, 11
failure [rate of], 18, 23–26, 28–29, 37, 50
infection risk and, 108, 109, 261–264
and spermicides, 50, 261–269
and user age, 37
See also Vaginal methods
Staphylococci, 92, 110
microbicides and, 275, 279, 281
toxic shock syndrome and, 121
STD sequelae, 108
See also Sexually transmitted disease
Sterilization, 328
demographics and, 11
female, 66
Streptococci, 92, 95, 96
Surfactants, 147
See also Spermicides
Surveys
and bias, 18–19, 18–22
contraceptive failure and, 18
Symphysis pubis, 77
Syphilis, 95, 203, 207

Tampons
oxygen tension and, 110
vaginal flora and, 95
Testosterone, 184

Torulopsis glabrata, 96
Toxic shock syndrome, 83
  barrier contraceptives and, 105, 106, 109
  and staph organism, 121
Transudation, vaginal, 83
Treponema pallidum, 107, 203
  See also Syphilis
Trichomonas vaginalis, 96, 203, 207, 240, 241
  barrier contraceptives and, 105, 106, 108, 249–253
  female condom and, 249–253
  and HIV, 204, 229
  microbicides and, 107, 268, 277, 281, 284
  polyherbal cream and, 277, 279, 281, 284
Tubal infertility, 108, 227

Ulcers, genital, 106, 203
  and HIV, 204, 228
  nonoxynol-9 and, 247, 264
Unmarried individuals, sexually active and use of barrier methods, 9–10, 15
Ureaplasma, 93
Ureaplasma urealyticum, 96
Urethra, 77, 79, 80
Urethritis, 226
  health care costs of, 229–230
Urinary tract infection
  barrier contraceptives and, 105, 106, 109
Urogenital diaphragm, 77, 79
Urogenital sinus, 77, 78
Uterus, 77, 79
  disease/infections, 227–228

Vagina, 77, 79, 80, 222
  anatomy of, 77–86, 214
  bacterial flora of, 82–83, 91–97, 102–103, 110–112, 114–115, 153, 228, 277, 355
  cervical/glandular secretions and, 83, 91, 110, 111, 150–153, 154–159, 173–181
  embryologic development of, 77–78
  estrogens and, 81, 93, 94
  hormonal milieu of, 110–112
  immunoglobulin secretions and, 111, 115
  infections and, 105, 106–107, 108, 110, 114, 115, 153, 207, 209, 228, 249–253, 277–285, 354–358
  innervation of, 79
  microbiology of, 91–97, 102–103, 121, 217–218, 225, 226–227, 230–233, 278–279
  oxygen tension in, 110
  physiology of, 82–86, 110–112
  pregnancy and, 93, 94–97
  spermicides and, 95, 102, 110, 114–115, 137–144, 147–158, 161–162, 218, 261–269, 271–275, 277–282, 283–285, 354–358
  systemic disease and, 111–112
  trauma and, 83, 95
Vaginal barrier methods
  acceptability issues and, 58–63, 113–114, 147–157, 163–169, 323–331, 354–356
  and availability, 15
  cervical mucus and, 83, 150–153, 154–159, 173–181, 183
  demographic use of, 5–10, 11, 14, 15, 186–195, 255–259, 261–269, 275, 277–278
  development issues for, 114–115, 181, 183, 185–196, 197–200, 323–331, 339–350, 354–358
  failure [rate of], 18, 22–28, 29, 32–43, 195, 257
  infection and, 105–116, 208–210, 232, 245–249, 261–269, 281, 283–285, 353–356, 360–361
  insertion-drugs/drug-devices, 137–146, 147–157, 163–169, 277–285, 342, 344–345, 354–358
  prevalence of, 15, 209
  types of, 3, 209, 255, 259–263, 339–350
  and user age, 32–43, 188
  See also Cervical cap; Diaphragm; Spermicides; Sponge, contraceptive

Vaginal Contraceptive Film (VCF)
acceptability issues and, 59, 66
demographic use of, 43
failure [rate of], 18, 27–28, 29, 43
and user age, 43
Vaginitis, 82–83, 93, 96, 309
abnormal discharge and, 277
nonoxynol-9 and, 263
Vaginosis, 82–83
barrier contraceptives and, 105, 106, 109
and HIV, 204
hydrogen peroxide and, 92
Vasoactive intestinal peptide (VIP), 83
Vasocongestion, sexual response and, 83
Vestibular bulbs, 77
Virucides, 145, 146, 153
acceptability issues and, 64
nonoxynol-9, and HIV, 106, 215, 216–217
polyherbal, 273–275, 277–285
Viruses
antibiotics and, 204
and pH, 217–218
seminal plasma and, 223
STDs and, 203, 204, 228, 360
vaginal microflora and, 96–97, 277
Vulva, 207, 208
Vulvitis, 109, 115, 309
nonoxynol-9 and, 264

Warts, genital, 204
Women, sexually active
and age, 9–10, 14–15
contraception demographics and, 13–14
male condom use and, 22
and microbicide developments, 271–275, 277–282, 283–285
and monkeys, 214, 222
preventative behavior and, 13, 209–210, 341, 347–348, 360–361
and use of barrier methods, 9–10, 209–210, 249–253
vaginal discharge and, 277
and vaginal pH, 217–218
See also Sexually transmitted disease

Yeast infections, 96
Lactobacillus and, 111
microbicides and, 107, 354
nonoxynol-9 and, 247